Published for
OXFORD INTERNATIONAL AQA EXAMINATIONS

W0044041

International GCSE
COMBINED SCIENCES
Biology

Ann Fullick
Editor: Lawrie Ryan

OXFORD
UNIVERSITY PRESS

OXFORD
UNIVERSITY PRESS

Great Clarendon Street, Oxford, OX2 6DP, United Kingdom

Oxford University Press is a department of the University of Oxford. It furthers the University's objective of excellence in research, scholarship, and education by publishing worldwide. Oxford is a registered trade mark of Oxford University Press in the UK and in certain other countries.

British Library Cataloguing in Publication Data

Data available

978-0-19-840793-5

12

Paper used in the production of this book is a natural, recyclable product made from wood grown in sustainable forests. The manufacturing process conforms to the environmental regulations of the country of origin.

Printed and bound by CPI Group (UK) Ltd, Croydon, CR0 4YY

Acknowledgements

The publishers would like to thank the following for permissions to use their photographs:

Cover: Denise Allison Coyle/Shutterstock

p2l: Wim van Egmond/Visuals Unlimited/Corbis; **p2r:** Steve Gschmeissner/Science Photo Library; **p3:** iStockphoto; **p4t:** Eye of Science/Science Photo Library; **p4b:** Scott Camazine/Alamy; **p5:** Science Photo Library; **p8:** Eric Grave/Science Photo Library; **p17:** ImageBroker/Imagebroker/FLPA; **p21:** Science Photo Library; **p23:** Eye of Science/Science Photo Library; **p25:** Lewis Whyld/PA Archive/Press Association Images; **p26:** Science Photo Library; **p29:** Juniors Bildarchiv GmbH/Alamy Stock Photo; **p33:** iStockphoto; **p34:** Eye of Science/Science Photo Library; **p36:** Erik Dreyer/Getty Images; **p37:** Paul Bock/Alamy; **p40:** Science Photo Library; **p44:** St Bartholomew's Hospital/Science Photo Library; **p45:** National Cancer Institute/Science Photo Library; **p48r:** Image Source/Alamy; **p48l:** Maximilian Stock Ltd/Science Photo Library; **p49:** Dorling Kindersley/Getty Images; **p50:** J.C. Revy, ISM/Science Photo Library; **p51:** Martyn F. Chillmaid/Science Photo Library; **p52:** Ingo Arndt/Minden Pictures/FLPA; **p55:** Trevor Clifford Photography/Science Photo Library; **p57:** Dr P. Marazzi/Science Photo Library; **p58l:** Manfred Kage/Science Photo Library; **p58r:** Dr. Richard Kessel & Dr. Gene; **p62l:** Philippe Lissac/Godong/Corbis; **p62r:** Christian Larue/fotolia.com; **p65:** Shutterstock; **p66t:** Anthony Short; **p67t:** Frans Lanting Studio/Alamy Stock Photo; **p67b & p68t:** Anthony Short; **p68b:** Wayne R Bilenduke; **p69t:** Jeffrey Lepore/Science Photo Library; **p69b:** Tony Short; **p70t:** Arturo de Frias photography/Getty Images; **p70b:** Eugene Sergeev; **p71t:** Natiure Picutre Library/Kim Taylor; **p71b:** Andrew Rutherford -www.flickr.com/photos/arutherford1/Getty Images; **p72t:** Shutterstock; **p72b:** Thomas Imo/Alamy Stock Photo; **p73l:** Mike Goldwater/Alamy Stock Photo; **p73r:** Roger Coulam/Getty Images; **p74l:** Shutterstock; **p74r:** Thomas Marent/Minden Pictures/Getty Images; **p75:** Ann Fullick; **p76:** Gavin Rodgers/Rex Features; **p77:** Gorilla/fotolia.com; **p78t:** Gavin Hellier/Getty Images; **p78b:** iStockphoto; **p81:** Steve Gschmeissner/Science Photo Library; **p82t:** Lea Paterson/Science Photo Library; **p82b:** Scott Camazine/Science Photo Library; **p83:** Shutterstock; **p86t:** Steve Gschmeissner/Science Photo Library; **p86b:** Dr. Harold Fisher, Visuals Unlimited/Science Photo Library; **p87:** Karl Schoendorfer/Rex Features; **p88t:** Scott Camazine/Alamy; **p88b:** Eye of Science/Science Photo Library; **p90 & p92t:** Shutterstock; **p92b:** St Mary's Hospital Medical School/Science Photo Library; **p93:** CC Studio/Science Photo Library; **p95:** iStockphoto; **p96:** CDC/Science Photo Library; **p99:** Christian Darkin/Alamy Stock Photo; **p100:** SAPS; **p103:** Fuse/Getty Images; **p104:** iStockphoto; **p105t:** Cordelia Molloy/Science Photo Library; **p105b:** FLPA/Chris Mattison; **p107:** FLPA/Wayne Hutchinson; **p110t:** iStockphoto; **p110b:** Steve Gschmeissner/Science Photo Library; **p112:** iStockphoto; **p113t:** Mat Hayward/fotolia.com; **p113b:** iStockphoto; **p114t:** Shutterstock; **p114b:** Anthony Short; **p116t:** Corbis; **p116b:** CNRI/Science Photo Library; **p121t:** Perla Copernik/Alamy Stock Photo; **p121b:** Dietmar Temps/Shutterstock.com; **p122:** Zephyr/Science Photo Library; **p124:** Dr. Stanley Flegler, Visuals Unlimited/Science Photo Library; **p125l:** Look at Sciences/Science Photo Library; **p125r:** iStockphoto; **p130:** Roslin Roslin Institute/Press Association Images; **p133:** International Rice Research Institute (IRRI); **p134:** Courtesy Golden Rice Humanitarian Board. goldenrice.org; **p135l:** A&M University/Rex Features; **p135r:** Pat Sullivan/AP/Press Association Images; **p139t:** sodapix/Getty Images; **p139b:** English Heritage Photo Library; **p140t:** Phil Degginger/Alamy; **p140b:** Zoonar GmbH/Alamy; **p141:** NickR/fotolia.com; **p142:** ZSSD/Minden Pictures/FLPA; **p143t:** iStockphoto; **p143b:** Ulla Lohman; **p144:** iStockphoto; **p145r:** Norbert Wu/Minden Pictures/FLPA; **p145l:** Murton/Southampton Oceanography Centre/Science Photo Library; **p146:** Louise Murray/Science Photo Library; **p147t:** Anthony Short; **p147b:** FLPA/Mike Lane/Holt; **p148:** FLPA/Bob Gibbons; **p149l:** iStockphoto; **p149r:** David Hughes/fotolia.com; **p150t - p151t:** iStockphoto; **p151b:** Nature Picture Library/Phil Savoie; **p153t:** Roxana/fotolia.com; **p153m:** Lynwood Chase/Science Photo Library; **p153b:** iStockphoto; **p155t:** WLDavies/iStockphoto; **p155b:** coopder/iStock; **p156t:** David R. Frazier Photolibrary, Inc/Alamy; **p156b:** Westend61 - WEP/Getty Images; **p158t - p160:** iStockphoto; **p161:** Copper Age/Getty Images; **p162 & p163t:** iStockphoto; **p163b:** Steve Morgan/Alamy.

The publishers would like to thank the following for permissions to use their statistics:

p91: Open University; **p95:** Office of National Statistics.

Although we have made every effort to trace and contact all copyright holders before publication this has not been possible in all cases. If notified, the publisher will rectify any errors or omissions at the earliest opportunity.

Links to third party websites are provided by Oxford in good faith and for information only. Oxford disclaims any responsibility for the materials contained in any third party website referenced in this work.

Biology   Contents

How to use this book

This book has been written for you by experienced teachers and subject experts. It covers what you need to know for your exams and is packed full of features to help you achieve the very best that you can.

Figure 1 *Many diagrams are as important for you to learn as the text, so make sure you revise them carefully*

Key words are highlighted in the text. You can look them up in the glossary at the back of the book if you are not sure what they mean.

Required practical

This feature helps you to become familiar with key practicals. It may be a simple introduction, a reminder, or the basis for a practical in the classroom.

Summary questions

These questions give you the chance to test whether you have learnt and understood everything in the topic. If you get any wrong, go back and have another look. They are designed to be increasingly challenging.

And at the end of each chapter you will find …

Chapter summary questions

These will test you on what you have learnt throughout the whole chapter, helping you to work out what you have understood and where you need to go back and revise.

Practice questions

These questions are examples of the types of questions that you will answer in your actual exam, so you can get lots of practice during your course.

Ext Extension Tier material is shown with this label. This will not be included in the Core Tier exams.

Key points

At the end of the topic are the important points that you must remember. They can be used to help with revision and summarising your knowledge.

Learning objectives

Each topic begins with key statements that you should know by the end of the lesson.

Study tip

These hints give you important advice on things to remember and what to watch out for.

 ## Did you know …?

There are lots of interesting and often strange facts about science. This feature tells you about many of them.

∞ links

Links will tell you where you can find more information about what you are learning and how different topics link up.

Practical skills

During this course, you will develop your understanding of the scientific process and the skills associated with scientific enquiry. Practical work is an important part of the course as it develops these skills and in addition it reinforces concepts and knowledge developed during the course.

As part of this course, you are expected to undertake practical work in many topics and must carry out the three required practicals listed below:

Required practicals

1 Investigating how variables effect the rate of photosynthesis.
2 Investigating the effects of exercise on the human body.
3 Investigating the effect of disinfectants and antibiotics on uncontaminated cultures of microorganisms.

In Paper 2, you will be assessed on aspects of the practical skills listed below, and may be required to read and interpret information from scales given in diagrams and charts, present data in appropriate formats, design investigations, and evaluate information that is presented to you.

Designing a practical procedure

- Design a practical procedure to answer a question, solve a problem, or test a hypothesis.
- Comment on/evaluate plans for practical procedures.
- Select suitable apparatus for carrying out experiments accurately and safely.

Control

- Appreciate that, unless certain variables are controlled, experimental results may not be valid.
- Recognise the need to choose appropriate sample sizes, and study control groups where necessary.

Risk assessment

- Identify possible hazards in practical situations, the risks associated with these hazards, and methods of minimising the risks.

Collecting data

- Make and record observations and measurements with appropriate precision and record data collected in an appropriate format (such as a table, chart, or graph).

Analysing data

- Recognise and identify the cause of anomalous results and suggest what should be done about them.
- Appreciate when it is appropriate to calculate a mean, calculate a mean from a set of at least three results, and recognise when it is appropriate to ignore anomalous results in calculating a mean.
- Recognise and identify the causes of random errors and systematic errors.
- Recognise patterns in data, form hypotheses, and deduce relationships.
- Use and interpret tabular and graphical representations of data.

Making conclusions

- Draw conclusions that are consistent with the evidence obtained and support them with scientific explanations.

Evaluation

- Evaluate data, considering its repeatability, reproducibility, and validity in presenting and justifying conclusions.
- Evaluate methods of data collection and appreciate that the evidence obtained may not allow a conclusion to be made with confidence.
- Suggest ways of improving an investigation or practical procedure to obtain extra evidence to allow a conclusion to be made.

1.1 Animal and plant cells

Earth is covered with a great variety of living things. However, they all have one thing in common – they are all made up of cells. Most cells are very small and you can only see them using a microscope. Eggs are the biggest animal cells. Unfertilised ostrich eggs are the biggest of all – they have a mass of around 1.3 kg and you certainly don't need a microscope to see them! The **light microscopes** in schools may magnify things several hundred times. Scientists have found out even more about cells using **electron microscopes**. These can magnify objects more than a hundred thousand times.

Most of the organisms you see around you are **eukaryotes**. This includes all animals and plants. Many microorganisms are **prokaryotes**. You will compare **eukaryotic cells** and **prokaryotic cells** on page 5.

Animal cells – structure and function

All eukaryotic cells have some features in common. You can see these clearly in animal cells. Human cells have the same features as other animal cells, and so do the cells of most other living things.

- The **nucleus** – controls all the activities of the cell. It contains the **genes** on the **chromosomes** that carry the instructions for making the proteins needed to build new cells or new organisms.
- The **cytoplasm** – a liquid gel in which most of the chemical reactions needed for life take place, for example, the first stages of respiration.
- The **cell membrane** – controls the passage of substances such as glucose and mineral **ions** into the cell. It also controls the movement of substances such as urea or hormones out of the cell.
- The **mitochondria** – structures in the cytoplasm where oxygen is used and where most of the energy is released during respiration.
- **Ribosomes** – where **protein synthesis** takes place, making all the proteins needed in the cell.

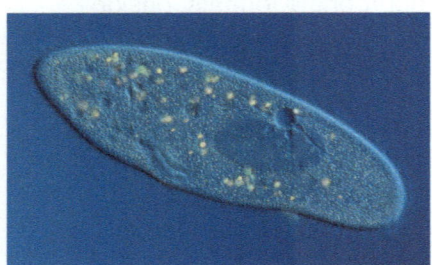

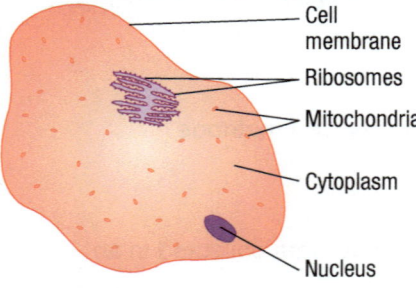

Cell membrane
Ribosomes
Mitochondria
Cytoplasm
Nucleus

Figure 1 *Diagrams of cells are much easier to understand than the real thing seen under a microscope. The top picture shows an animal cell magnified ×2000 times under an electron microscope. Below it is the way that a model animal cell is drawn to show the main features common to most living cells, including those in humans*

Plant cells – structure and function

Plants are very different organisms from animals. They make their own food by photosynthesis. They stay in one place, and do not move their whole bodies about from one place to another.

Plant cells have all the features of a typical animal cell, but they also contain features that are needed for their very different way of life. **Algae** are simple aquatic organisms. They also make their own food and have many similar features to plant cells. For centuries they were classified as plants, but now they are part of a different kingdom.

Figure 2 *Algal cells contain a nucleus and chloroplasts so they can photosynthesise*

All plant and **algal cells** have:

- a **cell wall** made of **cellulose** that strengthens the cell and gives it support.

Many (but not all) plant cells also have these other features:

- **Chloroplasts** are found in all the green parts of the plant. They are green because they contain the green substance **chlorophyll**. Chlorophyll absorbs light energy to make food by photosynthesis. Root cells do not have chloroplasts because they are underground and do not photosynthesise.
- A **permanent vacuole** is a space in the cytoplasm filled with cell sap. This is important for keeping the cells rigid to support the plant.

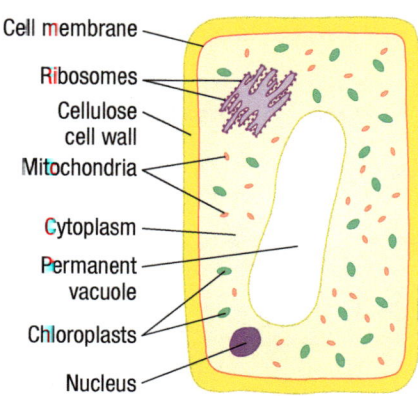

Cell membrane
Ribosomes
Cellulose cell wall
Mitochondria
Cytoplasm
Permanent vacuole
Chloroplasts
Nucleus

Figure 3 *A plant cell has many features in common with an animal cell, as well as other features that are unique to plants*

Practical

Looking at cells

Set up a microscope to look at plant cells, for example, from onions, *Elodea,* and/or algal cells. You should see the cell wall, the cytoplasm, and sometimes a vacuole. You will see chloroplasts in the *Elodea* and the algae, but not in the onion cells because they do not photosynthesise.

Figure 4 *Some of the common features of plant cells show up well under a light microscope. Here, the features are magnified ×40*

Summary questions

1 a List the main structures you would expect to find in an animal cell.
  b You would find all the things that are present in animal cells in a plant cell or algal cell, too. There are three extra features that may be found in plant cells but not in animal cells. What are they?
  c What are the main functions of these three extra structures?

2 Why are the nucleus and the mitochondria so important in all cells?

3 Chloroplasts are found in many plant cells but not in all of them. Give an example of plant cells without chloroplasts, and explain why they have none.

1.2 Eukaryotes and prokaryotes

Using units

1 km = 1000 m
1 m = 100 cm
1 cm = 10 mm
1 mm = 1000 μm (micrometres)

Bacteria are single-celled living organisms that are much smaller than animal and plant cells. Most bacteria are less than 1 μm in length. They are prokaryotic cells. You could fit hundreds of thousands of bacteria onto the full stop at the end of this sentence, so you can't see individual bacteria without a powerful microscope.

When you culture bacteria on an agar plate, you grow many millions of bacteria. This enables you to see the **bacterial colony** with your naked eye.

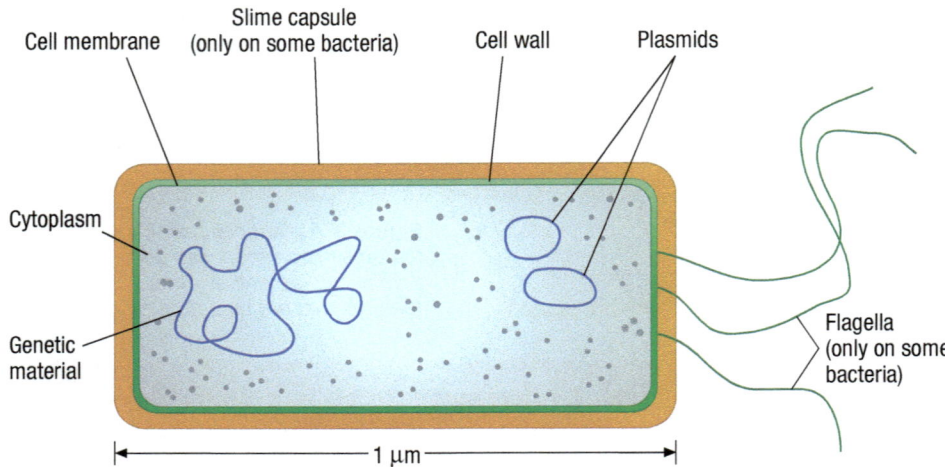

Figure 1 *Bacteria come in a variety of shapes, but they all have the same basic structure*

Bacterial cells

Each bacterium is a single cell. It is made up of cytoplasm surrounded by a membrane and a cell wall. Inside the bacterial cell is the **genetic material**. Unlike animal, plant, and algal cells, the genes are not contained in a nucleus. The long strand of **DNA** (the bacterial chromosome) is usually circular and is found free in the cytoplasm.

Many bacterial cells also contain **plasmids**, which are small, circular bits of DNA. These carry extra genetic information. Plasmids are widely used by scientists in the process of genetic engineering.

Some bacteria have specialised structures. A slime, capsule around the outside of the cell wall protects some bacteria from your immune system. Others have tiny whip-like threads called flagella to help them move around.

Although some bacteria cause disease, many are harmless. Some are actually really useful to humans. People use them to make foods like yoghurt and cheese. Others are used in sewage treatment and to make medicines. Bacteria are vital as decomposers in food chains and webs, and in natural cycles such as the carbon and nitrogen cycles. They are also an important part of a healthy gut.

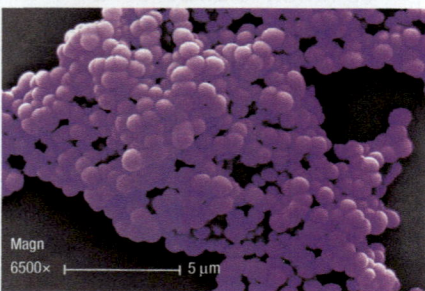

Figure 2 *Bacteria come in several different shapes and sizes. This helps us to identify them under the microscope*

Comparing eukaryotic cells with prokaryotic cells

As you have seen, all living things are made of cells. Prokaryotic organisms such as bacteria are often made of single cells. Eukaryotic organisms are often multicellular, including plants, fish, and people.

- **Prokaryotic cells** are smaller and simpler than eukaryotic cells. They are often an order of magnitude smaller than plant cells, for example. They do not have a nucleus or any other membrane-bound organelles. They have a cell wall, but it is *not* made of cellulose.
- **Eukaryotic cells** are larger and more complex that prokaryotic cells. They all have a nucleus, although a few types, such as red blood cells, lose their nucleus as they grow and mature. They have many different cell organelles, all surrounded by membranes. Plant cells have cell walls made of cellulose.

The main similarities and differences are summarised in the following table:

Prokaryotic cells	Eukaryotic cells
Very small cells often less than 5 μm	Bigger cells usually between 10 and 100 μm
No nucleus – loop of DNA	Membrane-bound nucleus containing DNA
No membrane-bound organelles	Membrane-bound organelles
Cell wall not made of cellulose	Animal cells have no cell wall, plant cells have cellulose cell wall
Some genes may be in separate circular structures called plasmids	No plasmids

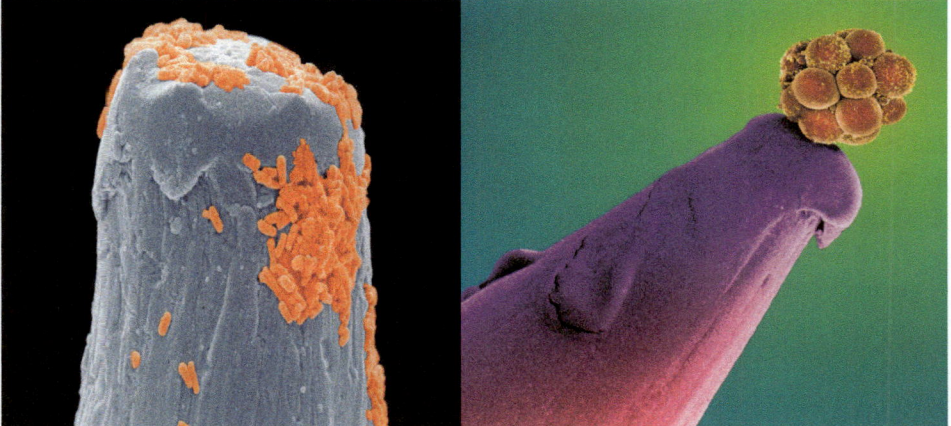

Figure 3 *These scanning electron micrographs show prokaryotic bacterial cells on the right and eukaryotic human cells on the left, balanced on the end of a pin. The bacteria are magnified more than ×1500, but the human cells are only magnified around ×300 – yet they are still much bigger*

Summary questions

1 a What is unusual about the genetic material in bacterial cells?
   b Which are bigger, bacterial cells or human cells?
   c What is the difference between the cell walls of plant cells and the cell walls of bacteria?

2 Explain how bacteria are both useful and/or damaging to people.

3 Make a table to compare the structures in animal, plant, algal, and bacterial cells.

?? Did you know …?

When making sourdough bread, a mixture of yeast and bacteria is used to provide a natural raising agent. This gives the bread its typical sharp taste.

Study tip

Be clear about the similarities and differences between animal, plant, algal, and bacterial cells.

Key points

- A bacterial cell consists of cytoplasm and a membrane surrounded by a cell wall. The genes are not in a distinct nucleus. Some of the genes are in circular structures called plasmids.
- Animal and plant cells are eukaryotic cells and bacterial cells are prokaryotic cells.

1.3

Specialised cells

Learning objectives

After this topic, you should know:

- that cells may be specialised to carry out a particular function.

∞ links

You can find out much more about the organisation of specialised cells into tissues, organs, and organ systems in 1.4 Tissues and organs and 1.5 Organ systems.

??? Did you know … ?

An adult human who is not overweight will typically have 30 to 50 billion fat cells in their body.

The smallest living organisms are single cells. They can carry out all of the functions of life. These functions range from feeding and respiration to excretion and reproduction.

Most organisms are bigger and are made up of lots of cells. Some of these cells become **specialised** in order to carry out particular jobs.

When a cell becomes specialised, its structure is adapted to suit the particular job it does. As a result, specialised cells often look very different to a typical plant or animal cell. Sometimes cells become so specialised that they only have one function within the body. Examples of this include sperm cells, egg cells, red blood cells, and nerve cells. Some specialised cells, such as egg and sperm cells, work individually. Others are adapted to work as part of a tissue, an organ, or a whole organism.

Fat cells

If you eat more food than you need, your body makes fat and stores it in fat cells. The fat can be broken down and used to transfer energy when it is needed. Fat cells help animals, including humans, to survive when food is in short supply. Thousands of fat cells together form adipose tissue.

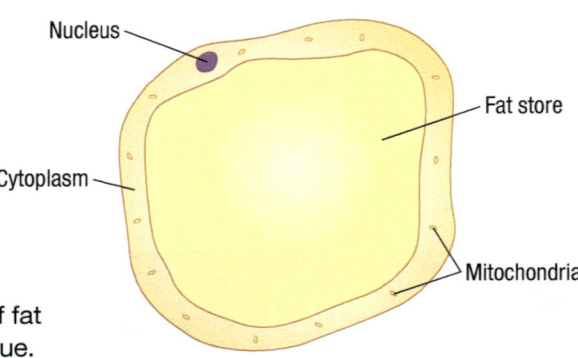

Fat cells have three main adaptations:

- They have a small amount of cytoplasm and large amounts of fat.
- They have few mitochondria as the cell needs very little energy.
- They can expand – a fat cell can end up 1000 times its original size as it fills up with fat.

Cone cells from human eye

There are cone cells in the light-sensitive layer of your eye (the retina). They make it possible for you to see in colour.

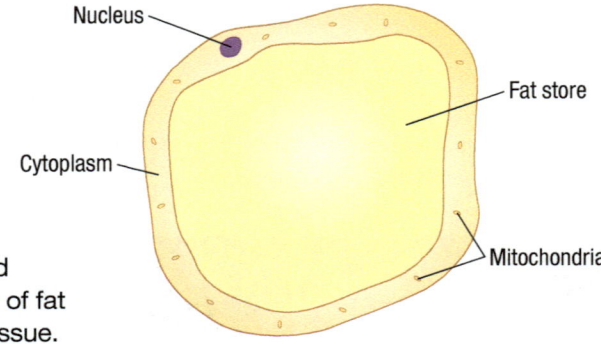

Cone cells have three main adaptations:

- The outer segment contains a special chemical, a visual pigment, which changes chemically in coloured light. It needs energy to change back to its original form. The visual pigments are based on the vitamin A in your diet.
- The middle segment is packed full of mitochondria. The mitochondria transfer the energy needed to reform the visual pigment. This lets you see continually in colour.
- The final part of the cone cell is made up of specialised synapses that connect to the **optic nerve**. When coloured light makes your visual pigment change, a nerve impulse is triggered. This makes its way along the optic nerve to your brain.

Root hair cells

You find **root hair cells** close to the tips of growing roots. Plants need to take in lots of water (and dissolved mineral ions). The root hair cells help them to take up water and mineral ions more efficiently. The water and mineral ions then pass easily across the root to the **xylem tissue**. The xylem tissue carries water and **mineral ions** up into the rest of the plant. Mineral ions are moved into the cell by active transport.

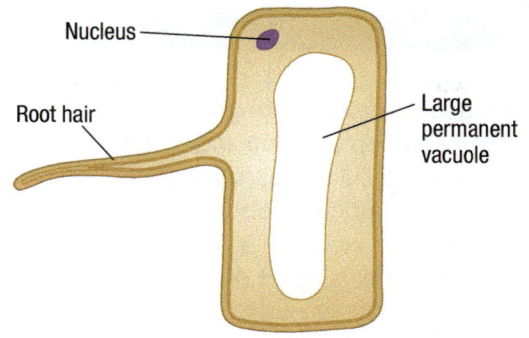

Nucleus

Root hair

Large permanent vacuole

Root hair cells have two main adaptations:
- The root hairs increase the surface area for water to move into the cell.
- The root hair cells have a large permanent vacuole that speeds up the movement of water by osmosis from the soil across the root hair cell.

Sperm cells

Sperm cells are usually released a long way from the egg they are going to fertilise. They contain the genetic information from the male parent. Depending on the type of animal, sperm cells need to move through water or the female reproductive system to reach an egg. Then they have to break into the egg.

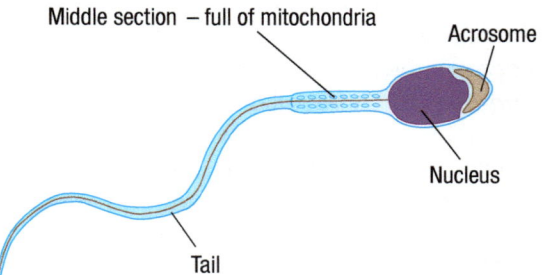

Middle section – full of mitochondria

Acrosome

Nucleus

Tail

Sperm cells have several adaptations to make all this possible:
- A long tail whips from side to side and helps move the sperm towards the egg.
- The middle section is full of mitochondria, which provide the energy for the tail to work.
- The acrosome stores digestive enzymes for breaking down the outer layers of the egg.
- A large nucleus contains the genetic information to be passed on.

⚭ links

You can find out more about osmosis in 1.7 Osmosis and about active transport in 1.8 Active transport. You will find out more about the specialised cells of the phloem and xylem in plants in 9.4 Transport systems in plants.

Summary questions

1 Make a table to explain how the structure of each cell discussed in this topic is adapted to its function.

2 a Muscle cells can contract (shorten) and are used to move the body around and also to move substances around your body. Muscle cells usually contain many mitochondria. Explain why this is an important adaptation.

  b The palisade cells are found near the top surface of a leaf. They contain many chloroplasts. Why is this an important adaptation?

3 Explain the types of features you would look for to decide on the function of an unknown specialised cell.

Key points

- Cells may be specialised to carry out a particular function.
- Examples of specialised cells are fat cells, cone cells, root hair cells, and sperm cells.
- Cells may be specialised to work as tissues, organs, or whole organisms.

1.4

Tissues and organs

Learning objectives

After this topic, you should know:

- how specialised cells become organised into tissues

- how several different tissues work together to form an organ.

Large **multicellular organisms** have to overcome problems linked to their size. They develop different ways of exchanging materials. During the development of a multicellular organism, cells **differentiate**. They become specialised to carry out particular jobs. For example, in animals, muscle cells have a different structure to blood cells and nerve cells. In plants, the cells where photosynthesis takes place are very different to root hair cells.

However, the adaptations of multicellular organisms go beyond specialised cells. Similar specialised cells are often found grouped together to form a tissue.

Tissues

A **tissue** is a group of cells with similar structure and function working together. **Muscular tissue** can contract to bring about movement. **Glandular tissue** contains secretory cells that can produce and **secrete** (release) substances such as enzymes and hormones. **Epithelial tissue** covers the outside of your body as well as your internal organs.

Plants have tissues too. **Epidermal tissues** cover the surfaces and protect them. **Palisade mesophyll** contains lots of chloroplasts and can carry out photosynthesis, whilst **spongy mesophyll** has some chloroplasts for photosynthesis but also has big air spaces and a large surface area to make the diffusion of gases easier. **Xylem** and **phloem** are the transport tissues in plants. They carry water and dissolved mineral ions from the roots up to the leaves and transport dissolved food from the leaves around the plant.

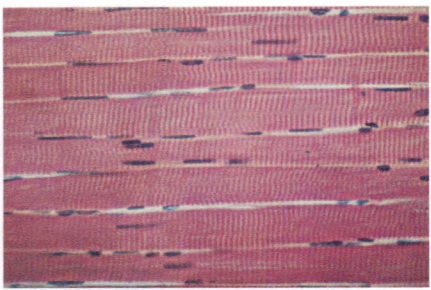

Figure 1 *Muscle tissue contracts to move your skeleton around*

Organs

Organs are made up of tissues. One organ can contain several tissues, all working together. For example, the stomach is an organ involved in the digestion of your food. It contains:

- muscular tissue, to churn the food and **digestive juices** together and move the contents through the digestive system
- glandular tissue, to produce the digestive juices that break down food
- epithelial tissue, which covers the inside and the outside of the organ.

links

For more information on specialised cells, see 1.3 Specialised cells.

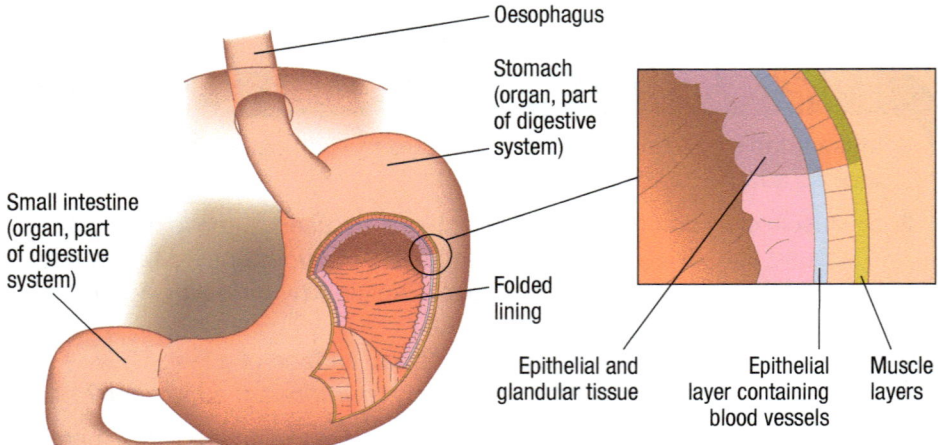

Oesophagus

Stomach (organ, part of digestive system)

Small intestine (organ, part of digestive system)

Folded lining

Epithelial and glandular tissue

Epithelial layer containing blood vessels

Muscle layers

Figure 2 *The stomach contains several different tissues, each with a different function in the organ*

The pancreas is an organ that has two important functions. It makes hormones to control our blood glucose, as well as some of the enzymes that digest our food. It contains two very different types of glandular tissue to produce these different secretions.

Plant organs

Animals are not the only organisms to have organs – plants do, too.

Plants have differentiated cells that form specialised tissues. Within the body of a plant, tissues such as the palisade and spongy mesophyll, xylem, and phloem are arranged to form organs. Each organ carries out its own particular functions.

Plant organs include the leaves, stems, and roots, each of which has a very specific job to do (Figure 3).

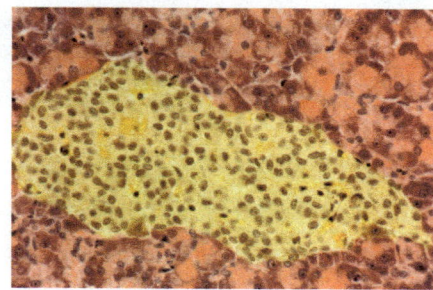

Figure 4 *The pancreas, showing two different types of glandular tissue*

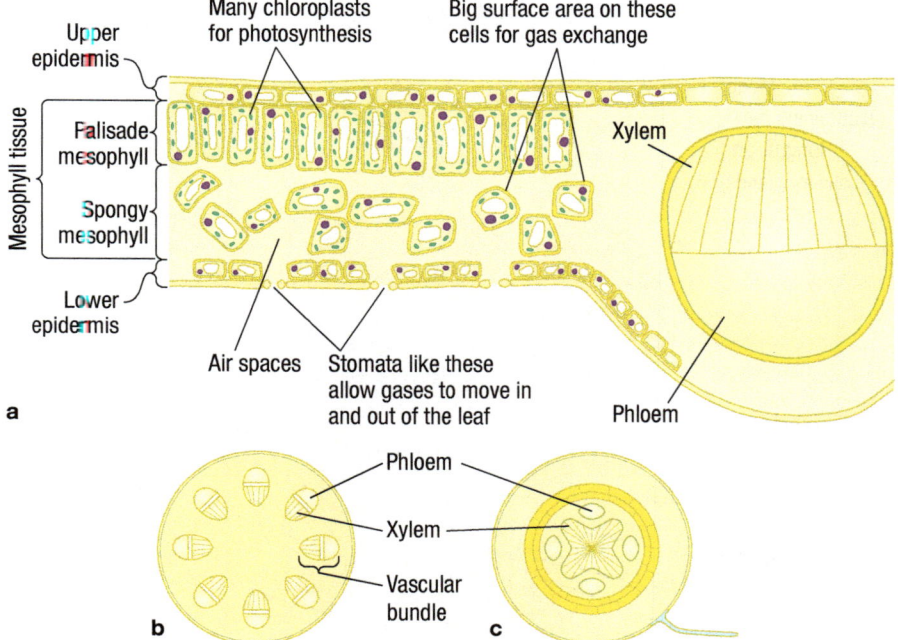

Figure 3 *Plants have specific tissues to carry out particular functions. They are arranged in organs such as **a** the leaf, **b** the stem, and **c** the roots*

To summarise, whether in a plant or an animal, an organ is a collection of different tissues working together to carry out important functions for the organism.

??? **Did you know … ?**

Some trees, such as the giant redwood, have trunks that are over 40 m tall. A plant cell is about 100 μm long. So the plant organ is 400 000 times bigger than the individual cells.

??? **Did you know … ?**

A human liver cell is about 10 μm (1×10^{-5} m) in diameter. A human liver is about 22 cm across. It contains a lot of liver cells!

Summary questions

1 a What is a tissue?
 b What is an organ?

2 State whether each of the following is a specialised cell, a tissue, or an organ and explain your answer:
 a sperm
 b kidney
 c stomach.

3 a Explain how the tissues in a leaf are arranged to form an effective organ for photosynthesis.
 b Explain how the stomach is adapted for its role in the digestion of food.

Key points

- A tissue is a group of cells with similar structure and function.

- Organs are made of tissues. One organ may contain several types of tissue.

- Animal organs include the stomach and the heart.

- Plant organs include stems, roots, and leaves.

1.5

Organ systems

After this topic, you should know:

- what makes up an organ system

- the main organs of the digestive system.

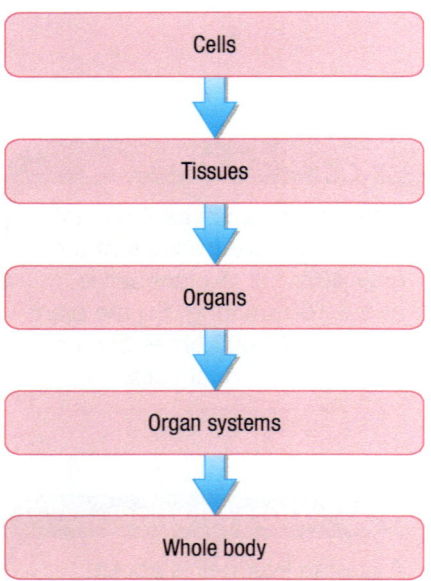

Figure 1 *Larger multicellular organisms have many levels of organisation*

A whole multicellular organism is made up of a number of **organ systems** working together. Organ systems are groups of organs that all work together to perform a particular function. The way one organ functions often depends on other organs in the system. The human digestive system is a good example of an organ system.

The digestive system

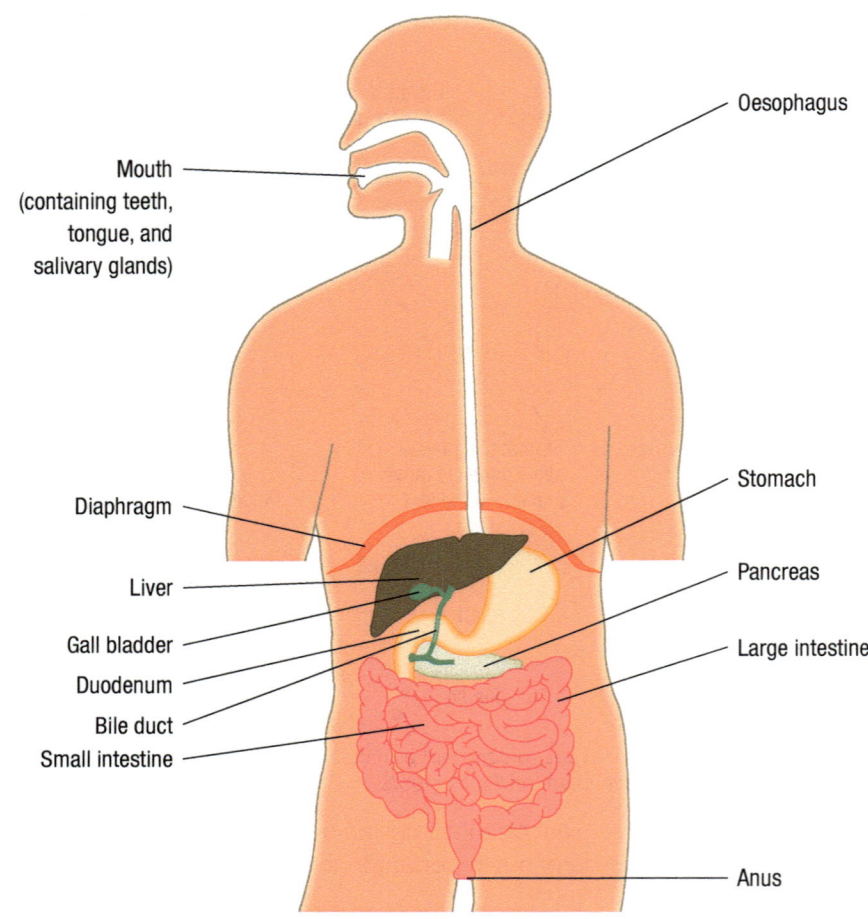

Figure 2 *The main organs of the human digestive system*

The **digestive system** is one of several organ systems that make it possible for mammals and other animals to exchange substances with the environment. The food you eat is made up of large insoluble **molecules**. Your body cannot take these molecules in and use them. They need to be digested (broken down) into smaller, soluble molecules that can be absorbed into your bloodstream and used by your cells. This is the function of your digestive system. The digestive system is made up of many different organs that all work together. They include:

Part of the digestive system	Function
Glands (e.g. salivary glands, the pancreas)	Produce digestive juices containing enzymes that chemically break down food molecules
Stomach	Where the digestion of protein in food takes place
Liver	Produces bile, which helps in the digestion of lipids (fats)
Small intestine	Where digestion takes place and where soluble food is absorbed into the bloodstream
Large intestine	Where water is absorbed from undigested food, producing faeces

??? Did you know … ?

The human digestive system is between 6 m and 9 m long. That is about 9 million times longer than an average human cell!

You will learn a lot more about the human digestive system and how it works in Chapter 5. Other human organ systems include:

- the breathing system, which allows the body to exchange the gases oxygen and carbon dioxide with the environment
- the cardiovascular system (the heart and the blood vessels), which carries substances to and from the cells of the body.

Plant organ systems

In small plants, the whole body works as an organ system with the roots, stems, and leaves working together to allow the plant to exchange substances with the environment. For example, water comes into the plant through the roots, is carried through the plant in the stems, and is lost into the environment again through the leaves. You will learn more about how plant organ systems work in Chapter 9.

links

You will find out more about the adaptations of the villi in the small intestine as an exchange surface in 5.6 Exchange in the gut, and about the role of the liver and bile in the digestion of food in 5.5 Making digestion efficient.

Figure 3 *The organs of a plant*

Summary questions

1 Match each organ (A–D) to its correct function (1–4):

A	Stem	1	Breaking down large insoluble molecules into smaller soluble molecules
B	Root	2	Photosynthesising in plants
C	Small intestine	3	Providing support in plants
D	Leaf	4	Anchoring plants and obtaining water and minerals from soil

2 Explain the difference between organs and organ systems, giving two examples.

3 Using the human digestive system as an example, explain how the organs in an organ system rely on each other to function properly.

Key points

- Organ systems are groups of organs that perform a particular function.
- The digestive system in a mammal is an example of a system where substances are exchanged with the environment.

1.6

Diffusion

After this topic, you should know:

- how diffusion takes place and why it is important in living organisms

- what affects the rate of diffusion.

Your cells need to take in substances such as glucose and oxygen for respiration. Cells also need to get rid of waste products and release chemicals that are needed elsewhere in your body. Dissolved substances and gases can move into and out of your cells across the cell membrane. One of the main ways in which they move is by **diffusion**.

Diffusion

Diffusion is the spreading out of the particles of a gas, or of any substance in solution (a **solute**). This results in the **net movement** (overall movement) of particles. The net movement is from an area of high concentration to an area of lower concentration. It takes place because of the random movement of the particles. The motion of the particles causes them to bump into each other, and this moves them all around.

Imagine a room containing a group of boys and a group of girls. If everyone closes their eyes and moves around briskly but randomly, children will bump into each other. They will scatter until the room contains a mixture of boys and girls. This gives you a good model of diffusion (Figure 1).

Figure 1 *The random movement of particles results in substances spreading out, or diffusing, from an area of higher concentration to an area of lower concentration*

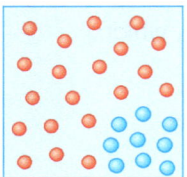 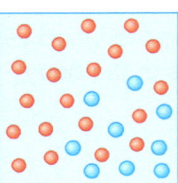 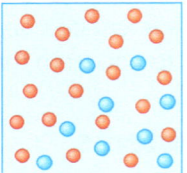 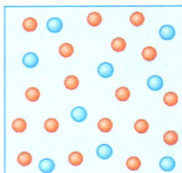

At the moment when the blue particles are added to the red particles they are not mixed at all	As the particles move randomly, the blue ones begin to mix with the red ones	As the particles move and spread out, they bump into each other. This helps them to keep spreading randomly	Eventually, the particles are completely mixed and diffusion is complete

Rate of diffusion

If there is a big difference in concentration between two areas, diffusion will take place quickly. Many particles will move randomly towards the area of low concentration. Only a few will move randomly in the other direction.

However, if there is only a small difference in concentration between two areas, the net movement by diffusion will be quite slow. The number of particles moving into the area of lower concentration by random movement will only be slightly more than the number of particles that are leaving the area.

net movement = particles moving in – particles moving out

In general, the greater the difference in concentration, the faster the rate of diffusion. This difference between two areas of concentration is called the **concentration gradient**. The bigger the difference, the steeper the concentration gradient and the faster the rate of diffusion. In other words, diffusion occurs *down* a concentration gradient (Figure 2).

Temperature also affects the rate of diffusion. An increase in temperature means the particles in a gas or a solution move around more quickly. When this happens, diffusion takes place more rapidly as the random movement of the particles speeds up.

Both types of particles can pass through this membrane – it is freely permeable

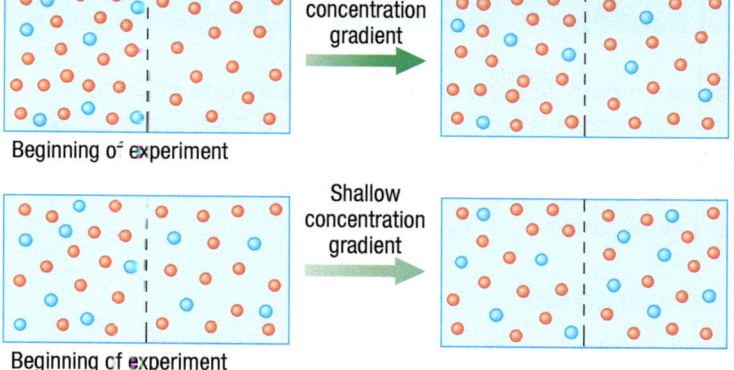

Beginning of experiment

Steep concentration gradient

Random movement means three blue particles have moved from left to right by diffusion

Beginning of experiment

Shallow concentration gradient

Four blue particles have moved from left to right as a result of random movement – but two have moved from right to left. There is a **net** movement of **two** particles to the right by diffusion

Figure 2 *This diagram shows the effect of concentration on the rate of diffusion. This is why so many body systems are adapted to maintain steep concentration gradients*

Diffusion in living organisms

Dissolved substances move into and out of your cells by diffusion across the cell membrane. These substances include **simple sugars** such as glucose, gases such as oxygen, and waste products such as urea from the breakdown of amino acids in your liver.

The oxygen you need for **respiration** passes from the air into your lungs. From the lungs it enters your red blood cells through the cell membranes by diffusion. The oxygen moves down a concentration gradient from a region of high oxygen concentration to a region of lower oxygen concentration. Oxygen then also moves by diffusion, down a concentration gradient, from the blood cells into the cells of the body where it is needed.

Carbon dioxide moves by diffusion, down a concentration gradient, from the body cells into the red blood cells and then into the air in the lungs in a similar way.

Individual cells may be adapted to make diffusion easier and more rapid. The most common adaptation is to increase the surface area of the cell membrane. Increasing the surface area means there is more room for diffusion to take place. By folding up the membrane of a cell, or the tissue lining an organ, the area over which diffusion can take place is greatly increased. Therefore the rate of diffusion is also greatly increased, so that much more of a substance moves by diffusion in a given time.

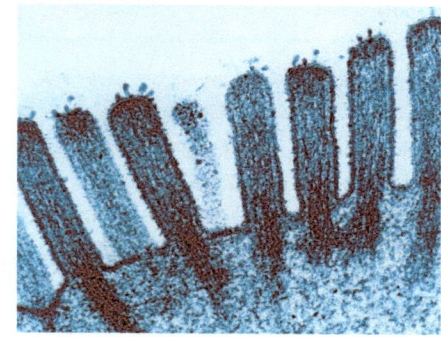

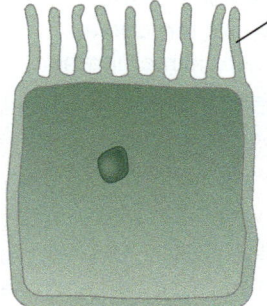

Infoldings of the cell membrane form microvilli, which increase the surface area of the cell

Figure 3 *An increase in the surface area of a cell membrane means diffusion can take place more quickly. This is an intestinal cell*

Key points

- Diffusion is the net movement of particles from an area where they are at a high concentration to an area where they are at a lower concentration, down a concentration gradient.

- The greater the difference in concentration, the faster the rate of diffusion.

- Dissolved substances such as glucose and gases such as oxygen move in and out of cells by diffusion.

Summary questions

1 Explain the process of diffusion in terms of the particles involved.

2 a Explain why diffusion takes place faster when there is an increase in temperature.
  b Explain in terms of diffusion why so many cells have folded membranes along at least one surface.

3 Explain the following statements in terms of diffusion:
  a Digested food products move from the inside of your gut into the bloodstream.
  b Carbon dioxide moves from the blood in the capillaries in your lungs to the air in the lungs.
  c Male moths can track down a mate from up to 3 miles away because of the special chemicals produced by the female.

1.7 Osmosis

Study tip

Remember, all particles can diffuse from an area of high concentration to an area of lower concentration, provided they are *soluble* and *small* enough to pass through the membrane. Osmosis refers only to the diffusion of water through a partially permeable membrane.

Diffusion takes place when particles can spread freely from one place to another. However, the solutions inside cells are separated from those outside by the cell membrane. This membrane does not let all types of particles through. Membranes that only let some types of particles through are called **partially permeable membranes**.

The process of osmosis

Partially permeable cell membranes let water move across them. Remember:

- A *dilute* solution of sugar contains a *high* concentration of water (the solvent). It has a *low* concentration of sugar (the solute).
- A *concentrated* sugar solution contains a relatively *low* concentration of water and a *high* concentration of sugar.

The cytoplasm of a cell is made up of chemicals dissolved in water inside a partially permeable bag of cell membrane. The cytoplasm contains a fairly concentrated solution of salts and sugars. Water moves from a dilute solution (with a high concentration of water molecules) to a concentrated solution (with fewer water molecules) across the membrane of the cell.

This special type of diffusion, where water moves across a partially permeable membrane, is called **osmosis**.

Practical

Investigating osmosis

You can make model cells using bags made of partially permeable membrane (Figure 1). You can see what happens to them if the concentrations of the solutions inside or outside the 'cells' change.

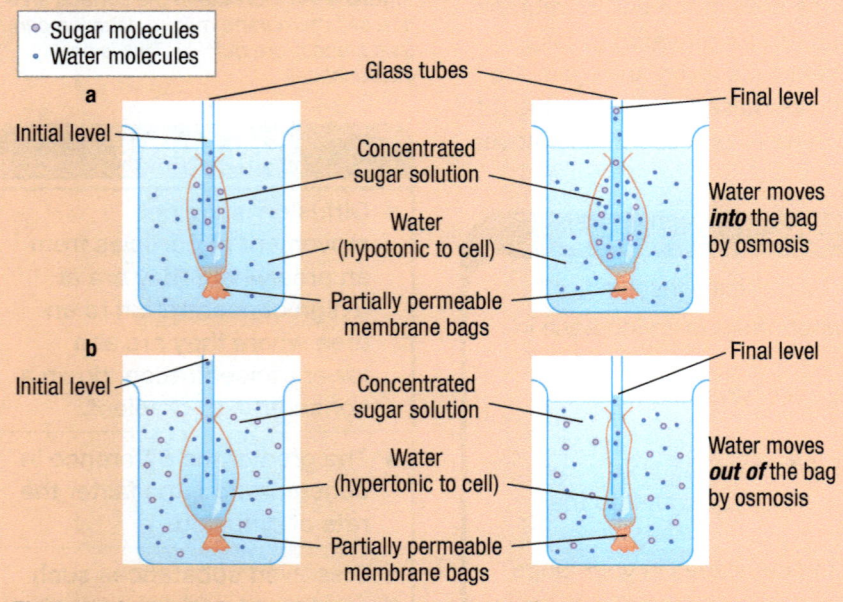

Figure 1 *A model of osmosis in* **a** *cell. In* **a** *the 'cell' contents are more concentrated than the surrounding solution. In* **b** *the 'cell' contents are less concentrated than the surrounding solution*

The concentration inside your body cells needs to stay the same for them to work properly. However, the concentration of the solutions outside your cells may be very different to the concentration inside them. This concentration gradient can cause water to move into or out of the cells by osmosis.

- If the concentration of solutes in the solution outside the cell is *lower* than the concentration inside the cell, the solution is **hypotonic** to the cell.
- If the concentration of solutes in the solution outside the cell is *the same* as the concentration inside the cell, the solution is **isotonic** to the cell.
- If the concentration of solutes in the solution outside the cell is *higher* than the concentration inside the cell, the solution is **hypertonic** to the cell.

Osmosis in animals

If a cell uses up water in its chemical reactions, the cytoplasm becomes more concentrated. The surrounding fluid becomes hypotonic and more water immediately moves in by osmosis.

If the cytoplasm becomes too dilute because more water is made in chemical reactions, the surrounding fluid becomes hypertonic and water leaves the cell by osmosis. So osmosis restores the balance in both cases.

However, osmosis can also cause big problems in animal cells. If the solution outside the cell becomes much more dilute than the cell contents (hypotonic), a lot of water will move into the cell by osmosis. The cell will swell and may burst.

If the solution outside the cell becomes more concentrated than the cell contents (hypertonic), water will move out of the cell by osmosis. The cytoplasm will become too concentrated and the cell will shrivel up. Then it can no longer survive.

Once you understand the effect osmosis can have on cells, the importance of maintaining constant internal conditions in the human body becomes clear.

Osmosis in plants

Plants rely on osmosis to support their stems and leaves. Water moves into plant cells by osmosis. This causes the vacuole to swell and press the cytoplasm against the plant cell walls. The pressure builds up until no more water can physically enter the cell – this pressure is known as turgor. Turgor pressure makes the cells hard and rigid, which in turn keeps the leaves and stems of the plant rigid and firm.

Plants need the fluid surrounding the cells to always be hypotonic to the cytoplasm, with a lower concentration of solutes and a higher concentration of water than the plant cells themselves. This keeps water moving by osmosis in the right direction and the cells are turgid. If the solution surrounding the plant cells is hypertonic to (more concentrated than) the cell contents, water will leave the cells by osmosis. The cells will no longer be firm and swollen – they become flaccid (soft) as there is no pressure on the cell walls. At this point, the plant wilts as turgor no longer supports the plant tissues.

If more water is lost by osmosis, the vacuole and cytoplasm shrink, and eventually the cell membrane pulls away from the cell wall. This is plasmolysis. Plasmolysis is usually only seen in laboratory experiments. Plasmolysed cells die quickly unless the osmotic balance is restored.

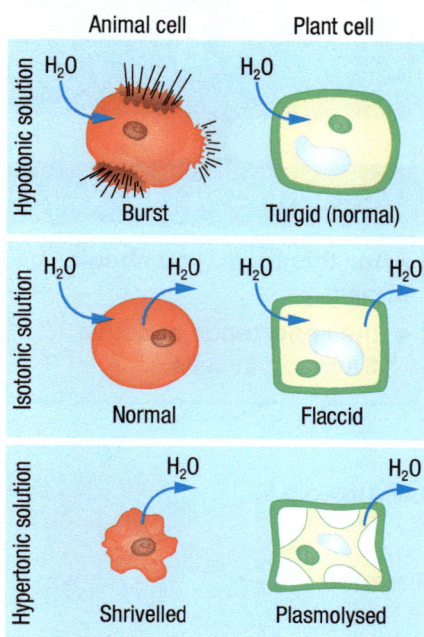

Figure 2 *Osmosis is important in all living organisms*

Study tip

When writing about osmosis, be careful to specify whether it is the concentration of water or solutes you are referring to. Simply saying 'higher concentration outside cell' will gain no marks!

Key points

- Osmosis is a special case of diffusion. It is the movement of water from a dilute to a more concentrated solution through a partially permeable membrane that allows water to pass through.
- Differences in the concentrations of solutions inside and outside a cell cause water to move into or out of the cell by osmosis.
- Osmosis is important to maintain turgor in plant cells. Animal cells can be damaged if the concentrations inside and outside the cells are not kept the same.

Summary questions

1 a What is the difference between osmosis and simple diffusion?
  b How does osmosis help to maintain the cytoplasm of plant and body cells at a specific concentration?

2 a Define the following terms:
    i isotonic solution   ii hypotonic solution   iii hypertonic solution.
  b Why is it so important for the cells of the human body that the solute concentration of the fluid surrounding the cells is kept as constant as possible?

3 Explain why osmosis is so important in the support systems of plants.

4 Animals that live in fresh water have a constant problem with their water balance. The single-celled organism called *Amoeba* has a special vacuole in its cell. It fills with water and then moves to the outside of the cell and bursts. A new vacuole starts forming straight away.
  Explain in terms of osmosis why the *Amoeba* needs one of these vacuoles.

1.8 Active transport

Cells need to move substances in and out. Water often moves across the cell boundaries by osmosis. Dissolved substances also need to move in and out of cells. There are two main ways in which this happens:

- Substances move by diffusion, down a concentration gradient. This must be in the right direction to be useful to the cells.
- Sometimes the substances needed by a cell have to be absorbed against a concentration gradient. This needs a special process called **active transport**.

Moving substances by active transport

Active transport allows cells to move substances from an area of low concentration to an area of high concentration. This movement is *against* the concentration gradient. As a result, cells can absorb ions from very dilute solutions. It also enables them to move substances, such as sugars and ions, from one place to another through the cell membranes.

It takes energy for the active transport system to carry a molecule across the membrane and then return to its original position (Figure 1). The energy for active transport comes from cellular respiration. Scientists have shown in a number of different cells that the rate of respiration and the rate of active transport are closely linked (Figure 2).

In other words, if a cell is releasing plenty of energy, it can carry out lots of active transport. Examples include root hair cells and the cells lining your gut. Cells involved in a lot of active transport usually have many mitochondria to provide the energy they need.

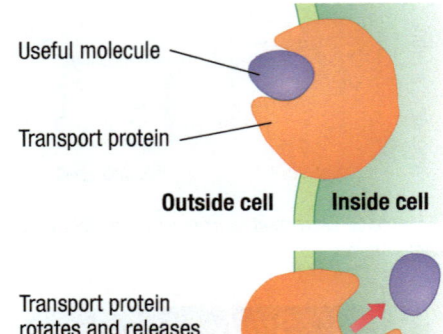

Useful molecule

Transport protein

Outside cell   **Inside cell**

Transport protein rotates and releases molecule inside cell (using energy)

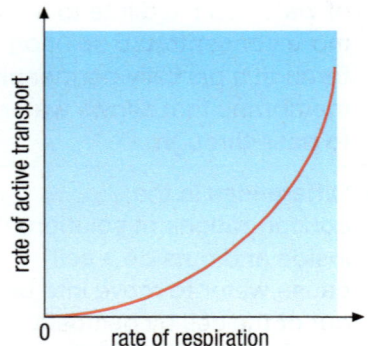

Transport protein rotates back again (often using energy)

Figure 1 *Active transport uses energy to move substances against a concentration gradient*

rate of active transport

0   rate of respiration

Figure 2 *The rate of active transport depends on the rate of respiration*

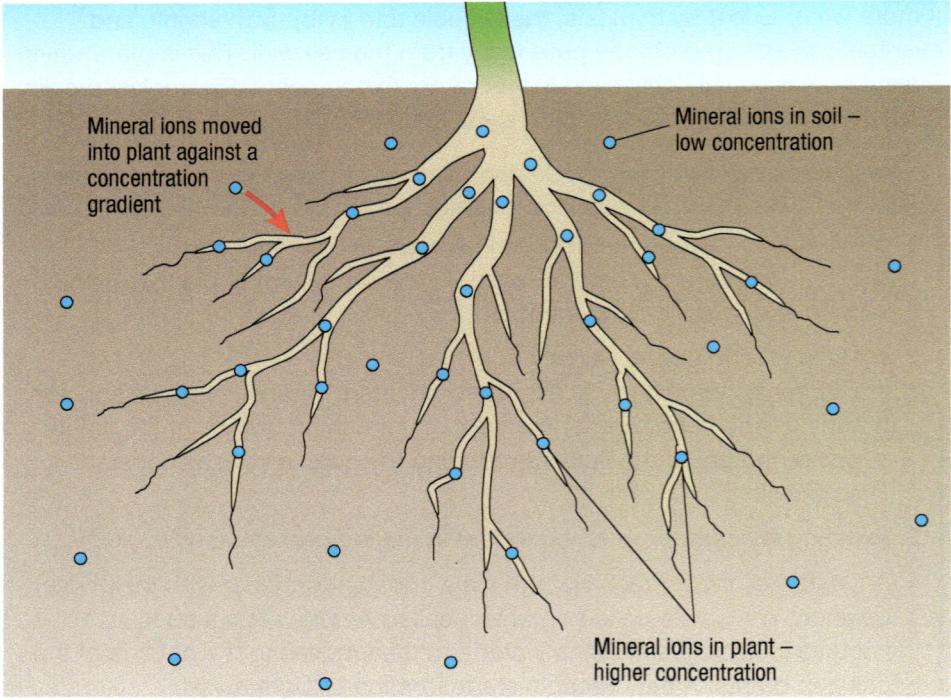

Mineral ions moved into plant against a concentration gradient

Mineral ions in soil – low concentration

Mineral ions in plant – higher concentration

Figure 3 *Plants use energy from respiration in active transport to move mineral ions from the soil into the roots against a concentration gradient*

The importance of active transport

Active transport is widely used in cells. There are some situations in which it is particularly important. For example, mineral ions in the soil, such as nitrate ions, are usually found in very dilute solutions. These solutions are more dilute than the solution within the plant cells. By using active transport, plants can absorb these mineral ions, even though it is against a concentration gradient (Figure 3).

Sugars, such as the simple sugar glucose, are actively absorbed out of your gut and **kidney tubules** into your blood. This is often done against a large concentration gradient.

Figure 4 *Some crocodiles have special salt glands in their tongues. These remove excess salt from the body against the concentration gradient by active transport. That's why members of the crocodile species* Crocodylus porosus *can live in estuaries and even in the sea*

 Did you know …?

People with cystic fibrosis (see 10.7 Inherited conditions in humans) have thick, sticky mucus in their lungs, guts, and reproductive systems. This is the result of a mutation affecting a protein involved in the active transport system of the mucus-producing cells.

Summary questions

1 Explain how active transport works in a cell.

2 a How does active transport differ from diffusion and osmosis?
  b Why do cells that carry out a lot of active transport also usually have many mitochondria?

3 Explain fully why active transport is so important to:
  a marine birds such as albatrosses, which have special salt glands producing very salty liquid
  b plants.

⬭ links

You can find out more about the absorption of glucose in the gut in 5.4 The digestive system.

Study tip

Do not refer to movement *along* a concentration gradient. Always refer to movement as *down* a concentration gradient (from high to low) for diffusion or osmosis and *against* a concentration gradient (from low to high) for active transport.

Key points

- Substances are sometimes absorbed against a concentration gradient by active transport.

- Active transport uses energy from respiration.

- Cells can absorb ions from very dilute solutions (e.g., root hair cells), and actively absorb substances such as sugars and salt against a concentration gradient, using active transport (e.g., glucose may be absorbed from low concentrations in the small intestine and the kidney tubules).

Summary questions

1

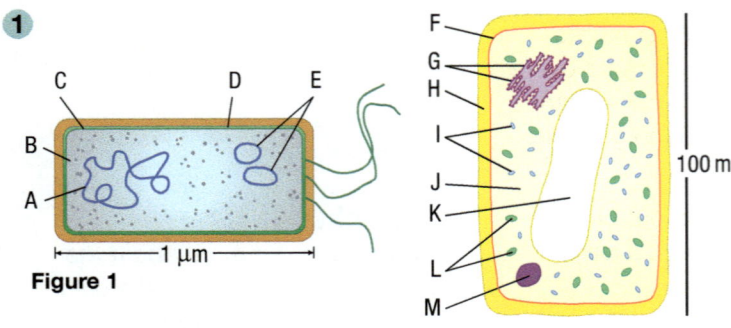

Figure 1

Figure 2

a Name the structures labelled A–E in the bacterial cell (Figure 1).

b Name the structures labelled F–M in the plant cell (Figure 2).

c Explain the similarities and differences between a bacterial cell and a plant cell.

d Explain the similarities and differences between a bacterial cell and a non-specialised animal cell.

2 a Produce a table to compare diffusion, osmosis, and active transport. Write a brief explanation of the advantages and disadvantages of all three processes in cells.

b In an experiment to investigate osmosis, two Visking tubing bags were set up with sugar solution inside the bags and water outside the bags. Bag A was kept at 20 °C and bag B was kept at 30 °C (Figure 3).

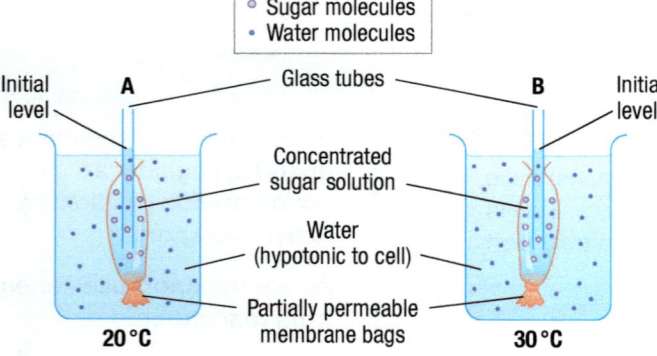

Figure 3

Describe what you would expect to happen and explain it in terms of osmosis and particle movements.

3 Plants have specialised cells, tissues, and organs, just as animals do.

a Give **three** examples of plant tissues.

b What are the main plant organs and what do they do?

c Which plant tissues are found in all of the main plant organs, and why?

4 Figure 4 shows the human digestive system.

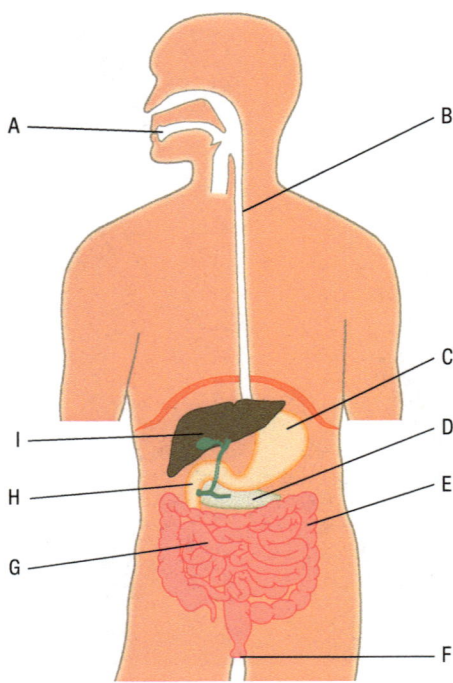

Figure 4

a What is an organ system?

b Make a table and name the parts of the human digestive system labelled A–I in Figure 4. Give the function of each organ in the digestive system as a whole.

c Select **two** examples of individual tissues that you would find in an organ of the digestive system and explain how they are specialised for their role.

Practice questions

1 Figure 1 shows a typical plant, animal, and bacterial cell.

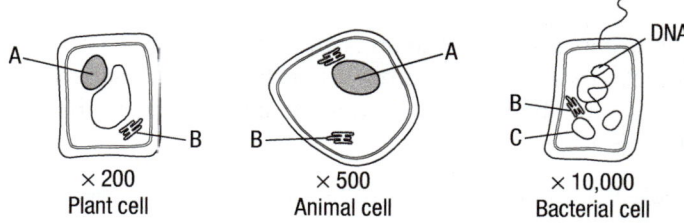

×200
Plant cell

×500
Animal cell

×10,000
Bacterial cell

Figure 1

a Name the structures A, B, and C. (4)

b Describe the function of the nucleus in cells. (2)

c Which parts of a bacterial cell carry out the same function as the nucleus does in plant and animal cells? (2)

d Calculate the length of each cell in µm (1 µm = 0.001 mm). Use the formula:
length on diagram = real length × magnification
i Plant cell ....
ii Animal cell ....
iii Bacterial cell .... (3)

e Ribosomes make protein molecules for the cell.

Suggest two possible uses for these protein molecules. (2)

f Suggest why there are no mitochondria in bacterial cells. (1)

2 The two plant cells in Figure 2 are very different because they have been specialised to carry out different functions.

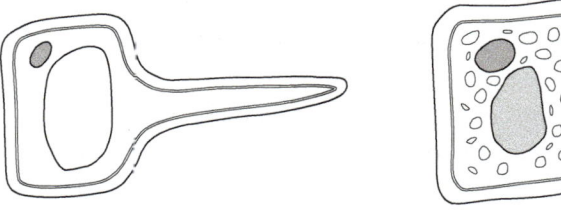

Root hair cell

Mesophyll cell

Figure 2

a Describe the function of each cell. (5)

b *In this question you will be assessed on using good English, organising information clearly, and using specialist terms where appropriate.*

Compare and contrast the structures of these two cells in relation to their function. (6)

3 The plant in pot A has been adequately watered and is healthy. The plant in pot B has not been given enough water and is wilting.

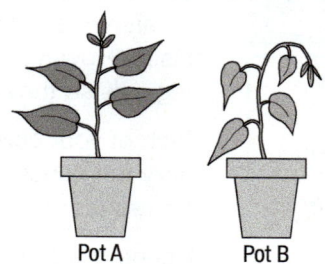

Pot A   Pot B

Figure 3

a Draw and label a stem cell from each plant.

Use the correct term to describe the state of each cell. (4)

b Explain in detail why the plant in pot A remains upright whilst the plant in pot B does not. (3)

4 Some scientists wish to grow mouse skin cells in tissue culture. They know that the tissue culture liquid must have the same concentration of salts and sugars as the cytoplasm of the cell.

They made solutions at four different concentrations and added some mouse skin cells.

After 24 hours they removed some cells and observed them under a microscope.

The following table shows the results.

Test number	1	2	3	4	Control
Concentration of salts in mol/dm³	0.24	0.26	0.28	0.30	Fresh cell from mouse
Appearance of cells					

a Name the term that describes a solution that has the same concentration as the contents of the cells. (1)

b Which concentration of salts is suitable for the tissue culture liquid? (1)

c *In this question you will be assessed on using good English, organising information clearly, and using specialist terms where appropriate.*

The concentration in test 1 is not suitable as it has damaged the cells.

Explain why the solution in test 1 has damaged the cells. (6)

Cell division, growth, and differentiation

2.1

⚭ links

For more information on alleles, look at Chapter 10 Variation and inheritance.

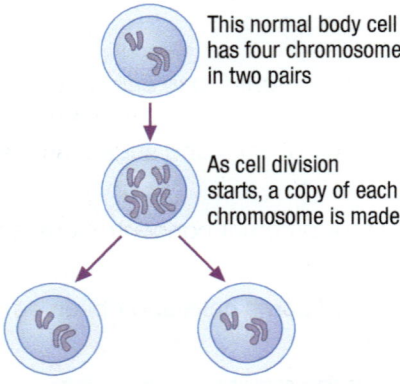

This normal body cell has four chromosomes in two pairs

As cell division starts, a copy of each chromosome is made

The cell divides in two to form two daughter cells. Each daughter cell has a nucleus containing four chromosomes identical to the ones in the original parent cell.

Figure 1 *Two identical cells are formed by the simple division that takes place during mitosis. This cell is shown with only two pairs of chromosomes rather than 23*

New cells are needed for an organism, or part of an organism, to grow. They are also needed to replace cells that become worn out and to repair damaged tissue. However, the new cells must have the same genetic information as the originals so they can do the same job.

Each of your cells has a nucleus that contains chromosomes. Chromosomes carry the genes that contain the instructions for making both new cells and all the tissues and organs needed to make an entire new you.

A gene is a small packet of information that controls a characteristic, or part of a characteristic, of your body. It is a section of DNA. Different forms of the same gene are known as **alleles**. Different alleles of the same gene may result in different characteristics. For example, there is a gene which determines whether or not you have dimples – one allele gives dimples, another allele gives no dimples. The genes are grouped together on chromosomes. A chromosome may carry several hundred or even thousands of genes.

You have 46 chromosomes in the nucleus of your body cells. They are arranged in 23 pairs. One of each pair is inherited from your father and one from your mother. Your sex cells (gametes) have only one of each pair of chromosomes.

Mitosis

The cell division in normal body cells produces two identical cells and is called **mitosis**. As a result of mitosis, all your body cells have the same chromosomes. This means that they have the same genetic information. Mitosis produces the additional cells needed for growth or replacement.

In asexual reproduction, the cells of the offspring are produced by mitosis from the cells of their parent. This is why they contain exactly the same alleles as their parent, with no genetic variation.

How does mitosis work? Before a cell divides, it produces new identical copies of the chromosomes in the nucleus. Then the cell divides once to form two genetically identical cells (Figure 1).

In some parts of an animal or plant, cell division like this carries on rapidly all the time. Your skin is a good example. You constantly lose cells from the skin's surface, and make new cells to replace them. In fact, about 300 million body cells die every minute, so mitosis is very important.

Practical

Observing mitosis

View a special preparation of a growing root tip under a microscope. You should be able to see the different stages of mitosis as they are taking place. Use Figure 2 for reference.

● Describe your observations of mitosis.

Differentiation

In the early development of animal and plant embryos, the cells are unspecialised. Each one of them (known as a **stem cell**) can become any type of cell that is needed.

In many animals, the cells become specialised very early in life. By the time a human baby is born, most of its cells are specialised. They will all do a particular job, such as liver cells, skin cells, or muscle cells. They have differentiated. Some of their genes have been switched on and others have been switched off.

This means that when, for example, a muscle cell divides by mitosis, it can only form more muscle cells. So, in a mature (adult) animal, cell division is mainly restricted. It is needed for the repair of damaged tissue and to replace worn-out cells, because in most adult cells, differentiation has already occurred. Specialised cells can divide by mitosis, but they only form the same sort of cell. Some differentiated cells, such as red blood cells, cannot divide so **adult stem cells** replace dead or damaged cells.

In contrast, most plant cells are able to differentiate all through their lives. Undifferentiated cells are formed at active regions in the stems and roots. These regions of actively dividing plant cells are known as the meristems. In these areas, mitosis takes place almost continuously.

Plants keep growing all through their lives at these growing points. The plant cells produced don't differentiate until they are in their final position in the plant. Even then, the differentiation isn't permanent. You can move a plant cell from one part of a plant to another. There it can redifferentiate and become a completely different type of cell. You can't do that with animal cells – once a muscle cell, always a muscle cell.

Producing identical offspring is known as **cloning**. You can produce huge numbers of identical plant clones from a tiny piece of leaf tissue. This is because, in the right conditions, a plant cell will become unspecialised and undergo mitosis many times. Each of these undifferentiated cells will produce more cells by mitosis. Given different conditions, these will then differentiate to form a tiny new plant. The new plant will be identical to the original parent.

It is difficult to clone animals because animal cells differentiate permanently, early in embryo development. The cells can't change back. Animal clones can only be made by cloning embryos, although by using very specialised techniques the nuclei of adult cells can be used to make an embryo.

links

You learned about the results of differentiation in 1.3 Specialised cells and 1.4 Tissues and organs.

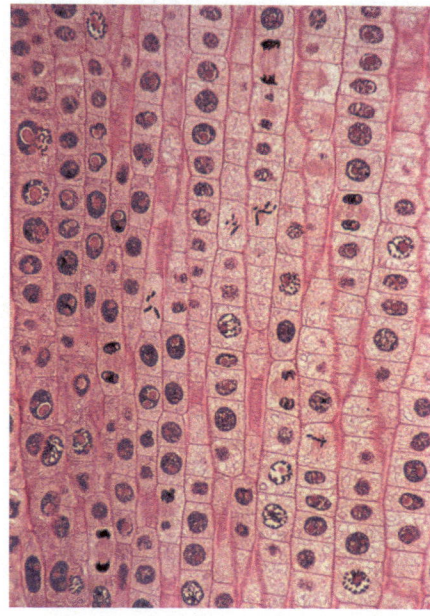

Figure 2 *The undifferentiated cells in this onion root tip are dividing rapidly. You can see mitosis taking place, with the chromosomes in different positions as the cells divide*

Study tip

Cells produced by mitosis are genetically identical.

Key points

- In body cells, chromosomes are found in pairs.
- Body cells divide by mitosis to produce more identical cells for growth, repair, and replacement, or in some cases asexual reproduction.
- In plant cells, mitosis takes place throughout life in the meristems found in the shoot and root tips.
- Most types of animal cell differentiate at an early stage of development. Many plant cells can differentiate throughout their life.

Summary questions

1 Define the following terms:
   a chromosome   b gene   c allele.

2 a Explain why the chromosome number must stay the same when the cells divide to make other normal body cells.
   b Explain clearly what happens during mitosis and why it is so important in the body.

3 a What is differentiation, and why is it important in living organisms?
   b How does differentiation differ in animal and plant cells?
   c Explain how this difference affects the cloning of plants and animals.

2.2 Cell division in sexual reproduction

Mitosis takes place all the time, in tissues all over your body. Yet there is another type of cell division that takes place only in the reproductive organs of animals and plants. In humans, these are the ovaries and the testes. **Meiosis** results in sex cells, called **gametes**, with only half the original number of chromosomes.

Meiosis

The female gametes, the egg cells or **ova**, are made in the ovaries. The male gametes, or sperm, are made in the testes.

The gametes are formed by meiosis. In meiosis, the chromosome number is reduced by half in the following way:

● When a cell divides to form gametes, the chromosomes (the genetic information) are copied so there are four sets of chromosomes instead of the normal two sets. This is very similar to mitosis.

● The cell then divides twice in quick succession to form four gametes, each with a single set of chromosomes (Figure 1).

Each gamete that is produced is slightly different from all the others. They contain random mixtures of the original chromosome pairs. This introduces variation.

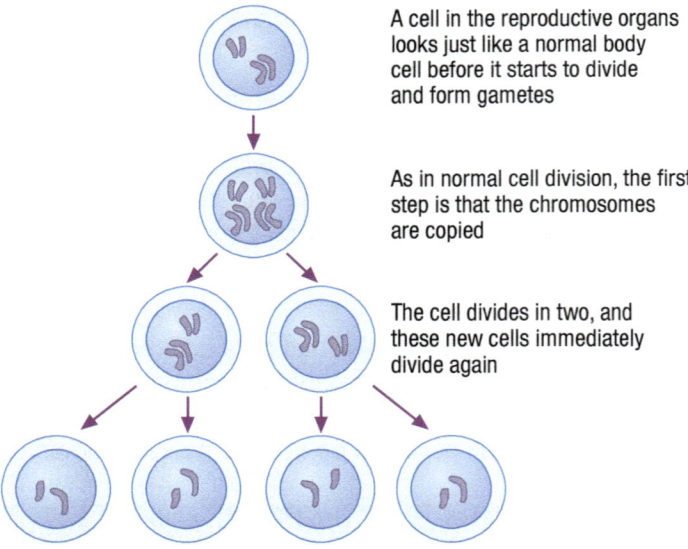

A cell in the reproductive organs looks just like a normal body cell before it starts to divide and form gametes

As in normal cell division, the first step is that the chromosomes are copied

The cell divides in two, and these new cells immediately divide again

This gives four sex cells, each with a single set of chromosomes – in this case two instead of the original four

Figure 1 *The formation of sex cells in the ovaries and testes involves meiosis to halve the chromosome number. The original cell is shown with only two pairs of chromosomes, to make it easier to follow what is happening*

Fertilisation

More variation is added when fertilisation takes place. Each sex cell has a single set of chromosomes. When two sex cells join during fertilisation, the single new cell formed has a full set of chromosomes. In humans, the egg cell (ovum) has 23 chromosomes and so does the sperm. When they join together, they produce a single new body cell with 46 chromosomes in 23 pairs – the correct number of chromosomes for all humans.

The combination of genes on the chromosomes of every newly fertilised ovum is unique. Once fertilisation is complete, the unique new cell begins to divide by mitosis to form a new individual. As the organism develops, the cells differentiate to form different kinds of cells. Mitosis will continue long after the fetus is fully developed and the baby is born. It is still important for adults for the repair of the body and replacement of worn-out cells.

In fact, about 80% of fertilised eggs never make it to become a live baby – about 50% never even implant into the lining of the womb.

Variation

The differences between asexual and sexual reproduction are reflected in the different types of cell division involved:

- In asexual reproduction, the offspring are produced from the parent cells as a result of mitosis. They contain exactly the same chromosomes and the same genes as their parents. There is little variation in the genetic material.
- In sexual reproduction, the gametes are produced by meiosis in the sex organs of the parents. This introduces variation as each gamete is different. Then, when the gametes fuse, one of each pair of chromosomes, and so one of each pair of genes, comes from each parent.

The combination of genes in the new pair of chromosomes will contain alleles from each parent. This also helps to produce variation in the characteristics of the offspring.

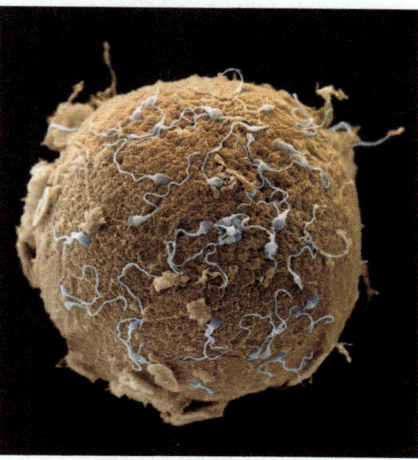

Figure 2 *At the moment of fertilisation, the chromosomes of the two gametes are combined. The new cell has a complete set of chromosomes, like any other body cell. This new cell will then grow and divide by mitosis to form a new individual*

Study tip

Learn to spell mitosis and meiosis.
Remember their meanings:
Mitosis – **m**aking **i**dentical **t**wo.
Meiosis – **m**aking **e**ggs (and sperm).

Key points

- Cells in the reproductive organs divide by meiosis to form gametes (sex cells).
- Body cells have two sets of chromosomes – gametes have only one set.
- In meiosis, the genetic material is copied and then the cell divides twice to form four gametes, each with a single set of chromosomes.
- Sexual reproduction gives rise to variation because genetic information from two parents is combined.

Summary questions

1. a How many pairs of chromosomes are there in a normal human body cell?
   b How many chromosomes are there in a human sperm cell?
   c How many chromosomes are there in a fertilised human egg cell?

2. Sexual reproduction results in variation. Explain clearly exactly how this comes about.

3. a What is the name of the special type of cell division that produces gametes in the reproductive organs? Describe clearly what happens to the chromosomes in this process.
   b Where in your body would this type of cell division take place?
   c Explain why this type of cell division is so important in sexual reproduction.

2.3 Stem cells

Learning objectives

After this topic, you should know:

● how stem cells are different from other body cells

● why scientists hope that it will be possible to use stem cells to treat a number of health problems.

The function of stem cells

An egg and a sperm cell fuse to form a **zygote**, a single new cell. That cell divides and becomes a hollow ball of cells – the embryo. The inner cells of this ball are the stem cells. Stem cells differentiate to form the specialised cells of your body that make up your various tissues and organs. They will eventually produce every type of cell in your body.

Even when you are an adult, some of your stem cells remain. Your bone marrow is a good source of stem cells. Scientists now think there may be a tiny number of stem cells in most of the different tissues in your body. This includes your blood, brain, muscles, and liver.

Many of your differentiated cells can divide to replace themselves. However, some tissues cannot do this. The stem cells can stay in these tissues for many years. They are only needed if the cells are injured or affected by disease. Then they start dividing to replace the different types of damaged cell.

Using stem cells

Many people suffer and even die because parts of their body stop working properly. For example, spinal injuries can cause paralysis. That's because if the spinal nerves are damaged, they do not repair themselves. Millions of people would benefit if doctors could replace damaged body parts.

In 1998, there was a breakthrough. Two American scientists managed to culture human **embryonic stem cells**. These were capable of forming other types of cell.

Scientists hope that the embryonic stem cells can be encouraged to grow into almost any of the different types of cell needed in the body. For example, scientists in the USA have grown nerve cells from embryonic stem cells. In rats, these have been used to reconnect damaged spinal nerves. The rats regained some movement of their legs. In 2010, the first trials using nerve cells grown from embryonic stem cells were carried out in humans. The nerve cells were injected into the spinal cords of patients with new, severe spinal cord injuries. These first trials were to make sure that the technique is safe. The scientists and doctors hope it will not be long before they can use stem cells to help people who have been paralysed to walk again.

Scientists might also be able to grow whole new organs from embryonic stem cells. These could then be used in transplant surgery (Figure 1). Scientists in Edinburgh have already grown functioning kidney structures using stem cells from amniotic fluid.

Doctors in the USA and the UK are carrying out trials to see if embryonic stem cells can be used to treat common causes of blindness. They have to discover first if it is safe to inject stem cells into the eyes, and then find out whether the stem cells can restore sight. Conditions ranging from infertility to dementia could eventually be treated using stem cells.

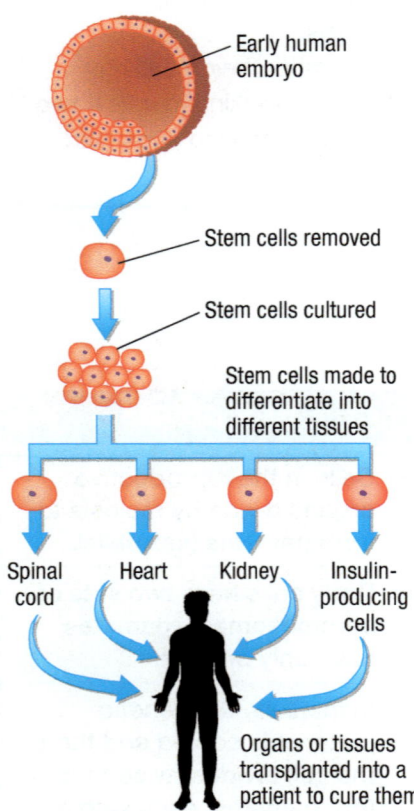

Early human embryo

Stem cells removed

Stem cells cultured

Stem cells made to differentiate into different tissues

Spinal cord   Heart   Kidney   Insulin-producing cells

Organs or tissues transplanted into a patient to cure them

Figure 1 *This shows one way in which scientists hope embryonic stem cells might be formed into adult cells and used as human treatments in the future*

Problems with stem cells

Many embryonic stem cells come from aborted embryos. Others come from spare embryos in fertility treatment. This raises ethical issues. There are people, including many religious groups, who feel this is wrong. They question the use of a potential human being as a source of cells, even to cure others. Some people feel that, as the embryo cannot give permission, using it is a violation of its human rights.

In addition, progress in developing therapies using stem cells has been relatively slow, although in fact scientists have been working with them for less than 20 years. There is some concern that embryonic stem cells might cause cancer if they are used to treat sick people. This has certainly been seen in mice. Furthermore, making stem cells is slow, difficult, expensive and hard to control.

The future of stem cell research

Scientists have found embryonic stem cells in the umbilical cord blood of newborn babies and even in the amniotic fluid that surrounds the fetus as it grows. These sources of embryonic stem cells may help to overcome some of the ethical concerns.

Scientists are also finding ways of growing the **adult stem cells** found in bone marrow and some other tissues. So far, they can only develop into a limited range of cell types. However, this is another possible way of avoiding the controversial use of embryonic tissue. Adult stem cells have been used successfully to treat some forms of heart disease and to grow some new organs such as **tracheas** (windpipes).

The area of stem cell research known as **therapeutic cloning** could be very useful. However, it is proving very difficult. It involves using the nuclei from the cells of an adult to produce a cloned early embryo of themselves. This would provide a source of perfectly matched embryonic stem cells. In theory, these could then be used to grow new organs for the original donor. The new organs would not be rejected by the body because they have been made from the body's own cells.

Most people remain excited by the possibilities of embryonic stem cell use in treating many diseases. At the moment, after years of relatively slow progress, hopes are high again that stem cells will change the future of medicine. People don't know how many of these hopes will be fulfilled – only time will tell.

Figure 2 *In 2010, Ciaran Finn-Lynch was the first child to be given a life-saving new windpipe grown using his own stem cells. His recovery wasn't easy, but a year later he was back at school*

Summary questions

1 a What is the difference between a stem cell and a normal body cell?
   b What are the different types of stem cells?

2 a What are the advantages of using stem cells to treat diseases?
   b Suggest three areas in which the use of stem cells could provide valuable medical treatments.
   c Explain how successful stem cell treatments have been so far.

3 a What are the difficulties with stem cell research?
   b How are scientists hoping to overcome ethical objections to using embryonic stem cells in their research?

Key points

- Cells from human embryos and adult bone marrow, called stem cells, can be made to differentiate into many different types of human cell.

- Stem cells have the potential to treat previously incurable conditions. Scientists may be able to grow nerve cells to cure paralysis or grow whole new organs for people who need them.

Summary questions

1 a What is mitosis?

b Explain, using diagrams, what takes place when a cell divides by mitosis.

c Mitosis is very important during the development of a baby from a fertilised egg. It is also important throughout life. Why?

2 What is meiosis, and where does it take place?

3 a Why is meiosis so important?

b Explain, using labelled diagrams, what takes place when a cell divides by meiosis.

4

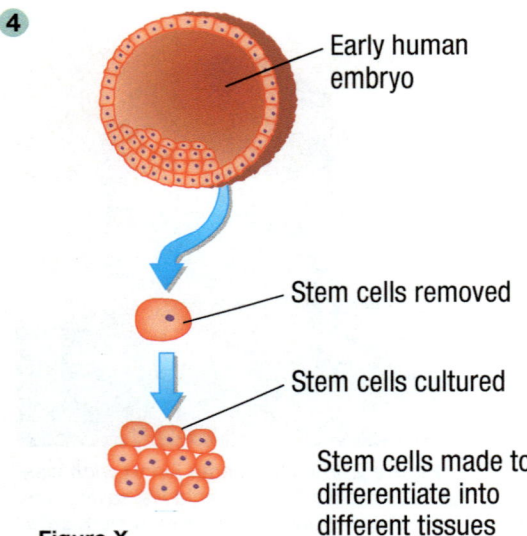

Early human embryo

Stem cells removed

Stem cells cultured

Stem cells made to differentiate into different tissues

Figure X

Fig X shows one way in which stem cells can be produced for possible medical treatments.

a What are stem cells?

b It is hoped that many different medical problems may be cured using stem cells. Explain how this might work.

c There are some ethical issues associated with the use of embryonic stem cells. Explain the arguments both for and against their use.

5

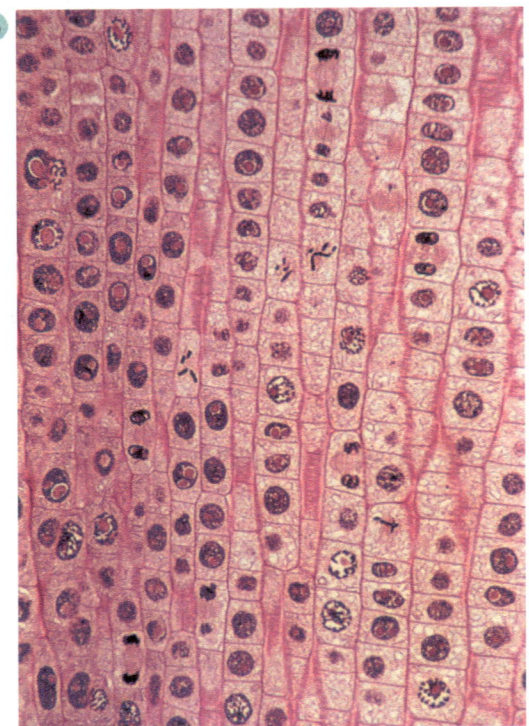

Figure Y

a Fig Y shows the cells in an onion root tip. What process is taking place in these cells?

b What is this region of the plant called?

c Explain how the process taking place in fig Y differs from the way the gametes of the plant are formed in the sex organs?

Practice questions

1 List A contains words about genetic information in cells. List B contains explanations of the words.

Match each word in list A with the correct explanation in list B.

List A	List B
Chromosome	a different form of one gene
Allele	a cell with a single set of chromosomes
Gene	a structure carrying a large number of genes
Nucleus	a section of genetic material coding for one characteristic
Gamete	the part of a cell that contains the genetic material

(5)

2 Cells from organism Z have three pairs of chromosomes.

  a Copy and complete stages C and D of the diagram below to show the chromosomes at each stage of meiosis.

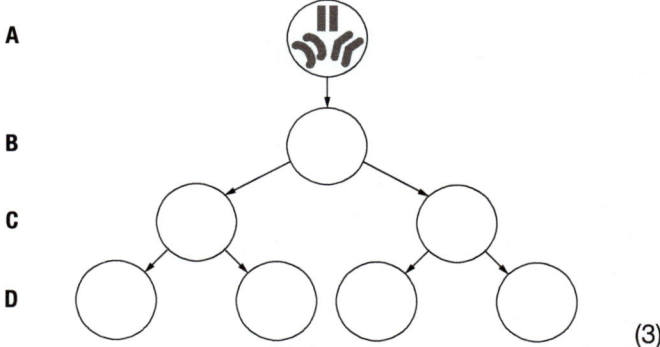

(3)

  b Name the cells formed at stage D. (1)

  c Complete the diagram below to show a cell from organism Z, which has been formed by mitosis.

(1)

3 The numbers **i** to **v** represent stages in the life cycle of a sunflower plant.

  i Pollen grains from one sunflower fuse with the ova of another sunflower at fertilisation.

  ii A fertilised ovum develops into a seed.

  iii The seed germinates. It develops a radicle (early root) and a plumule (early shoot).

  iv The seed grows a shoot and a single root.

  v The shoot develops into a stem and leaves. The single root forms a complex root system.

  a Put a number from the stages **i** to **v** in each box to represent the stage at which you would find:

Differentiated cells	
Cells with a single set of chromosomes	
Undifferentiated cells	
Cells dividing rapidly by mitosis	
An embryo	

(5)

  b The following terms describe the organisation within a plant or animal:

organ   cell   organism
tissue   organ system

Rearrange these words in increasing order of complexity. (2)

  c Use words from the list in **b** above to describe each of these parts of the sunflower plant:
  i phloem
  ii stem
  iii root hair cell
  iv water transport system
  v sunflower plant.

Exchanging materials

After this topic, you should know:

● how the surface area to volume ratio varies depending on the size of an organism

● why large multicellular organisms need special systems for exchanging materials with the environment.

For many single-celled organisms, diffusion, osmosis, and active transport are all that is needed to exchange materials with their environment. A single-celled organism such as *Amoeba* has a relatively large surface area compared with the volume of the cell. This is known as the surface area to volume ratio. So, for example, in *Amoeba* the diffusion distances are small enough for all the oxygen needed by the organism to be moved into the cell from the surrounding water by simple diffusion. The carbon dioxide produced during metabolism can be removed in the same way.

Surface area to volume ratio

The surface area to volume ratio is very important in biology. It makes a big difference to the way in which animals can exchange substances with the environment. Surface area to volume ratio is also important when you consider how energy is transferred to and from the surroundings by living organisms, and how water evaporates from the surfaces of plants and animals.

∞ links

You will use the idea of surface area to volume ratio when you study the adaptations of animals and plants for living in a variety of different habitats in 12.4 Adapt and survive through to 12.7 Competition in animals.

Worked example

1 cm
1 cm
1 cm
SA : V ratio = 6 : 1

3 cm
3 cm
3 cm
SA : V ratio = 54 : 27 = 2 : 1

Surface area to volume ratio

The ratio of surface area to volume falls as objects get bigger. You can see this clearly in the diagram. In a small object, the surface area to volume (SA : V) ratio is relatively large. This means that the diffusion distances are short and that simple diffusion is sufficient for the exchange of materials.

As organisms get bigger, the surface area to volume ratio falls. As the distances between the centre of the organism and the surface get bigger, simple diffusion is no longer enough to exchange materials between the cells and the environment.

Getting bigger

As living organisms get bigger and more complex, their surface area to volume ratio gets smaller. This makes it increasingly difficult to exchange materials quickly enough with the outside world:

● Gases and food molecules can no longer reach every cell inside the organism by simple diffusion.

● Metabolic waste cannot be removed fast enough to avoid poisoning the cells.

So in many larger organisms, there are special surfaces where the exchange of materials takes place. These surfaces are adapted to be as effective as possible. You can find them in humans, in other animals, and in plants.

Adaptations for exchanging materials

There are various adaptations to make the process of exchange more efficient. The effectiveness of an **exchange surface** can be increased by:

Study tip

Active transport requires energy. You will remember this if you think about mineral ions having to *push* into the root hair cells *against* a concentration gradient.

- having a large surface area that provides a big area over which exchange can take place
- being thin, which provides a short diffusion path
- having an efficient blood supply (in animals), which moves the diffusing substances away and maintains a concentration (diffusion) gradient
- being **ventilated** (in animals), which makes gaseous exchange more efficient by maintaining steep concentration gradients.

Different organisms have very different adaptations for the exchange of materials, such as the leaves of a plant, the gills of a fish, and the kidneys of a desert rat. For example, scientists have recently discovered that the common musk turtle has a specially adapted tongue (see Figure 1).

Its tongue is covered in tiny buds that greatly increase the surface area. Its tongue also has a good blood supply. These turtles don't just use their tongue for eating – they use it for **gaseous exchange**, too. The buds on the tongue absorb oxygen dissolved in the water that passes over them. Most turtles have to surface regularly for air. However, the musk turtle's tongue is so effective at gaseous exchange that it can stay underwater for months at a time.

Examples of adaptations

Many of your own organ systems are specialised for exchanging materials. This is because the human surface area to volume ratio is so low that without specialised organs the cells inside your body cannot possibly get the food and oxygen they need, or get rid of the waste they produce quickly enough.

One of these exchange systems is your breathing system, particularly your lungs. Air is moved into and out of your lungs when you breathe, ventilating millions of tiny air sacs called **alveoli**. The alveoli have an enormous surface area, thin walls made of flattened cells, and a very rich blood supply, all making your lungs very effective for gas exchange.

Animals are not the only organisms that need effective gas and solute exchange systems. Even the smallest plants need specialised exchange systems to get the water, mineral ions, and carbon dioxide they need. Plants need to be able to take in plenty of water and dissolved mineral ions through their root systems. The roots have a very large surface area that is increased still more by root hair cells. Water is constantly moved away from the roots in the transpiration stream, which maintains a steep concentration gradient into the cells.

Plant leaves are also modified to make gaseous and solute exchange as effective as possible. Flat, thin leaves, the presence of air spaces in the spongy mesophyll tissue, and the **stomata** all help to provide a big surface area. They also maintain a steep concentration gradient for the diffusion of substances needed by plants, such as water and carbon dioxide.

Figure 1 *The common musk turtle has a very unusual tongue, which is adapted for gaseous exchange*

∞ links

You will find much more detail about the alveoli of the lungs as a site of gas exchange in 3.2 Breathing and gas exchange in the lungs.

∞ links

You can see how active transport and osmosis are important in plant roots in 1.8 Active transport and 9.4 Transport systems in plants.

∞ links

You can find out more about the adaptations of plant leaves for diffusion and the exchange of materials in 9.4 Transport systems in plants.

Key points

- Single-celled organisms have a relatively large surface area to volume ratio, so all necessary exchanges with the environment take place over this surface.
- In multicellular organisms, many organs are specialised with effective exchange surfaces.
- Exchange surfaces usually have a large surface area and thin walls, which provide short diffusion distances. In animals, exchange surfaces will have an efficient blood supply or, for gaseous exchange, will be ventilated.

Summary questions

1. Explain clearly how the surface area to volume ratio of an organism affects the way in which it exchanges materials with the environment.
2. a How does the tongue of a musk turtle differ from the tongues of most reptiles?
   b How does this adaptation help musk turtles survive?
3. a Summarise the adaptations you would expect to see in effective exchange surfaces, and explain the importance of each adaptation.
   b Explain how three different exchange surfaces show some or all of these adaptations to help them operate efficiently.

3.2

Breathing and gas exchange in the lungs

For a gas exchange system to work efficiently, you need a steep concentration gradient. Humans are like many big, complex mammals in that they move air in and out of their lungs regularly. By changing the composition of the air in the lungs, they maintain a steep concentration gradient for both oxygen diffusing into and carbon dioxide diffusing out of the blood. This is known as ventilating the lungs or **breathing**. It takes place in a specially adapted **respiratory (breathing) system**.

The respiratory (breathing) system

Your lungs are found in the upper part of your body – in your chest, or **thorax**. They are protected by your bony ribcage. They are separated from the digestive organs beneath (in your **abdomen**) by the **diaphragm**. The diaphragm is a strong sheet of muscle. The job of your breathing system is to move air in and out of your lungs. The lungs provide an efficient surface for gas exchange in the alveoli (Figure 1).

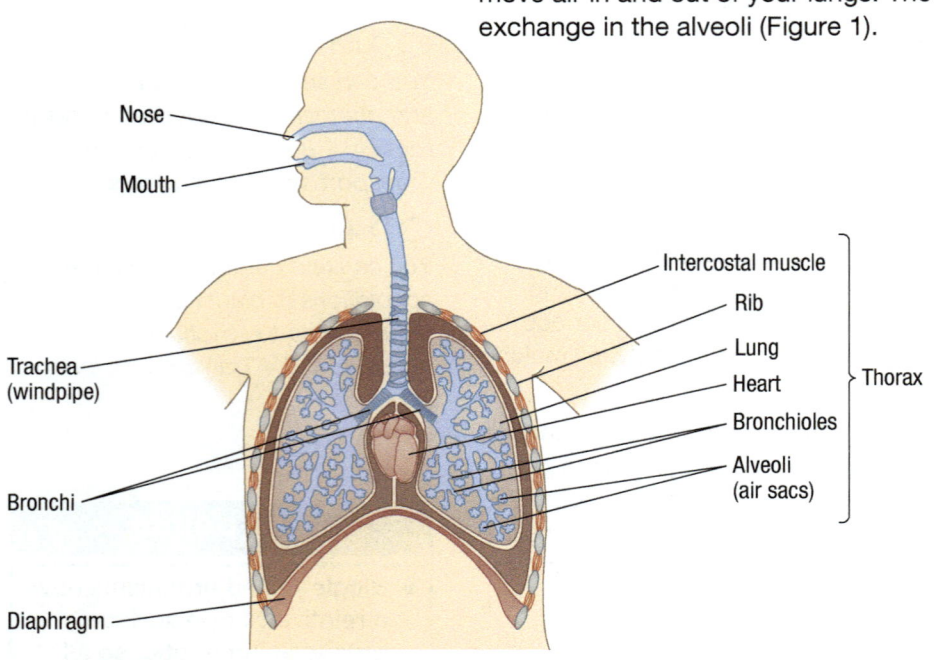

Nose
Mouth
Intercostal muscle
Rib
Lung
Heart   Thorax
Trachea (windpipe)
Bronchioles
Alveoli (air sacs)
Bronchi
Diaphragm

Figure 1 *The breathing system supplies your body with vital oxygen and removes waste carbon dioxide*

⬭ links

You can find out more about diffusion and concentration gradients in 1.6 Diffusion.

Table 1 *The composition of inhaled and exhaled air (~ means approximately)*

Atmospheric gas	% of air breathed in	% of air breathed out
nitrogen	~80	~80
oxygen	20	~16
carbon dioxide	0.04	~4

Moving air in and out of the lungs

Ventilation of the lungs is brought about by movements of your ribcage and diaphragm. You can see and feel the movements of your ribcage, but not your diaphragm.

When you breathe in, your **intercostal muscles** contract, which pulls your ribs upwards and outwards. At the same time, your diaphragm muscles contract, which flattens your diaphragm from its normal domed shape. These two movements *increase* the volume of your thorax. Because the same amount of gas is now inside a much bigger space, the pressure inside your thorax drops. Pressure inside the thorax is now lower than the pressure of air outside your body. As a result, air moves into your lungs, pushed in by atmospheric pressure.

When the intercostal muscles relax, your ribs drop down and in again. When the diaphragm relaxes, it curves back up into your thorax, resuming its domed shape. As a result, the volume of your thorax gets smaller again. This increases the pressure inside the chest so the air is squeezed and forced out of the lungs. That is how you breathe out (Figure 2).

When you breathe in, oxygen-rich air moves into your lungs. This maintains a steep concentration gradient with the blood. As a result, oxygen continually diffuses into your bloodstream through the gas exchange surfaces of your alveoli. Breathing out removes carbon dioxide-rich air from the lungs. This maintains a concentration gradient so carbon dioxide can continually diffuse out of the bloodstream into the air in the lungs.

Adaptations of the alveoli

Your lungs are specially adapted to make gas exchange more efficient. They are made up of clusters of alveoli which provide a very large surface area. This is important in order to achieve the most effective diffusion of oxygen and carbon dioxide.

The alveoli also have a rich supply of blood **capillaries**. This maintains a concentration gradient in both directions – the blood coming to the lungs is always relatively low in oxygen and high in carbon dioxide compared with the inhaled air.

As a result, gas exchange takes place along the steepest concentration gradients possible. This makes the exchange rapid and effective. The layer of cells between the air in the lungs and the blood in the capillaries is also very thin. This allows diffusion to take place over the shortest possible distance (see Figure 3).

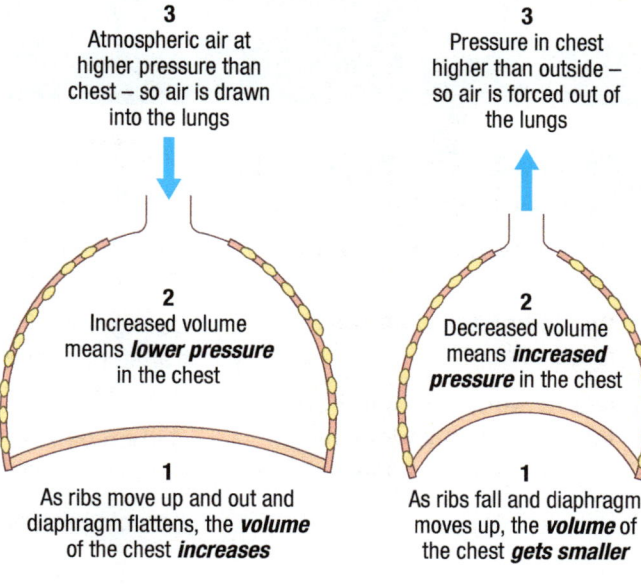

3
Atmospheric air at higher pressure than chest – so air is drawn into the lungs

2
Increased volume means *lower pressure* in the chest

1
As ribs move up and out and diaphragm flattens, the *volume* of the chest *increases*

Breathing in

3
Pressure in chest higher than outside – so air is forced out of the lungs

2
Decreased volume means *increased pressure* in the chest

1
As ribs fall and diaphragm moves up, the *volume* of the chest *gets smaller*

Breathing out

Figure 2 *Ventilation of the lungs*

Direction of blood flow
Oxygen
Carbon dioxide
Air

Air in

Air out

Oxygen moves into blood by diffusion

Carbon dioxide passes out of blood by diffusion

An alveolus

Ventilation moves air in and out and helps maintain a steep diffusion gradient

Very thin walls of the alveolus and the capillary (a single cell thick) provide a short diffusion distance between the air and the blood

Good blood supply maintains concentration gradient for diffusion by removing oxygen and bringing lots of carbon dioxide

Figure 3 *The alveoli are adapted so gas exchange can take place as efficiently as possible in the lungs*

??? Did you know … ?

If all the alveoli in your lungs were spread out flat, they would have a surface area about the size of 20 table tennis tables.

Summary questions

1 Explain clearly how air is moved into and out of your lungs.

2 What is meant by the term gaseous exchange, and why is it so important in your body?

3 a Draw a bar chart to show the difference in composition between the air you breathe in and the air you breathe out (use Table 1 on Page 32).

   b People often say you breathe in oxygen and breathe out carbon dioxide. Use your bar chart to explain why this is wrong.

   c Explain how your respiratory system is adapted to make gaseous exchange as efficient as possible.

Key points

- The lungs are in your thorax. They are protected by your ribcage and separated from your abdomen by the diaphragm.

- The intercostal muscles contract to move your ribs up and out as the diaphragm flattens, increasing the volume of your thorax. The pressure decreases and air moves into your lungs.

- The intercostal muscles relax and the ribs move down and in as the diaphragm domes up, decreasing the volume of your thorax. The pressure increases and air is forced out of your lungs.

- The alveoli provide a very large surface area and have a rich supply of blood capillaries. This means gases can diffuse into and out of the blood as efficiently as possible.

3.3 Aerobic respiration

One of the most important enzyme-controlled processes in living things is aerobic respiration. It takes place all the time in plant and animal cells.

Your digestive system, lungs, and circulation all work to provide your cells with the glucose and oxygen they need for respiration.

During aerobic respiration, glucose (a sugar) reacts with oxygen. This reaction transfers energy that your cells can use. This energy is vital for everything that goes on in your body.

Carbon dioxide and water are produced as waste products of the reaction. The process is called aerobic respiration because it uses oxygen from the air.

Aerobic respiration can be represented by the equation:

glucose + oxygen → carbon dioxide + water

$$C_6H_{12}O_6 + 6O_2 \rightarrow 6CO_2 + 6H_2O$$

Did you know … ?

The average energy needs of a teenage boy are around 11 510 kJ every day – but teenage girls only need 8830 kJ a day. This is partly because, on average, girls are smaller than boys, but it is also because boys have more muscle cells, which means more mitochondria demanding fuel for aerobic respiration.

Practical

Investigating respiration

Animals, plants and microorganisms all respire. It is possible to show that cellular respiration is taking place. You can either deprive a living organism of the things it needs to respire, or show that waste products are produced from the reaction.

Depriving a living thing of food and/or oxygen would kill it. This would be unethical, so scientists concentrate on the products of respiration. Carbon dioxide is the easiest to identify. You can also measure the energy released to the surroundings.

Limewater goes cloudy when carbon dioxide bubbles through it. The higher the concentration of carbon dioxide, the quicker the limewater goes cloudy. This gives us an easy way of showing that carbon dioxide has been produced. You can also look for a rise in temperature to show that energy is being released during respiration.

- Plan an ethical investigation into aerobic respiration in living organisms.

Mitochondria – the site of respiration

Aerobic respiration involves lots of chemical reactions. Each reaction is controlled by a different enzyme. Most of these reactions take place in the mitochondria of your cells.

Mitochondria are tiny rod-shaped parts (organelles) that are found in almost all plant and animal cells as well as in fungi and algal cells. They have a folded inner membrane that provides a large surface area for the enzymes involved in aerobic respiration.

The number of mitochondria in a cell shows you how active the cell is.

Outer membrane

Folded inner membrane gives a large surface area where the enzymes involved in aerobic respiration are found

Figure 1 *Mitochondria are the powerhouses that transfer energy for all the functions of your cells*

Reasons for respiration

The energy transferred during respiration may be used by the organism in a variety of ways:

- Living cells need energy to carry out the basic functions of life. They build up large molecules from smaller ones to make new cell material. Much of the energy transferred in respiration is used for these 'building' activities (synthesis reactions).
- In animals, energy transferred from respiration is used to make muscles contract. Muscles are working all the time in your body. Even when you sleep, your heart beats, you breathe, and your gut churns. All muscular activities use energy.
- Mammals and birds maintain a constant internal body temperature almost regardless of the temperature of their surroundings. Doing this uses energy transferred from respiration. So on cold days you will shiver, and the energy transferred from your active muscles warms your body.
- In plants, the energy transferred from respiration is used to move mineral ions such as nitrates from the soil into root hair cells by active transport. It is also used to convert sugars, nitrates, and other nutrients into amino acids, which are then built up into proteins.

Figure 2 *When the weather is cold, birds like this robin need a lot of energy from respiration just to keep warm. Giving them extra food supplies during the winter can therefore mean the difference between life and death*

∞ links

You can find out more about active transport and the movement of mineral ions into root hair cells in 1.8 Active transport, about the use of energy in Chapter 13 Ecology, and about warming and temperature control in 7.2 Controlling body temperature.

Study tip

Make sure you know the word equation and balanced symbol equation for aerobic respiration. Remember that aerobic respiration takes place in the mitochondria.

Key points

- Aerobic respiration involves chemical reactions that use oxygen and glucose and transfer energy.
- Most of the reactions in aerobic respiration take place inside the mitochondria.
- The energy transferred during aerobic respiration may be used for building up large molecules, for muscle contraction in animals, for maintaining body temperature when it is cold in birds and mammals, and in plants for taking in nitrates from the soil and building up amino acids and then proteins.

Summary questions

1   a Give the word equation for aerobic respiration.
    b Give the balanced symbol equation for aerobic respiration.
    c Why do muscle cells have many mitochondria whilst fat cells have very few?

2   You need a regular supply of food to provide energy for your cells. If you don't get enough to eat, you become thin and stop growing. You don't want to move around and you start to feel cold.
    a What are the three main uses of the energy released in your body during aerobic respiration?
    b How does this explain the symptoms of starvation described above?

3   Plan an experiment to show that during aerobic respiration:
    a oxygen is taken up   b carbon dioxide is released.

3.4 The effect of exercise on the body

Figure 1 *All the work done by your muscles is based on these special protein fibres, which use energy transferred from respiration to contract*

Your muscles need a lot of energy. They move you around and help support your body against gravity. Your heart is made of muscle and pumps blood around your body. The movement of food along your gut depends on muscles, too.

Muscle tissue is made up of protein fibres that contract when they are supplied with energy transferred from respiration. Muscle fibres need a lot of energy to contract. They contain many mitochondria to carry out aerobic respiration and transfer the energy needed.

Muscle fibres usually occur in big blocks or groups known as muscles, which contract to cause movement. They then relax, which allows other muscles to work.

Your muscles also store glucose as the carbohydrate **glycogen**. Glycogen can be converted rapidly back to glucose to use during exercise. The glucose is used in aerobic respiration to provide the energy to make your muscles contract:

$$\text{glucose} + \text{oxygen} \rightarrow \text{carbon dioxide} + \text{water}$$
$$C_6H_{12}O_6 + 6O_2 \rightarrow 6CO_2 + 6H_2O$$

The response to exercise

Even when you are not moving about, your muscles use up a certain amount of oxygen and glucose. However, when you begin to exercise, many muscles start contracting harder and faster. As a result, they need more glucose and oxygen to supply their energy needs. During exercise the muscles also produce increased amounts of carbon dioxide. This needs to be removed for muscles to keep working effectively.

During exercise, when muscular activity increases, several changes take place in your body:

- Your heart rate increases and the arteries supplying blood to your muscles dilate (widen). These changes increase the blood flow to your exercising muscles. This in turn increases the rate of supply of oxygen and glucose to the muscles. It also increases the rate at which carbon dioxide is removed from the muscles.
- Your breathing rate increases and you breathe more deeply. This means you breathe more often and also bring more air into your lungs each time you breathe in. The rate at which oxygen is brought into your body and picked up by your red blood cells is increased, and this oxygen is carried to your exercising muscles. The increased breathing rate also means that carbon dioxide can be removed more quickly from the blood in the lungs and breathed out.
- Glycogen stored in the muscles is converted back to glucose, to supply the cells with the fuel they need for increased cellular respiration.

In this way, the heart rate and breathing rate increase during exercise to supply the muscles with what they need and to remove the extra waste produced. Cellular respiration increases to supply the muscle cells with the energy they need to contract during exercise. The increase in your breathing and heart rate is needed to keep up with the demands of the cells.

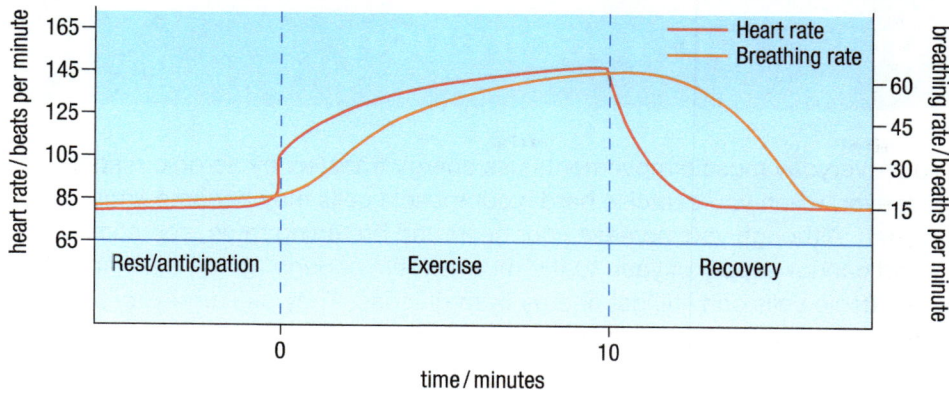

Figure 2 *The changes measured in the heart and breathing rate before, during, and after a period of exercise*

	Unfit person	Fit person
Amount of blood pumped out of the heart during each beat at rest (cm³)	64	80
Volume of the heart at rest (cm³)	120	140
Resting breathing rate (breaths/min)	14	12
Resting heart rate (beats/min)	72	63

Figure 3 *The heart and lung functions of a fit person and an unfit person at rest*

Required practical

The effect of exercise on the body

Investigate the effect of exercise on the human body using yourself and your classmates as experimental organisms. Plan and carry out your own investigation – here are some ideas to get you started:

1 You will need to find your resting breathing rate and heart rate. Take at least three readings and find the **mean**.

2 Exercise gently for a set period of time, for example, 2 minutes.

3 Record your heart and breathing rates at regular intervals after exercise until they return to resting rates.

4 *Either* vary the intensity of your exercise – you could try moderate and/or hard exercise *or* vary the length of your period of exercise.

5 Record all your results and produce graphs to show the data.

6 Put the data from the whole class together. Look for patterns and give biological reasons to explain your findings.

Summary questions

1 a What is glycogen?

  b Why do you think muscles contain a store of glycogen but most other tissues of the body do not?

2 Using Figure 2 and Figure 3, describe the effect of exercise on the heart rate and the breathing rate of an unfit person. Explain fully why these changes happen, and what you would expect to see if the person doing the exercise was fitter.

3 Plan an investigation into the fitness levels of your classmates. Describe how you might carry out this investigation, and explain what you would expect the results to be.

Study tip

A common mistake is to think that breathing becomes faster *because* the heart is beating faster. This is wrong – the two responses are not linked together. They are controlled separately. They are two different responses to the need of the muscles for more glucose and oxygen for respiration when they are working hard.

Key points

- During exercise the human body needs to react to the increased demand for energy from the muscle cells

- Body responses to exercise include:
  - an increase in the heart rate, increasing blood flow to the muscles
  - an increase in the breathing rate and depth of breathing
  - glycogen stores in the muscles are converted to glucose for cellular respiration.

- These responses act to increase the rate of supply of glucose and oxygen to the muscles and the rate of removal of carbon dioxide from the muscles.

3.5 Anaerobic respiration

Learning objectives

After this topic, you should know:

● why less energy is transferred by anaerobic respiration than by aerobic respiration

● what is meant by an oxygen debt

● that anaerobic respiration takes place in lots of different organisms, including bacteria and fungi.

Figure 1 *Training hard is the simplest way to avoid anaerobic respiration. When you are fit, you can get oxygen to your muscles and remove carbon dioxide more efficiently*

Practical

Making lactic acid

Repeat a single action many times. For example, you could step up and down, lift a weight, or clench and unclench your fist. You will soon feel the effect of a build-up of lactic acid in your muscles.

● How can you tell when your muscles have started to respire anaerobically?

Your everyday muscle movements use energy transferr by aerobic respiration. However, when you exercise hard, your muscle cells may become short of oxygen. Although you increase your heart and breathing rates, sometimes the blood cannot supply oxygen to the muscles fast enough. When this happens, the muscle cells can still get energy from glucose. They use **anaerobic respiration**, which takes place without oxygen.

In anaerobic respiration, glucose is not broken down completely. It produces **lactic acid** instead of carbon dioxide and water, and transfers a smaller amount of energy to the cells.

If you are fit, your heart and lungs will be able to keep a good supply of oxygen going to your muscles whilst you exercise. If you are unfit, your muscles will run short of oxygen much sooner.

Muscle fatigue

Using your muscle fibres vigorously for a long time can make them fatigued and they stop contracting efficiently. For example, repeated movements can soon lead to anaerobic respiration in your muscles – particularly if you're not used to the exercise.

One cause of this muscle fatigue is the build-up of lactic acid, which is made by anaerobic respiration in the muscle cells. Blood flowing through the muscles eventually removes the lactic acid.

Anaerobic respiration is not as efficient as aerobic respiration because the glucose molecules are not broken down completely. Since the breakdown of glucose is incomplete, far less energy is transferred than during aerobic respiration.

The end product of anaerobic respiration is lactic acid. This leads to the transfer of a small amount of energy, instead of the carbon dioxide and water plus lots of energy transferred by aerobic respiration.

Anaerobic respiration can be represented by the equation:

$$\text{glucose} \rightarrow \text{lactic acid}$$
$$C_6H_{12}O_6 \rightarrow 2C_3H_6O_3$$

Ext

Oxygen debt

If you have been exercising hard, you often carry on puffing and panting for some time after you stop. The length of time for which you remain out of breath depends on how fit you are. So why do you carry on breathing faster and more deeply when you have stopped using your muscles?

The waste lactic acid you produce during anaerobic respiration is a problem. You cannot simply get rid of lactic acid by breathing it out, as you can with carbon dioxide. As a result, when the exercise is over, lactic acid has to be broken down to produce carbon dioxide and water. This needs oxygen.

The amount of oxygen needed to break down the lactic acid to carbon dioxide and water is known as the **oxygen debt**.

Ext After running a race, your heart rate and breathing rate stay high to supply the extra oxygen needed to pay off the oxygen debt. The bigger the debt (the larger the amount of lactic acid), the longer you will puff and pant!

Oxygen debt repayment can be represented by the equation:

lactic acid + oxygen → carbon dioxide + water

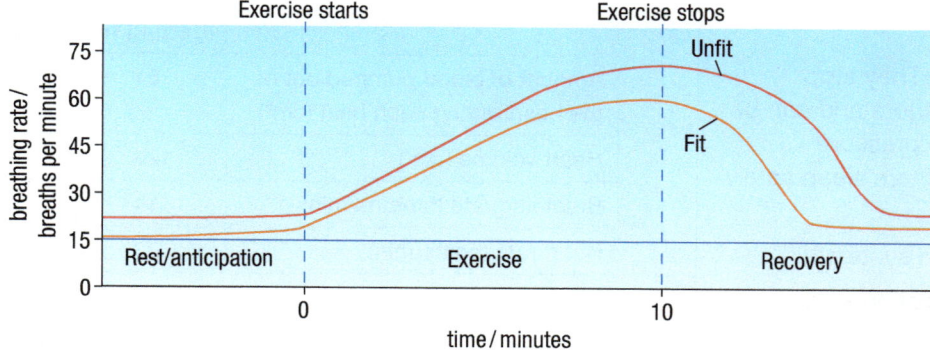

Figure 2 *Everyone gets an oxygen debt if they exercise hard, but if you are fit you can pay it off faster*

Anaerobic respiration in other organisms

Humans and other animals are not the only living organisms that can respire anaerobically. Plants and microorganisms can also respire without oxygen. This allows them to survive in environments with low oxygen levels. However, when plant cells respire anaerobically they do not form lactic acid – they form ethanol and carbon dioxide. Some microorganisms form lactic acid during anaerobic respiration – the bacteria used to produce yoghurts, for example. Other microorganisms, such as yeast, form ethanol and carbon dioxide. People have made use of the products of anaerobic respiration for thousands of years in the production of bread and alcoholic drinks. Relatively small amounts of energy are transferred during anaerobic respiration in bacteria and fungi:

$$\text{glucose} \rightarrow \text{ethanol} + \text{carbon dioxide}$$
$$C_6H_{12}O_6 \rightarrow 2C_2H_5OH + 2CO_2$$

Study tip

Note that energy is not an actual product in a chemical reaction – so do not include it when answering questions requiring you to write a word equation or balanced symbol equation.

Summary questions

1 If you exercise very hard or for a long time, your muscles begin to ache and do not work as effectively. Explain why.

2 If you exercise vigorously, you often puff and pant for some time after you stop. Explain what is happening.

3 a What is anaerobic respiration?
  b Explain how anaerobic respiration differs between animals, plants, and microorganisms. In each case, give the word and balanced symbol equations for what is happening, and explain the benefits to the organism of being able to respire in this way.

Figure 3 *Anaerobic respiration in yeast cells produces ethanol and carbon dioxide*

Key points

- Anaerobic respiration is respiration without oxygen. When this takes place in muscle cells, glucose is incompletely broken down to form lactic acid.

- During long periods of exercise muscles become fatigued and stop contracting efficiently. One cause is the build-up of lactic acid in the muscles. Blood flowing through the muscles eventually removes the lactic acid.

- The anaerobic breakdown of glucose releases less energy than aerobic respiration.

- After exercise, oxygen is still needed to break down the lactic acid that has built up. The amount of oxygen needed is known as the oxygen debt.

- Anaerobic respiration in plant cells and some microorganisms results in the production of ethanol and carbon dioxide.

Summary questions

1 a How are the lungs adapted to allow the exchange of oxygen and carbon dioxide between the air and the blood?

b How is air moved in and out of the lungs and how does this ventilation of the lungs make gaseous exchange more efficient?

2 Some people suffer from sleep apnoea. They stop breathing in their sleep, which disturbs them and can be dangerous. A nasal intermittent positive pressure ventilation system can be used to help them sleep safely through the night.

a What is a positive pressure ventilation system?

b Explain how a positive pressure ventilation system differs from normal breathing.

c What are the advantages of a system like this over a negative pressure ventilation system?

3 Some students investigated the process of cellular respiration. They set up three vacuum flasks. One contained live, soaked peas. One contained dry peas. One contained peas which had been soaked and then boiled. They took daily observations of the temperature in each flask for a week. The results are shown in the table.

Day	Room temperature (°C)	Temperature in flask A containing live, soaked peas (°C)	Temperature in flask B containing dry peas (°C)	Temperature in flask C containing soaked, boiled peas (°C)
1	20.0	20.0	20.0	20.0
2	20.0	20.5	20.0	20.0
3	20.0	21.0	20.0	20.0
4	20.0	21.5	20.0	20.0
5	20.0	22.0	20.0	20.0
6	20.0	22.2	20.0	20.5
7	20.0	22.5	20.0	21.0

a Plot a graph to show these results.

b Explain the results in flask A containing the live, soaked peas.

c Why were the results in flask B the same as the room temperature readings?

d Why was the room temperature in the lab recorded every day?

e Look at the results for flask C.
  i Why is the temperature at 20 °C for the first five days?
  ii After five days the temperature increases. Suggest **two** possible explanations for the temperature increase.

4 It is often said that taking regular exercise and getting fit is good for your heart and your lungs. The following table shows the effect of getting fit on the heart and lungs of one person.

	Before getting fit	After getting fit
Amount of blood pumped out of the heart during each beat (cm³)	64	80
Heart volume (cm³)	120	140
Breathing rate (breaths/min)	14	12
Heart rate (beats/min)	72	63

a The table shows the effect of getting fit on the heart and lungs of one person. Display the data in the table in four bar charts.

b Use the information in your bar charts to help you explain exactly what effect increased fitness has on:
  i your heart
  ii your lungs.

5 a What is aerobic respiration?

b What is anaerobic respiration?

c How does anaerobic respiration differ between a human cell and a yeast cell?

d Define the term oxygen debt.

e Explain the difference in the responses to exercise of a fit and an unfit individual in terms of their muscles, heart and lung function, and oxygen debt.

6 Athletes want to be able to use their muscles aerobically for as long as possible when they compete. They train to develop their heart and lungs. Many athletes also train at altitude. There is less oxygen in the air at altitude so your body makes more red blood cells, which helps to avoid oxygen debt. Some athletes remove some of their own blood, store it, and then transfuse it back into their system just before a competition. This is called blood doping and it is illegal. Other athletes use hormones to stimulate the growth of extra red blood cells. This is also illegal.

a Why do athletes want to be able to use their muscles aerobically for as long as possible?

b How does developing more red blood cells by training at altitude help athletic performance?

c How does blood doping help performance?

d Explain in detail what happens to the muscles if the body cannot supply glucose and oxygen quickly enough when they are working hard.

Practice questions

1 Figure 1 is a diagram of the human respiratory system.

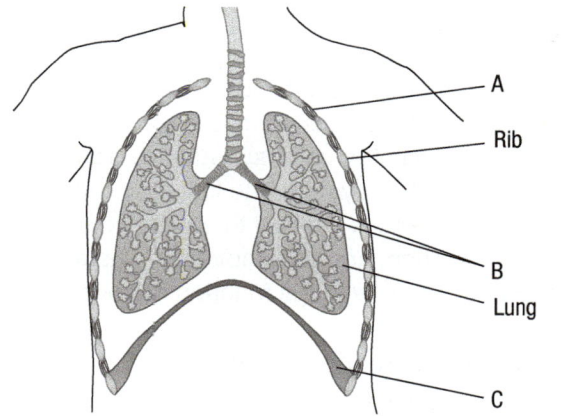

Figure 1

a Names the structures labelled A–C. (3)

b Describe the sequence of events that occurs so that air is drawn into the lungs during inhalation (breathing in). (5)

2 Figure 2 shows an alveolus and a blood capillary.

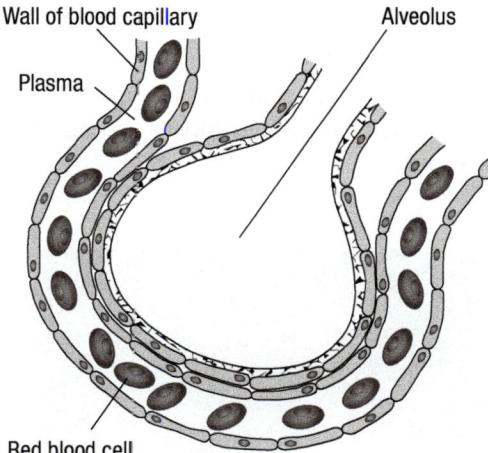

Figure 2

a Give **two** features seen in Figure 2 that increase the diffusion of oxygen from the air in the alveolus into the blood. Explain how each feature increases the rate of diffusion. (4)

b A steep diffusion gradient also increases the rate of oxygen diffusion. Give **two** ways in which a steep diffusion gradient is maintained between the air in the alveolus and the blood in the capillary. (2)

c A constant supply of oxygen is required for all body cells for aerobic respiration.

  i Copy and complete the word and balanced symbol equations for aerobic respiration:

    .... + oxygen → carbon dioxide + ...

    .... + $6O_2$ → ... CO_2 + ... (4)

  ii Name the cell parts where aerobic respiration takes place. (1)

  iii Muscle cells contain a large number of these cell parts. Why? (2)

3 *In this question you will be assessed on using good English, organising information clearly, and using specialist terms where appropriate.*

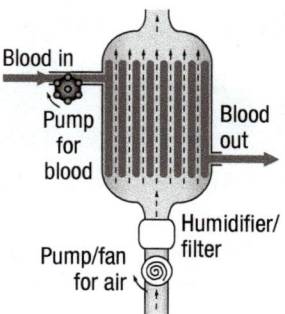

Figure 3

Figure 3 shows a design for an artificial lung.

Many people with lung disease are confined to a wheelchair or are unable to do much exercise. Scientists hope that a portable artificial lung, the size of a spectacle case, can be developed. This device might replace the need for lung transplants and allow patients to live a more normal life.

When scientists design an artificial lung, what features of a normal lung must they copy? Suggest the advantages of the artificial lung compared to a lung transplant. (6)

4 Yeast is used to make bread. Sealed inside the dough, the yeast cells must respire anaerobically.

a Give the word equation for the anaerobic respiration in yeast cells. (3)

b Suggest which product of anaerobic respiration causes the dough to rise. Give a reason. (2)

c Food scientists investigated the anaerobic respiration of yeast in bread dough. They made a dough mixture and divided it into five measuring cylinders. They subjected each measuring cylinder to a different temperature for 15 minutes.

The following table shows their results.

Temperature of measuring cylinder (°C)	0	20	40	60	80
Volume of dough at 0 minutes (cm³)	22	21	21	22	21
Volume of dough at 15 minutes (cm³)	22	27	44	23	21

  i Suggest the best temperature at which to leave bread dough to rise before it is baked. (1)

  ii Use scientific knowledge and understanding to explain why the investigation gave the results seen in the table. (5)

4.1 The circulatory system and the heart

Learning objectives

After this topic, you should know:

- how substances are transported to and from cells
- the function of the heart.

⚭ links

To find out more about how digested food gets into the transport system, see 5.6 Exchange in the gut.

To find out more about how oxygen and carbon dioxide enter or leave the blood, see 3.2 Breathing and gas exchange in the lungs.

To learn how oxygen is used in the cells and how carbon dioxide is produced, read 3.3 Aerobic respiration.

You are made up of billions of cells, and most of them are a long way from a direct source of food or oxygen. This means that direct diffusion is not enough to supply cells in multicellular organisms such as humans, whose surface area to volume ratio is small. A **transport system** is vital to carry substances from where they come into your body (e.g., the digestive system, the lungs) to the cells where they are needed. This transport system is needed to supply your body cells with glucose and oxygen for respiration, and to remove the waste materials that are the by-products of respiration. This is the function of your **circulatory system**. It has three parts:

- the **blood vessels** (the tubes that carry blood around your body)
- the **heart** (which pumps blood around your body)
- the **blood** (the liquid that carries substances around your body).

The heart as a pump

Your heart is the organ that pumps blood around your body. It is made up of two pumps (for the double circulation) that beat together about 70 times each minute. The walls of your heart are almost entirely muscle. This muscle is supplied with oxygen by the **coronary arteries** (Figure 2).

The structure of the human heart is perfectly adapted for pumping blood to your lungs and your body. The two sides of the heart fill and empty at the same time. This gives a strong, coordinated heartbeat.

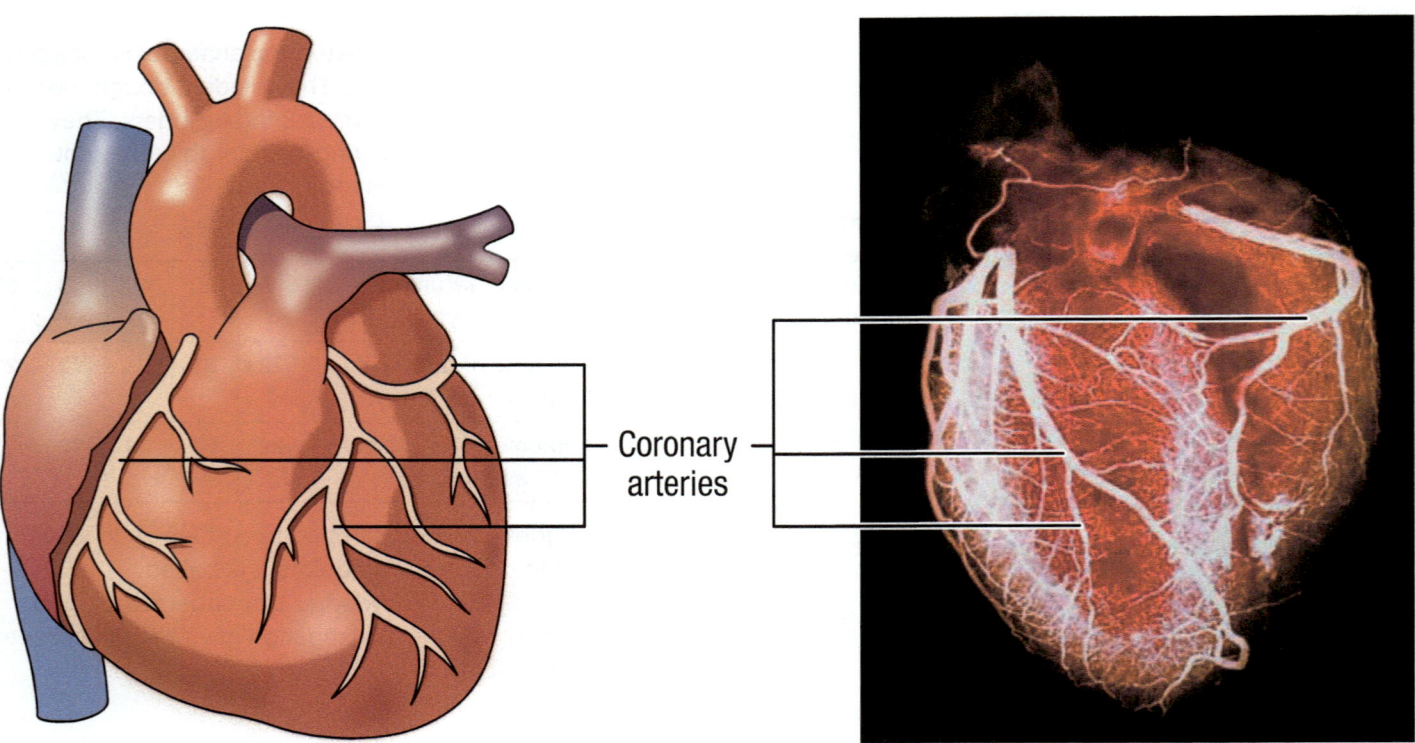

Coronary arteries

Figure 2 The muscles of the heart work hard, so they need a good supply of oxygen and glucose. This is supplied by the blood in the coronary arteries

Blood enters the top chambers of your heart (the **atria**). The blood coming into the right atrium from the **vena cava** is **deoxygenated** blood from your body. The blood coming into the left atrium in the **pulmonary vein** is oxygenated blood from your lungs. The atria contract together and force blood down into the **ventricles**. Valves close to stop blood flowing backwards out of the heart.

- The ventricles contract and force blood out of the heart.
- The right ventricle forces deoxygenated blood to the lungs in the **pulmonary artery**.
- The left ventricle pumps oxygenated blood around the body in a big artery called the **aorta**.

As blood is pumped into the pulmonary artery and the aorta, valves close to prevent backflow of blood to the heart. They make sure the blood flows in the right direction.

The muscle wall of the left ventricle is noticeably thicker than the wall of the right ventricle. This allows the left ventricle to develop much more pressure than the right. This higher pressure is needed as the blood leaving the left ventricle travels through the arterial system all over your body, whilst the blood leaving the right ventricle moves only through the pulmonary arteries to your lungs.

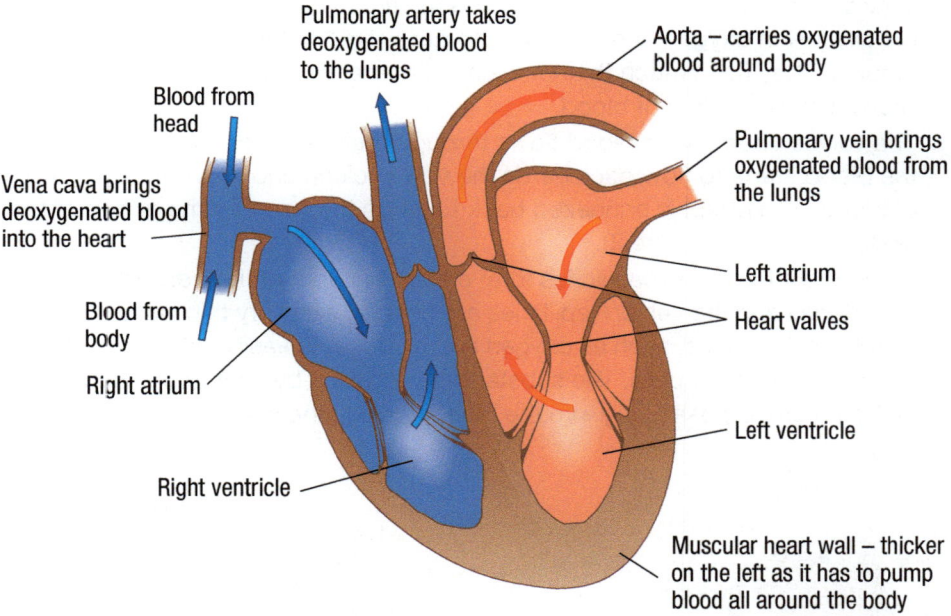

Pulmonary artery takes deoxygenated blood to the lungs

Aorta – carries oxygenated blood around body

Blood from head

Pulmonary vein brings oxygenated blood from the lungs

Vena cava brings deoxygenated blood into the heart

Left atrium

Heart valves

Blood from body

Right atrium

Left ventricle

Right ventricle

Muscular heart wall – thicker on the left as it has to pump blood all around the body

Figure 3 *The structure of the heart – you always label the diagram as if looking at the heart in a person who is facing you*

Summary questions

1 Explain carefully why people need a blood circulation system.

2 Blood in the arteries is usually bright red because it is full of oxygen. This is not true of the blood in the pulmonary arteries. Why not?

3 **a** Describe how the heart pumps blood around the body.

  b Explain the importance of the following in making the heart an effective pump in the circulatory system of the body:
  i heart valves
  ii coronary arteries
  iii the thickened muscular wall of the left ventricle.

Study tip

Remember:

Arteries carry blood **a**way from the heart and veins carry blood back to the heart. This applies to the circulation system of the lungs as well!

??? Did you know … ?

The noise of the heartbeat that you can hear through a stethoscope is actually the sound of the valves of the heart closing to prevent the blood from flowing backwards.

Study tip

Remember:

- the heart has *four* chambers
- ventricles pump blood *out* of the heart.

Key points

- The circulatory system transfers substances to and from the body cells. It consists of the blood vessels, the heart, and the blood.

- The heart is an organ that pumps blood around the body.

- The valves prevent backflow, ensuring that blood flows in the right direction through the heart.

4.2

The blood vessels

After this topic, you should know:

● how the blood flows round the body

● that there are different types of blood vessels

● why valves are important

● the importance of a double circulatory system.

 Did you know ... ?

No cell in your body is more than 0.05 mm away from a capillary.

The substances transported in the blood need to reach the individual cells. Every cell in your body is within 0.05 mm of a capillary – the tiniest blood vessels in your circulatory system.

The blood vessels

Blood is carried around your body in three main types of blood vessels, each adapted for a different function:

● Your arteries carry blood away from your heart to the organs of your body. This blood is usually bright red oxygenated blood. The arteries stretch as the blood is forced through them and go back into shape afterwards. You can feel this as a pulse where the arteries run close to the skin's surface (e.g., at your wrist). Arteries have thick walls containing muscle and elastic fibres, and small lumens (spaces in the middle through which blood flows). As the blood in the arteries is under pressure, it is very dangerous if an artery is cut, because the blood will spurt out rapidly every time the heart beats.

● Your veins carry blood from the organs towards your heart. This blood is usually low in oxygen and so is a deep purple-red colour. Veins do not have a pulse. They have much thinner walls than arteries and often have **valves** to prevent the backflow of blood.

● The valves open as the blood flows through them towards the heart, but if the blood starts to flow backwards the valves close and prevent a backflow of blood. The blood is squeezed back towards the heart by the action of the skeletal muscles (Figure 2).

● Throughout the body, capillaries form a huge network of tiny vessels linking the arteries and the veins. Capillaries are narrow, with very thin walls. This enables substances, such as oxygen and glucose, to easily diffuse out of your blood and into your cells. The substances produced by your cells, such as carbon dioxide, pass easily into the blood through the walls of the capillaries.

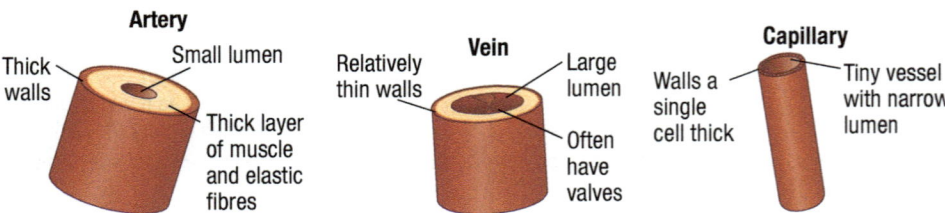

Figure 1 *The three main types of blood vessels*

Practical

Blood flow

You can practise finding your pulse in the arteries that run close to the surface of the body in your wrist and in your neck.

You can find the valves in the veins in your hands, wrists, and forearms and see how the valves prevent the blood flowing backwards.

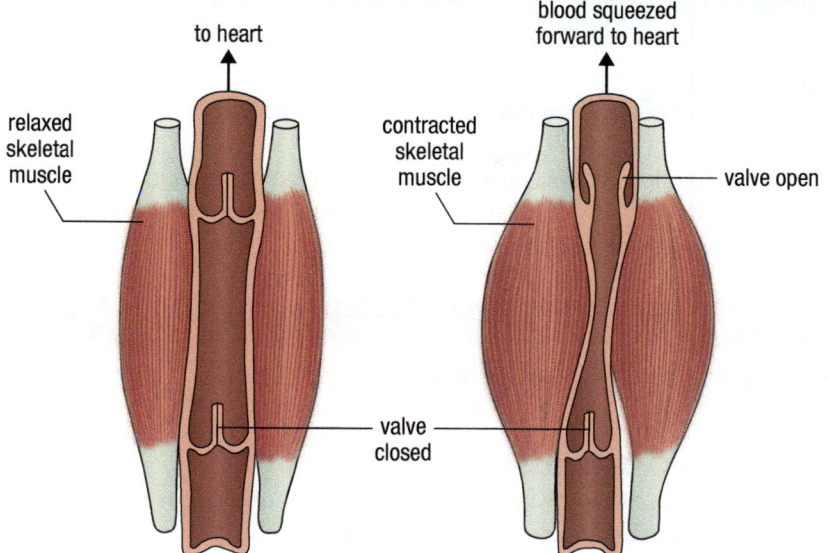

Figure 2 *How the valves and the muscles between them ensure that blood is moved from the body towards the heart.*

In your circulatory system, arteries carry blood away from your heart to the organs of the body. Blood returns to your heart in the veins. The two are linked by the capillary network.

Double circulation

In humans and other mammals the blood vessels are arranged into a **double circulatory system**.

- One transport system carries blood from your heart to your lungs and back again. This allows oxygen and carbon dioxide to be exchanged with the air in the lungs.
- The other transport system carries blood from your heart to all other organs of your body and back again.

A double circulation like this is vital in warm-blooded, active animals such as humans. It makes our circulatory system very efficient. Fully oxygenated blood returns to the heart from the lungs. This blood can then be sent off to different parts of the body at high pressure, so more areas of your body can receive fully oxygenated blood quickly.

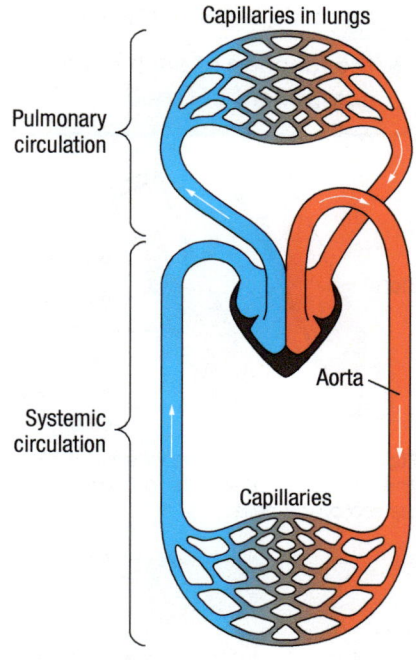

Figure 3 *The two separate circulation systems supply the lungs and the rest of the body*

Key points

- Blood flows around the body via the blood vessels. The main types of blood vessels are arteries, veins, and capillaries.
- Substances diffuse in and out of the blood in the capillaries.
- The valves prevent backflow, ensuring that blood flows in the right direction.
- Human beings have a double circulatory system.

Summary questions

1 Describe the following blood vessels:
   a artery
   b vein
   c capillaries.
2 a Draw a diagram that explains the way in which the arteries, veins, and capillaries are linked to each other and to the heart.
   b Label the diagram, and explain what is happening in the capillaries.

4.3 Transport in the blood

Learning objectives

After this topic, you should know:

- that blood is made up of many different components
- the function of each main component of blood.

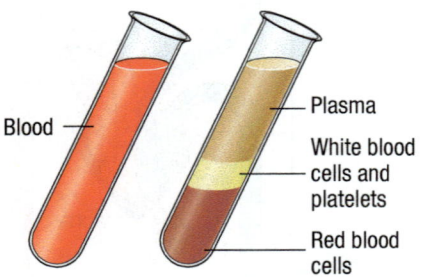

Figure 1 *The main components of blood. The red colour of your blood comes from the red blood cells*

Blood

Plasma

White blood cells and platelets

Red blood cells

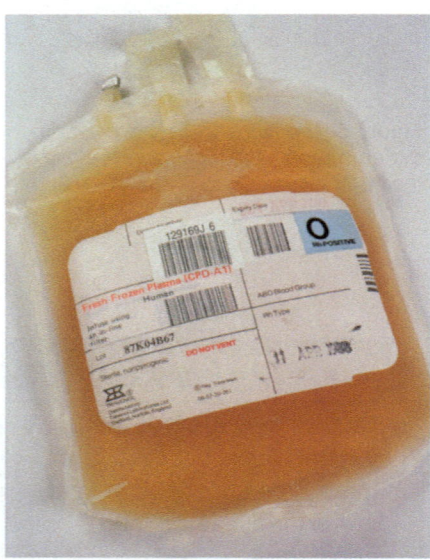

Figure 2 *Blood plasma is a yellow liquid that transports everything you need – and need to get rid of – around your body*

> **??? Did you know …?**
> There are more red blood cells than any other type of blood cell in your body – about 5 million in each cubic millimetre of your blood.

Your blood is a unique fluid based on a liquid called **plasma**. Plasma carries **red blood cells**, **white blood cells**, and **platelets** suspended in it. It also carries many dissolved substances around your body.

> **??? Did you know …?**
> The average person has between 4.7 and 5.0 litres of blood.

Blood plasma as a transport medium

Your blood plasma is a yellow liquid. Plasma transports all of your blood cells and some other substances around your body:

- Waste carbon dioxide produced in the organs of the body is carried to the lungs in the plasma, to be breathed out.
- **Urea** is carried to your kidneys. Urea is a waste product formed in your liver from the breakdown of proteins. It travels dissolved in the plasma from the liver to the kidneys. In the kidneys, the urea is removed from your blood to form **urine**.
- All the small, soluble products of digestion pass into the blood from your small intestine. These food molecules are carried in the plasma around your body to the other organs and individual cells.

Red blood cells

Red blood cells pick up oxygen from your lungs. They carry the oxygen to the organs, tissues, and cells where it is needed. These blood cells have adaptations that make them very efficient at their job:

- They have a very unusual shape – they are **biconcave discs**. This means that they are concave (pushed in) on both sides. This gives the cells an increased surface area over which the diffusion of oxygen can take place.
- Red blood cells are packed full of a special red **pigment** called **haemoglobin** that can carry oxygen.
- They do not have a nucleus. This makes more space to pack in molecules of haemoglobin.

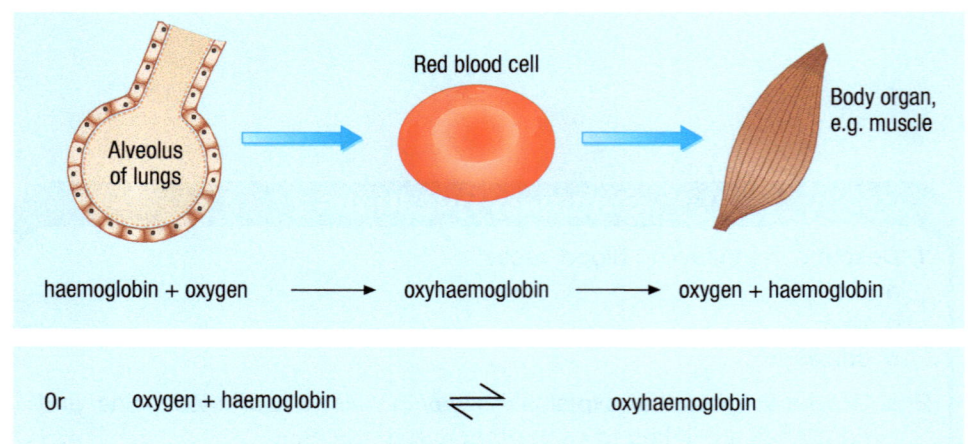

Figure 3 *The reversible reaction between oxygen and haemoglobin makes life as you know it possible by carrying oxygen to all the places where it is needed*

In the lungs where there is a high concentration of oxygen, haemoglobin reacts with oxygen to form bright red **oxyhaemoglobin**. In other organs, where the concentration of oxygen is lower, the oxyhaemoglobin splits up. It forms purple-red haemoglobin and oxygen, which diffuses into the cells where it is needed.

White blood cells

White blood cells are much bigger than red blood cells and there are fewer of them. They have a nucleus and form part of the body's defence system against harmful microorganisms. Some white blood cells form antibodies against microorganisms. Some form antitoxins against poisons made by microorganisms. Yet others (phagocytes) engulf and digest invading bacteria and viruses.

Platelets

Platelets are small fragments of cells that have no nucleus. They are very important in helping the blood to clot at the site of a wound. Blood clotting is a series of enzyme-controlled reactions that result in the change of fibrinogen into fibrin. This produces a network (or web) of protein fibres. The fibres then capture lots of red blood cells and more platelets to form a jelly-like clot. This stops you bleeding to death. The clot dries and hardens to form a scab. The scab protects the new skin as it grows and stops bacteria getting into your body through the wound.

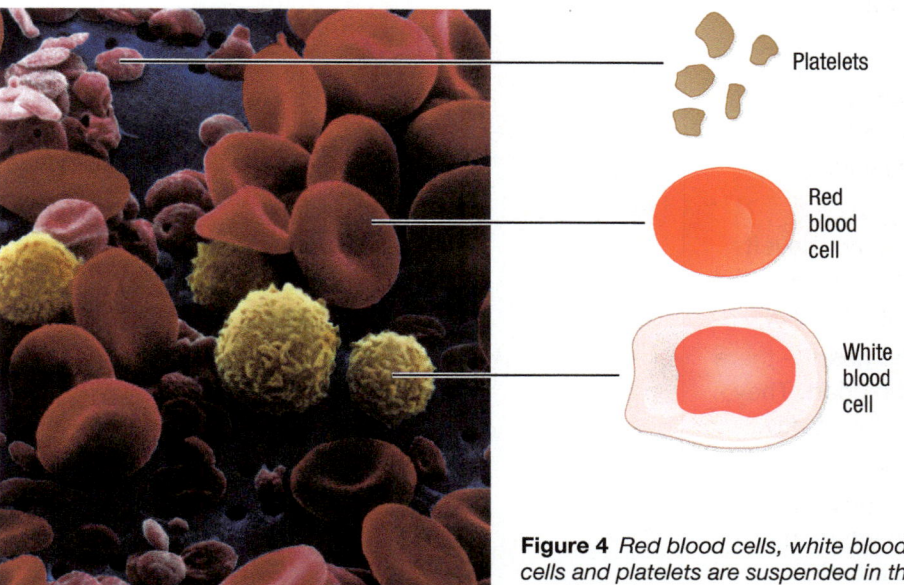

Platelets

Red blood cell

White blood cell

Figure 4 *Red blood cells, white blood, cells and platelets are suspended in the blood plasma*

Key points

- Your blood plasma, and the blood cells suspended in it, transport dissolved food molecules, carbon dioxide, and urea.

- Your red blood cells carry oxygen from your lungs to the organs of the body.

- Red blood cells are adapted to carry oxygen by being biconcave, which provides a bigger surface area, by containing haemoglobin, and by having no nucleus so more haemoglobin can fit in.

- White blood cells are part of the defence system of the body.

- Platelets are cell fragments involved in blood clotting.

- Blood clotting involves a series of enzyme-controlled reactions that turn fibrinogen to fibrin to form a network of fibres and a scab.

Summary questions

1 State three functions of the blood.

2 a Why is it not accurate to describe blood as a red liquid?
  b What actually makes blood red?
  c Give three important functions of blood plasma.

3 Explain carefully the main ways in which blood helps you to avoid infection, including a description of the parts of the blood involved.

Summary questions

1 Figure 1 is a diagram of the human heart.

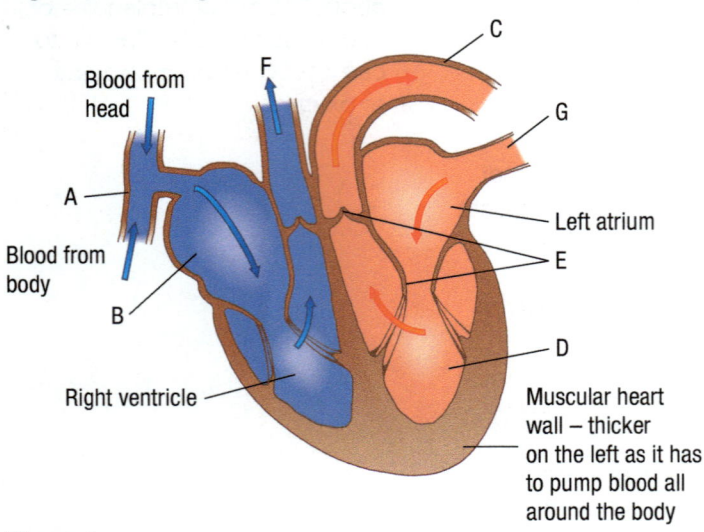

Blood from head

C

F

G

A

Left atrium

Blood from body

E

B

D

Right ventricle

Muscular heart wall – thicker on the left as it has to pump blood all around the body

Figure 1

a Name the parts labelled A–G.

b Describe how the blood flows through the heart from the point where it enters the heart from the head and body.

c Compare the structure and function of the arteries, veins, and capillaries.

2 Here are descriptions of three heart problems. In each case, use what you know about the heart and the circulatory system to explain the problems caused by the condition.

a The valve that stops blood flowing back into the left ventricle of the heart after it has been pumped into the aorta becomes weak and floppy and begins to leak.

b Some babies are born with a 'hole in the heart' – there is a gap in the central dividing wall of the heart. They may look blue in colour and have very little energy.

c The coronary arteries supplying blood to the heart muscle itself may become clogged with fatty material. The person affected may get chest pain when they exercise or they may even have a heart attack.

3 Exchanging materials with the outside world by diffusion is vital for most living organisms. Give four different adaptations that are found in living organisms to make this more efficient. For each adaptation, explain how it makes the exchange process more efficient, and give at least one example of where this adaptation is seen.

4 a Name three functions of the blood

b Copy and complete this table on the main components of the blood

	Plasma	Red blood cells	White blood cells	Platelets
Structure				
Function/ functions				

5 Write a paragraph to explain what happens when the blood clots and why blood clotting is so important

Practice questions

1 Figure 1 shows a vertical section through the heart.

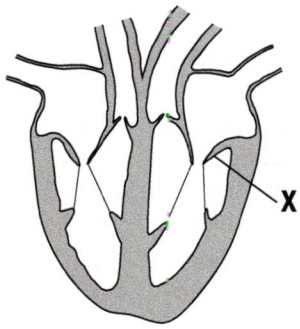

Figure 1

a Copy and complete sentences **i–iv** using the following words:

> *aorta*   *left atrium*   *pulmonary artery*
> *pulmonary vein*   *right atrium*   *vena cava*

   i Blood returning from the body enters the .... of the heart.
   ii The heart pumps blood to the lungs via the ....
   iii Blood returns from the lungs to the heart via the ....
   iv The heart pumps blood to the rest of the body via the .... (4)

b What is the function of the structure labelled **X** in Figure 1? (1)

2 Blood contains plasma, red blood cells, white blood cells, and platelets.

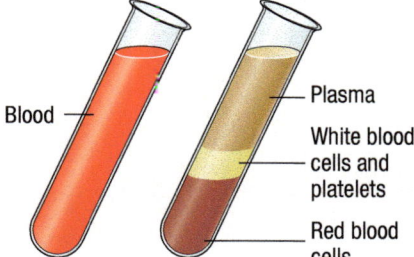

Blood —
Plasma
White blood cells and platelets
Red blood cells

Figure 2

the tube showing the different parts of the blood only.

a The blood is made up of several different types of cells - why does it look like a red liquid if you cut yourself (2)

b State the function of white blood cells. (1)

c State the function of platelets. (1)

d Describe how oxygen is moved from the lungs to the tissues. (3)

e Plasma transports dissolved substances from one part of the body to another. Name **two** of these substances. Explain where they are transported to and why. (4)

3

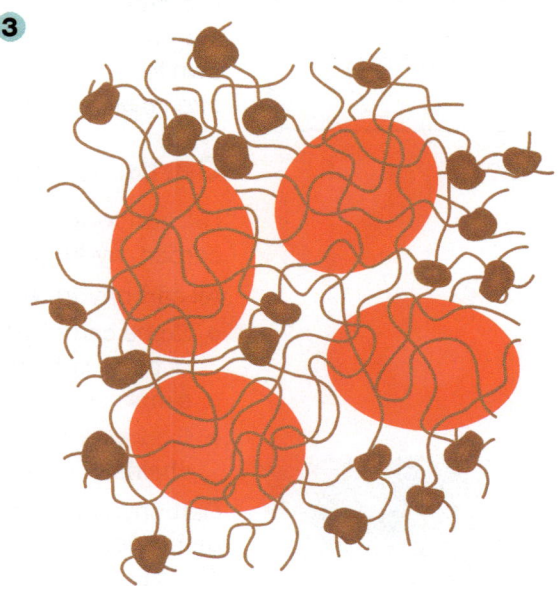

Figure 3

Fig 3 shows a magnified diagram of a blood clot.

a Why is the clotting of the blood so important? (3)

b What structures are labelled A, B and C? (3)

c Describe the process of the blood clotting. (3)

4

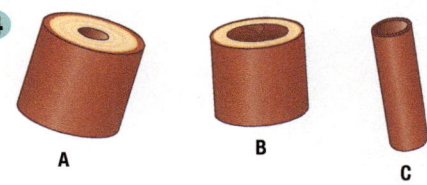

A   B   C

Figure 4

a Name the three main types of blood vessel labelled A, B and C (3)

b For each type of blood vessel explain how its structure is related to its function in the body (6)

Carbohydrates, lipids, and proteins

The food you take into your body is made up of big, insoluble molecules that are of no use to your cells. Your digestive system breaks these molecules down into small, soluble molecules that can be taken into your cells. Inside your cells, the digested food is built up into large, useful molecules again. Carbohydrates, lipids, and proteins are the main compounds that make up the structure of a cell. They are vital components in the balanced diet of any organism which cannot make its own food. Carbohydrates, lipids, and proteins are all large molecules that are often made up of smaller molecules joined together.

Carbohydrates

Carbohydrates provide us with energy. They contain the chemical elements carbon, hydrogen, and oxygen.

All carbohydrates are made up of units of sugar:

- Some carbohydrates contain only one or two units of sugar. The best known of these is glucose, $C_6H_{12}O_6$. These small carbohydrates are referred to as simple sugars.
- **Complex carbohydrates** such as starch and cellulose are made up of long chains of simple sugar units bonded together (Figure 1).

Carbohydrate-rich foods include bread, potatoes, rice, and pasta. Much of the carbohydrate food you take into your body will be broken down to form glucose. The glucose is used in respiration to provide energy for your cells. Carbohydrates in the form of cellulose are a very important support material in plants.

Figure 1 *Carbohydrates are all based on simple sugar units*

Lipids

Lipids are fats (solids) and oils (liquids). They are the most efficient energy store in your body and an important source of energy in your diet. Combined with other molecules, lipids are very important in your cell membranes, as hormones, and in your nervous system. Like carbohydrates, lipids are made up of carbon, hydrogen, and oxygen. All lipids are insoluble in water.

Lipids are made up of three molecules of fatty acids joined to a molecule of glycerol (Figure 2). The glycerol is always the same, but the fatty acids vary. It is the different fatty acids that cause some lipids to be solid fats and others to be liquid oils. Lipid-rich food includes all the oils, such as olive oil and corn oil, as well as butter, margarine, cheese, and cream.

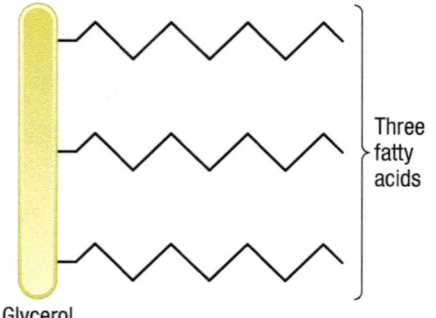

Figure 2 *It is the combination of fatty acids joined to the glycerol molecule that affect the melting point of a lipid*

Proteins

Proteins are used for building up the cells and tissues of your body as well as all your enzymes. They are made up of the elements carbon, hydrogen, oxygen, and nitrogen. Protein-rich foods include meat, fish, pulses, and cheese.

A protein molecule is made up of long chains of small units called **amino acids** (Figure 3). These amino acids are joined together into long chains to produce different proteins.

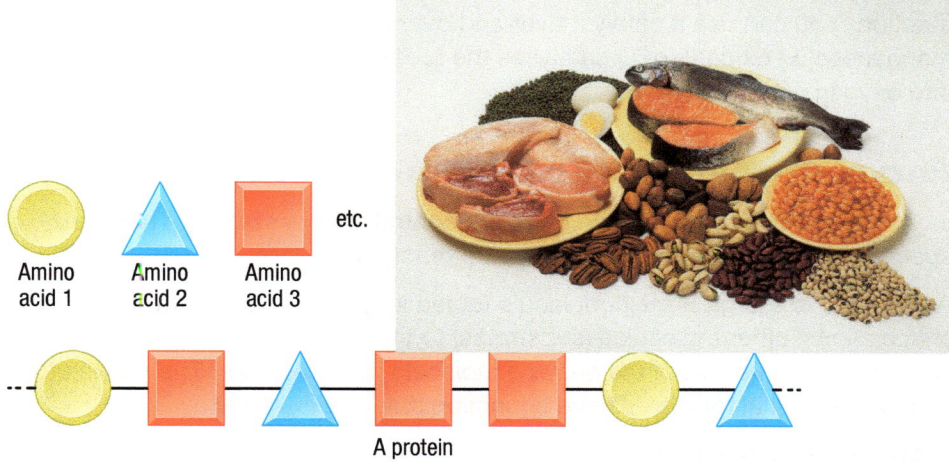

Figure 3 *Amino acids are the building blocks of proteins. They can join in an almost endless variety of ways to produce different proteins*

The long chains of amino acids that make up a protein are folded, coiled, and twisted to make specific 3-D shapes. It is these specific shapes that enable other molecules to fit into the protein. The bonds that hold the proteins in these 3-D shapes are very sensitive to temperature and pH, and can easily be broken. If this happens, the shape of the protein is lost and it may not function any more in your cells. The protein is **denatured**.

Proteins carry out many different functions in your body. They act as:
- structural components of tissues such as muscles and tendons
- hormones such as insulin
- antibodies, which destroy pathogens and are part of the immune system
- enzymes, which act as catalysts in the cells.

Summary questions

1 a What is a protein?
  b How are proteins used in the body?

2 Describe the main similarities and differences between the three main groups of chemicals (carbohydrates, proteins, and lipids) in the body.

3 How would you test a food sample to see if it contained:
  a starch
  b lipids?

4 a Explain why lipids can be either fats or oils.
  b Explain how simple sugars are related to complex carbohydrates.

Did you know …?

Between 15% and 16% of your body mass is protein. Protein is found in tissues ranging from your hair and nails to the muscles that move you around and the enzymes that control your body chemistry.

Practical

Food tests
You can identify the main food groups using standard food tests.
- *Carbohydrates* – Iodine test turns solution from yellowy-red to blue-black if starch is present. Benedict's test turns solution from blue to brick red on heating if a simple reducing sugar such as glucose is present.
- *Protein*: Biuret test turns solution from blue to purple if protein is present.
- *Lipids*: Ethanol (highly flammable, harmful) test gives a cloudy white layer if a lipid is present.

Safety: Wear eye protection.

Key points

- Simple sugars are carbohydrates that contain only one or two sugar units. Complex carbohydrates such as starch contain long chains of simple sugar units bonded together.

- Lipids consist of three molecules of fatty acids bonded to a molecule of glycerol.

- Protein molecules are made up of long chains of amino acids. These are folded to form specific shapes that are related to their different functions (e.g., as enzymes, antibodies, or hormones).

5.2 Catalysts and enzymes

In everyday life, you control the rate of chemical reactions all the time. For example, you increase the temperature of your oven to speed up chemical reactions when you cook. You place food in your fridge to slow down reactions that cause food to go off.

Sometimes special chemicals known as **catalysts** are used to speed up reactions. A catalyst speeds up a chemical reaction, but it is not used up in the reaction. You can use a catalyst over and over again. For example, manganese(IV) oxide (MnO_2) catalyses the breakdown of hydrogen peroxide into oxygen and water.

Enzymes – biological catalysts

In your body, chemical reaction rates are controlled by enzymes. These are special biological catalysts that speed up reactions.

An enzyme is a large protein molecule folded into a specific shape. This special shape allows other molecules (substrates) to fit into the enzyme protein. Scientists call this part of the enzyme molecule its **active site**. The shape of an enzyme is vital for the enzyme's function (the way it works).

Enzymes are involved in:

● building large molecules from lots of smaller ones

● changing one molecule into another

● breaking down large molecules into smaller ones.

Enzymes do not change a reaction in any way – they just make it happen faster. Different enzymes catalyse (speed up) specific types of reaction. In your body you need to build large molecules from smaller ones, such as making glycogen from glucose or proteins from amino acids. You need to change certain molecules into different ones, for example one sugar into another, such as glucose to fructose, and you need to break down large molecules into smaller ones, such as breaking down insoluble food molecules into small, soluble molecules, such as glucose. All these reactions are speeded up using enzymes.

Each of your cells can have a hundred or more chemical reactions going on within it at any one time. Each of the different types of reaction is controlled by a different specific enzyme. Enzymes deliver the control that makes it possible for your cell chemistry to work without one reaction interfering with another.

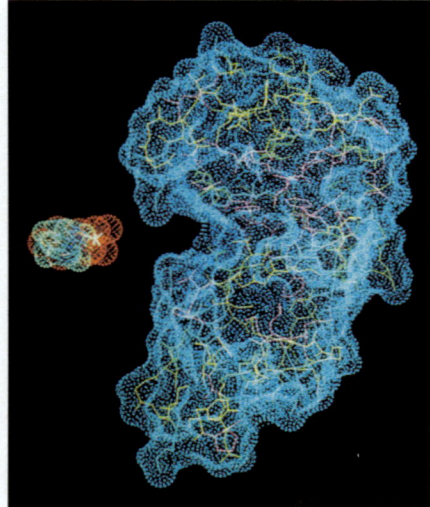

Figure 1 *This computer-generated model of an enzyme shows you the active site and the substrate molecule that fits into it*

?? **Did you know … ?**

For chemicals to react, they need to collide with sufficient energy to break the chemical bonds that hold the molecules together. Enzymes lower the energy needed to break the bonds, which is how they speed up the reactions (because a higher proportion of molecules have sufficient energy to react).

Practical

Breaking down hydrogen peroxide

You can investigate the impact of both an inorganic catalyst and an enzyme on the breakdown of 20 vol hydrogen peroxide solution into oxygen and water using:

a manganese(IV) oxide, and **b** raw liver or potato (both contain the enzyme catalase).

Your liver plays an important role in your body by breaking down toxins (poisons). Hydrogen peroxide is a poisonous compound that is often a waste product of reactions in cells. It is important that it is broken down into harmless oxygen and water quickly, before it causes any damage.

You can determine the rate of the reaction by measuring the volume of oxygen produced over time. A simple way to do a quick comparison between the inorganic catalyst and the enzyme is to add a drop of washing-up liquid to the hydrogen peroxide. Add the catalyst or the enzyme and measure how quickly the foam produced by the bubbles of oxygen gas rises up the test tube!

● Describe your observations and interpret the graph (Figure 2).

Safety: Wear eye protection. 20 'vol' hydrogen peroxide – irritant. Manganese(IV) oxide – harmful.

Figure 2 *The decomposition of hydrogen peroxide to oxygen and water goes much faster using a catalyst like manganese(IV) oxide. Raw liver contains the enzyme catalase, which speeds up the same reaction*

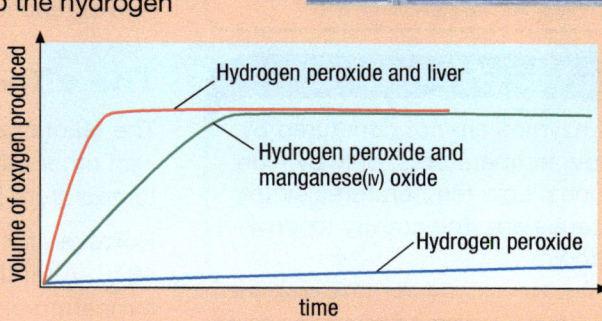

How do enzymes work?

The substrate (reactant) of the reaction to be catalysed fits into the active site of the enzyme. You can think of it like a lock and key. Once the substrate is in place, the enzyme and the substrate bind together.

The reaction then takes place rapidly and the products are released from the surface of the enzyme (Figure 3). Remember that enzymes can join small molecules together as well as break up large ones.

Study tip

Remember that the way an enzyme works depends on the shape of the active site that allows it to bind with the substrate.

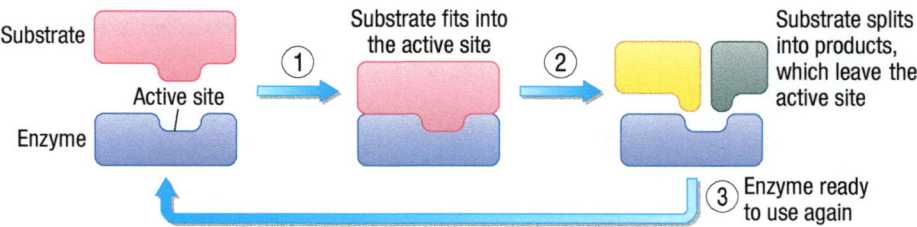

Figure 3 *Enzymes act as catalysts using the 'lock-and-key' mechanism shown here*

Summary questions

1 Define the following terms:
   a catalyst   **b** enzyme   **c** active site.

2 **a** What are enzymes made of?
   b Explain carefully how enzymes act to speed up reactions in your body.

3 **a** Give **three** clear examples of the types of reactions that are catalysed by enzymes.
   b Explain the importance of enzymes within cells.

Key points

● Catalysts increase the rate of chemical reactions without changing chemically themselves.

● Enzymes are biological catalysts.

● Enzymes are proteins. The amino acid chains are folded to form the active site, which matches the shape of a specific substrate.

● The substrate binds to the active site and the reaction is catalysed by the enzyme.

5.3 Factors affecting enzyme action

A container of milk left at the back of your fridge for a week or two will be disgusting. The milk will go off as enzymes in bacteria break down the protein structure.

Leave your milk in the sun for a day and the same thing happens – but much faster. Temperature affects the rate at which chemical reactions take place, even when they are controlled by biological catalysts.

Biological reactions are affected by the same factors as any other chemical reactions. These factors include concentration, temperature, and surface area. However, in living organisms, an increase in temperature only increases the rate of reaction up to a certain point.

The effect of temperature on enzyme action

The reactions that take place in cells happen at relatively low temperatures. As with other reactions, the rate of enzyme-controlled reactions increases as the temperature increases.

However, for most organisms this is only true up to temperatures of about 40 °C. After this, the protein structure of the enzyme is affected by the high temperature. The long amino acid chains begin to unravel and, as a result, the shape of the active site changes. The substrate will no longer fit in the active site. The enzyme has been denatured. It can no longer act as a catalyst, so the rate of the reaction drops dramatically. Most human enzymes work best at 37 °C, which is normal human body temperature.

Without enzymes, none of the reactions in your body would happen fast enough to keep you alive. This is why it is so dangerous if your temperature goes too high when you are ill. Once your body temperature reaches about 41 °C, your enzymes start to be denatured and you will soon die.

Learning objectives

After this topic, you should know:

- how temperature and pH affect enzyme action
- how digestive enzymes differ from most of the enzymes in your body.

Study tip

Enzymes are *not* denatured by low temperatures, only by high ones. Low temperatures simply cause enzyme activity to slow down.

Figure 1 *The magical light display of this comb jelly is caused by the action of an enzyme called luciferase*

??? Did you know … ?

Not all enzymes work best at around 40 °C. Bacteria living in hot springs survive at temperatures up to 80 °C and higher. On the other hand, some bacteria that live in the very cold, deep seas have enzymes that work effectively at 0 °C and below.

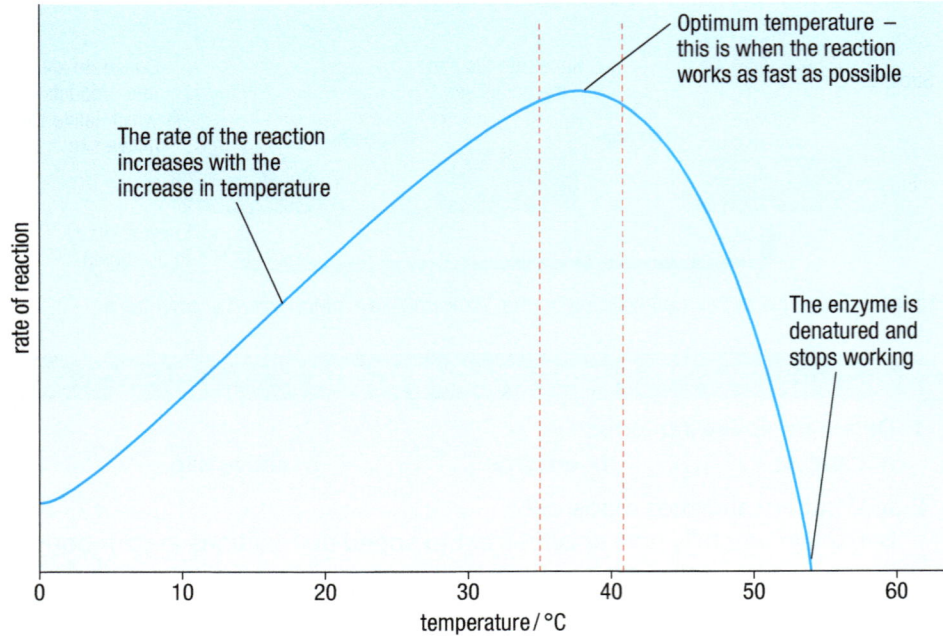

Figure 2 *The rate of an enzyme-controlled reaction increases as the temperature rises – but only until the protein structure of the enzyme breaks down*

Effect of pH on enzyme action

The shape of the active site of an enzyme comes from forces between the different parts of the protein molecule. These forces hold the folded chains in place. A change in the pH affects these forces. That's why it changes the shape of the molecule. As a result, the specific shape of the active site is lost, so the enzyme no longer acts as a catalyst.

Different enzymes work best at different pH levels. A change in the pH can stop them working completely.

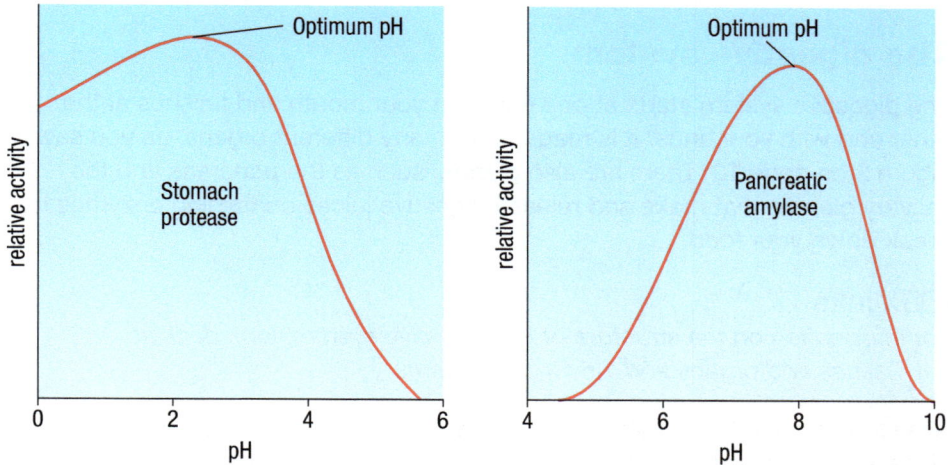

Figure 3 *These two digestive enzymes need very different pH levels to work at their maximum rate. The protease found in the stomach is mixed with hydrochloric acid, whilst pancreatic amylase is found in the first part of the small intestine along with alkaline bile*

The digestive enzymes

Most of your enzymes work *inside* the cells of your body, controlling the rate of the chemical reactions. Your digestive enzymes are different. They work *outside* your cells. They are produced by specialised cells in glands (such as the salivary glands and pancreas) and in the lining of your gut. The enzymes then pass out of these cells into the gut itself, where they come into contact with food molecules.

Your food is made up of large, insoluble molecules that your body cannot absorb. They need to be broken down or **digested** to form smaller, soluble molecules that can be absorbed and used by your cells. It is this chemical breakdown of your food which is controlled by digestive enzymes.

Different areas of the digestive system have different pH levels, which allow the enzymes in that region to work as efficiently as possible. For example, the mouth and small intestine are slightly alkaline, whilst the stomach has a low, acidic pH value.

Summary questions

1 Explain carefully, with the help of diagrams, how temperature affects enzyme-controlled reactions and why.

2 Look at Figure 3.
   a At which pH does the protease in the stomach work best?
   b At which pH does amylase work best?
   c What happens to the activity of the enzymes as the pH increases?
   d Explain why this change in activity happens.

⚭ links

You have already discovered many of the different organs in the human digestive system in 1.4 Tissues and organs and 1.5 Organ systems.

Study tip

Enzymes aren't killed (they are molecules, not living things) – so make sure that you use the term denatured.

Key points

- The shape of an enzyme is vital for the functioning of the enzyme.

- High temperatures denature the enzyme, changing the shape of the active site.

- Different enzymes work best at different pH values.

- Digestive enzymes are produced by specialised cells in glands and in the lining of the gut. The enzymes pass out of the cells into the gut, where they come into contact with food molecules and catalyse the breakdown of large molecules into smaller ones.

5.4

The digestive system

The food you take in and eat is made up of large, insoluble molecules, including starch (a carbohydrate), proteins, and fats. Your body cannot absorb and use these molecules, so they need to be broken down or digested to form smaller, soluble molecules. These can then be absorbed in your small intestine and used by your cells. This process of digestion takes place in your digestive system.

The digestive system

The digestive system starts at one end with your mouth and finishes at the other end with your anus. It is made up of many different organs, as you saw in Figure 2 on page 10. There are also glands, such as the pancreas and the salivary glands, that make and release digestive juices containing enzymes to break down your food.

⚙ links

For information on the structure of the digestive system, look back at 1.4 Tissues and organs and 1.5 Organ systems.

The stomach and the small intestine are the main organs where food is digested. Enzymes break down the large, insoluble food molecules into smaller, soluble ones.

Your small intestine is also where the soluble food molecules are absorbed into your blood. The digested food molecules are small enough to pass freely through the walls of the small intestine into the blood vessels by diffusion. They move in this direction because there is a very high concentration of food molecules in the gut and a much lower concentration in the blood. They move into the blood down a steep concentration gradient. Some substances are also moved from the gut into your blood by active transport. Villi and microvilli greatly increase the surface area of the small intestine so that the absorption of digested food is very efficient. Once absorbed, digested food molecules are transported in the bloodstream around your body.

The muscular walls of the gut squeeze the undigested food onwards into your large intestine. This is where water is absorbed from the undigested food into your blood. The material that remains makes up the bulk of your faeces. Faeces are stored and then pass out of your body through the anus back into the environment.

⚙ links

For more information on moving substances in and out of cells, see 1.6 Diffusion, 1.7 Osmosis, and 1.8 Active transport. For more information on the adaptations of the gut for absorption, see 5.6 Exchange in the gut.

Study tip

Learn the names of the parts of the digestive system. Make sure that you know the difference between the larger, lobed liver and the smaller, thinner pancreas.

Digestive enzymes

Most of your enzymes work *inside* the cells of your body, controlling the rate of chemical reactions. Your digestive enzymes are different. They work *outside* your cells. They are produced by specialised cells in glands (such as the salivary glands and pancreas), and in the lining of your gut.

The enzymes then pass out of these cells into the gut itself. Your gut is a hollow, muscular tube that squeezes your food. It helps to break up your food into small pieces with a large surface area for your enzymes to work on. It mixes your food with your digestive juices so that the enzymes come into contact with as much of the food as possible. The muscles of the gut move your food along from one area to the next.

⁇ Did you know … ?

When Alexis St Martin suffered a terrible gunshot wound in 1822, Dr William Beaumont managed to save his life. However, Alexis was left with a hole (or fistula) from his stomach to the outside world. Dr Beaumont then used this hole to find out what happened in Alexis's stomach as he digested food!

Digesting carbohydrates

Enzymes that break down carbohydrates are called carbohydrases. Starch is one of the most common carbohydrates that you eat. It is broken down into sugars in your mouth and small intestine. This reaction is catalysed by an enzyme called **amylase**.

Amylase is produced in your salivary glands, so the digestion of starch starts in your mouth. Amylase is also made in the pancreas and the small intestine. No digestion takes place inside the pancreas. All the enzymes made there flow into your small intestine, where most of the starch you eat is digested.

Practical

Investigating factors that affect the rate of digestion

The enzyme amylase breaks down starch into simple sugars.

Iodine may be used as an indicator for starch.

Set up test tubes containing the same volumes of starch solution and amylase in water baths at different temperatures ranging from 0 °C to 60 °C.

Sample the starch/amylase mixture from each temperature at regular intervals and record the time taken for the amylase to break down the starch at each temperature.

You can use the same basic experimental set-up to investigate the effect of pH on amylase activity. Keep all of the tubes at the same temperature but vary the pH of the tubes between pH 5 and pH 9.

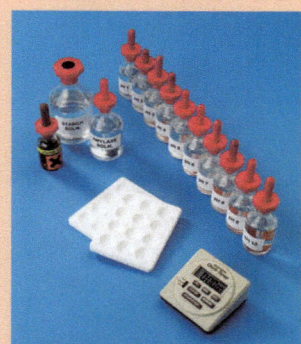

Figure 1 *Use spotting tiles with a drop of iodine solution in each well to test your samples for starch*

Digesting proteins

The breakdown of protein foods such as meat, fish and cheese into amino acids is catalysed by **protease** enzymes. Proteases are produced by your stomach, your pancreas, and your small intestine. The breakdown of proteins into amino acids takes place in your stomach and small intestine.

Digesting fats

The lipids (fats and oils) that you eat are broken down into fatty acids and glycerol in the small intestine. The reaction is catalysed by **lipase** enzymes, which are made in your pancreas and your small intestine. Again, the enzymes made in the pancreas are passed into the small intestine.

Once your food molecules have been completely digested into soluble glucose, amino acids, fatty acids, and glycerol, they leave your small intestine. They pass into your bloodstream to be carried around the body to the cells that need them.

Key points

- Digestion involves the breakdown of large, insoluble molecules into soluble substances that can be absorbed into the blood across the wall of the small intestine.

- Digestive enzymes are produced by specialised cells in glands and in the lining of the gut.

- Carbohdrases such as amylase catalyse the breakdown of carbohydrates such as starch to sugars.

- Proteases catalyse the breakdown of proteins to amino acids.

- Lipases catalyse the breakdown of lipids to fatty acids and glycerol.

Summary questions

1. Make a table that describes amylase, protease, and lipase. For each enzyme, show where it is made, which reaction it catalyses, and where it works in the gut.

2. Why is digestion of food so important? Explain your answer in terms of the molecules involved.

5.5

Making digestion efficient

Learning objectives

After this topic, you should know:

● the roles of hydrochloric acid and bile in making digestion more efficient.

Required practical (alternative)

Breaking down protein

You can see the effect of acid on the protease found in the stomach (called pepsin) quite simply. Set up three test tubes: one containing stomach protease, one containing hydrochloric acid, and one containing a mixture of the two. Keep them at body temperature in a water bath. Add a similar-sized chunk of meat to all three of them. Set up a webcam and watch for a few hours to see what happens.

● What conclusions can you make?

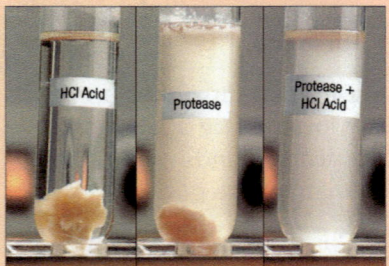

Figure 1 *These test tubes show clearly the importance of protein-digesting enzymes and hydrochloric acid in your stomach. Meat was added to each tube at the same time*

Safety: Wear eye protection (if HCl is stronger than 6.5 M then chemical splash-proof eye protection would be needed).

Your digestive system produces many enzymes that speed up the breakdown of the food you eat. As your body is kept at a fairly steady 37 °C, your enzymes have an optimum temperature that allows them to work as fast as possible.

Keeping the pH in your gut at optimum levels isn't that easy, because different enzymes work best at different pH levels. For example, the protease enzyme found in your stomach works best in acidic conditions, whilst the proteases made in your pancreas need alkaline conditions to work at their best. So, your body makes a variety of different chemicals that help to keep conditions ideal for your enzymes all the way through your gut.

Changing pH in the gut

You have around 35 million glands in the lining of your stomach. These secrete a protease enzyme to digest the protein you eat. This stomach protease works best in an acidic pH, so your stomach also produces a relatively concentrated solution of hydrochloric acid from the same glands. In fact, your stomach produces around 3 litres of hydrochloric acid a day! This acid allows your stomach protease enzymes to work very effectively. It also kills most of the bacteria that you take in with your food.

Your stomach also produces a thick layer of mucus. This coats your stomach walls and protects them from being digested by the acid and the enzymes.

After a few hours – depending on the size and the type of the meal you have eaten – your food leaves your stomach. It moves on into your small intestine. Some of the enzymes that catalyse digestion in your small intestine are made in your pancreas. Some are also made in the small intestine itself. They all work best in an alkaline environment.

The acidic liquid coming from your stomach needs to become an alkaline mix in your small intestine. Your liver makes a greeny-yellow alkaline liquid called **bile** so this can happen. Bile is stored in your gall bladder until it is needed.

As food comes into the small intestine from the stomach, bile is squirted onto it through the bile duct. The bile neutralises the acid that was added to the food in the stomach. This provides the alkaline conditions necessary for the enzymes in the small intestine to work most effectively.

Altering the surface area

It is very important for the enzymes of the gut to have the largest possible surface area of food to work on. This is not a problem with carbohydrates and proteins. However, the fats that you eat do not mix with all the watery liquids in your gut. They stay as large globules (like oil in water) that make it difficult for the lipase enzymes to act.

This is the second important function of bile – it **emulsifies** the fats in your food. This means that bile physically breaks up large drops of fat into smaller droplets. This provides a much bigger surface area of fats for the lipase enzymes to act upon. The larger surface area helps the lipase chemically break down the fats more quickly into fatty acids and glycerol.

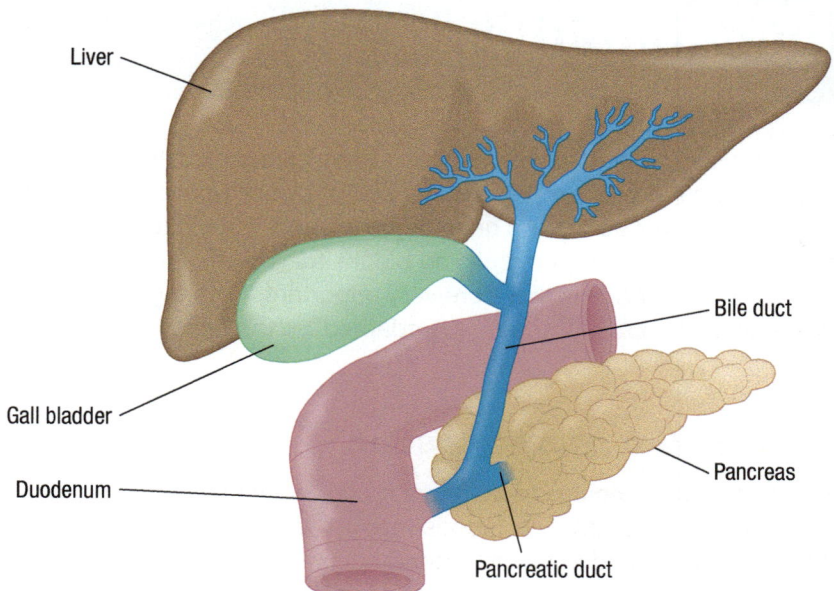

Liver

Bile duct

Gall bladder

Pancreas

Duodenum

Pancreatic duct

Figure 2 *Bile drains down small bile ducts in the liver. Most of it is stored in the gall bladder until it is needed*

Sometimes gall stones form, which can block the gall bladder and bile ducts. The stones can range from a few millimetres to several centimetres in diameter and can cause terrible pain (Figure 3). They can also stop bile being released onto the food in the small intestine, which reduces the efficiency of digestion.

Figure 3 *Gall stones can be very large and can cause extreme pain*

Summary questions

1 Look at Figure 1 opposite and Figure 3 on Page 61.
   a In what conditions does the protease from the stomach work best?
   b How does your body create the right pH in the stomach for this enzyme?
   c In what conditions do the proteases in the small intestine work best?
   d How does your body create the right pH in the small intestine for these enzymes?

2 Draw and label a diagram to explain how bile produces a big surface area for lipase to work on and explain why this is important.

3 Use everything you have learnt about digestion to describe the passage of a meal containing bread, butter, and egg through your digestive system from beginning to end.

For information on the sensitivity of enzymes to temperature and pH, look back at 5.3 Factors affecting enzyme action.

??? Did you know ... ?

If someone develops a stomach ulcer, the protective mucus is lost and acid production may increase. The lining of the stomach is then attacked by the acid and the protein-digesting enzymes, which is very painful.

links

For information on the sensitivity of enzymes to temperature and pH, look back at 5.3 Factors affecting enzyme action.

Study tip

Understand that:
Hydrochloric acid gives the stomach a low pH suitable for the protease secreted there to work efficiently.
Alkaline bile neutralises the acid and gives a high pH for the enzymes from the pancreas and small intestine to work well.
● Bile is **not** an enzyme as it does **not** break down fat molecules.
● Instead it emulsifies the fat into tiny droplets, which increases the surface area for lipase to act on, increasing the rate of digestion.

Key points

● The protease enzymes of the stomach work best in acid conditions. The stomach produces hydrochloric acid, which maintains a low pH.

● The enzymes made in the pancreas and the small intestine work best in alkaline conditions.

● Bile produced by the liver, stored in the gall bladder, and released through the bile duct neutralises acid and emulsifies fats.

5.6

Exchange in the gut

After this topic, you should know:

- how the small intestine is adapted to enable you to absorb food efficiently
- why villi are so important.

The food you eat is broken down in your gut. Large food molecules are digested by enzymes to give simple sugars (e.g., glucose), amino acids, fatty acids, and glycerol. Your body cells need these products of digestion to provide fuel for respiration and the building blocks for growth and repair. A successful exchange surface is therefore very important. Your digestive system, particularly the **small intestine**, is adapted for the effective exchange of solutes.

Absorption in the small intestine

For the digested food molecules to reach your cells, they must move from inside your small intestine into your bloodstream. They do this by a combination of diffusion and active transport.

The digested food molecules are small enough to pass freely through the walls of the small intestine into the blood vessels. They move in this direction because there is a very high concentration of food molecules in the gut and a much lower concentration in the blood. They move into the blood down a steep concentration gradient.

The lining of the small intestine is folded into thousands of tiny finger-like projections known as **villi** (singular: villus). These greatly increase the uptake of digested food by diffusion (Figure 1). Only a certain number of digested food molecules can diffuse over a given surface area of gut lining at any one time. Increasing the surface area means that there is more room for diffusion to take place (Figure 2).

Each individual villus is itself covered in many microscopic microvilli. This increases the surface area available for diffusion even more.

links

You will find information on glucose, amino acids, fatty acids, and glycerol in 5.4 'The digestive system'.

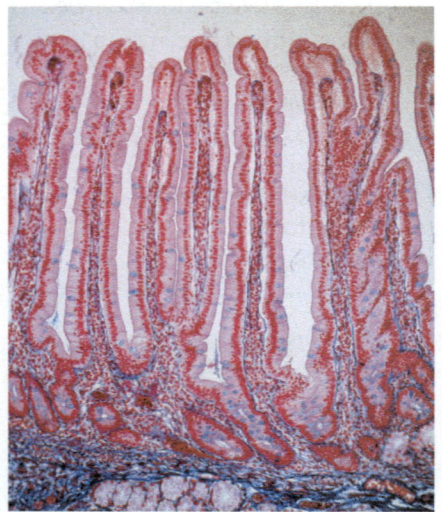

Light micrograph of a section through the villi of the small intestine

Scanning electron micrograph of the villi of the small intestine

Figure 1 *The villi of the small intestine increase the surface area available for diffusion many times over. This means you can absorb enough digested food to survive*

The lining of the small intestine has an excellent blood supply. This carries away the digested food molecules as soon as they have diffused from one side to the other. So, a steep concentration gradient is maintained all the time, from the inside of the intestine to the blood (see Figure 3). This in turn ensures that diffusion is as rapid and efficient as possible down the concentration gradient.

Active transport in the small intestine

Diffusion isn't the only way in which dissolved products of digestion move from the gut into the blood. As the time since your last meal increases, you will have

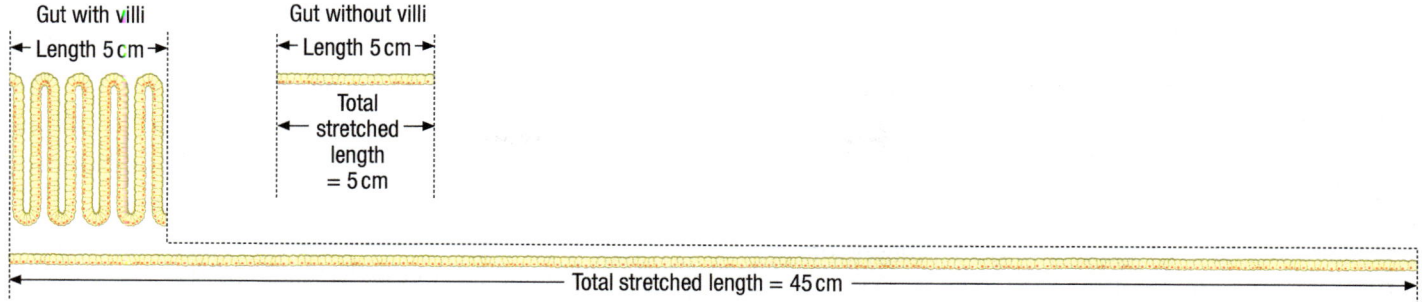

Figure 2 *The effect of folding on the available surface for exchange*

more dissolved food molecules in your blood than in your digestive system. Glucose and other dissolved food molecules are then moved from the small intestine into the blood by active transport. The digested food molecules have to move against the concentration gradient. This makes sure that none of the digested food is wasted and lost in your faeces.

??? Did you know ... ?

Although your gut is only around 7 metres long and a few centimetres wide, the way it is folded into your body along with the villi and microvilli give you a surface area for the absorption of digested products of between 200 and 300 m²!

Summary questions

1 In the following sentences, match each beginning (A, B, C or D) to its correct ending (1 to 4).

A	Food needs to be broken down into small soluble molecules ...	1	... by diffusion and active transport.
B	The villi are ...	2	... carry away the digested food to the cells and maintain a steep concentration gradient.
C	Food molecules move from the small intestine into the bloodstream ...	3	... so diffusion across the gut lining can take place.
D	The small intestine has a rich blood supply to ...	4	... finger-like projections in the lining of the small intestine that increase the surface area for diffusion.

2 Explain why a folded gut wall can absorb more nutrients than a flat one.

3 Coeliac disease is caused by gluten, a protein found in wheat, oats, and rye. Affected people react to gluten – the villi become flattened and the lining of the small intestine becomes damaged.
   a Why do you think people with untreated coeliac disease are often quite thin?
   b If someone with coeliac disease stops eating any food containing gluten, they will gradually gain weight and no longer suffer malnutrition. Suggest why this might be.

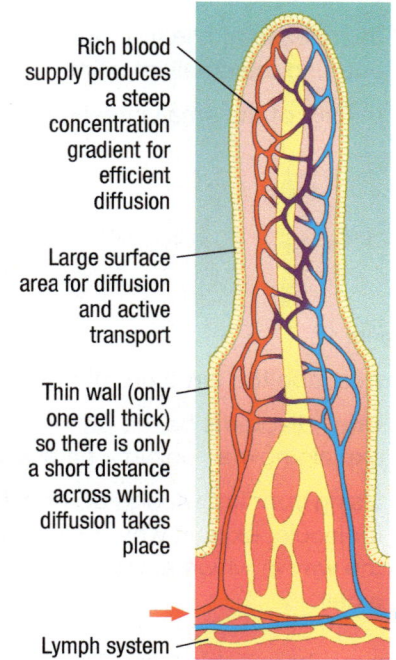

Rich blood supply produces a steep concentration gradient for efficient diffusion

Large surface area for diffusion and active transport

Thin wall (only one cell thick) so there is only a short distance across which diffusion takes place

Lymph system

Figure 3 *Thousands of finger-like projections in the wall of the small intestine – the villi – make it possible for all the digested food molecules to be transferred from your small intestine into your blood by diffusion and active transport*

Key points

- The villi in the small intestine provide a large surface area with an extensive network of blood capillaries.

- The villi mean the small intestine is well adapted as an exchange surface to absorb the products of digestion, both by diffusion and by active transport.

Summary questions

1 Imagine a meal of rice and chicken, with oil used to cook the chicken. Describe what happens to this meal in the digestive system after it is eaten until the remains are removed from the body.

2 Explain how each of the following adaptations makes digestion more efficient:

a the acid pH in the stomach and the alkaline pH in the small intestine

b the release of bile from the gall bladder onto the food entering the small intestine

c the villi of the small intestine.

3

Enzyme   Substrate

Figure 1

a Describe the structure of a protein.

b Draw diagrams based on Figure 1 to help you to explain how an enzyme catalyses a reaction.

c Using your knowledge of the structure of a protein molecule, explain why pH and temperature can affect the way in which a protein carries out its function in a cell.

4 The results in Table 1 and Table 2 come from a student who was investigating the breakdown of hydrogen peroxide using manganese(IV) oxide and grated raw potato.

Table 1 *Using manganese(IV) oxide*

Temperature (°C)	Time taken (s)
20	106
30	51
40	26
50	12

Table 2 *Using grated raw potato*

Temperature (°C)	Time taken (s)
20	114
30	96
40	80
50	120
60	no reaction

a Draw a graph of the results using manganese(IV) oxide.

b What do these results tell you about the effect of temperature on a catalysed reaction? Explain your observation.

c Draw a graph of the results when grated raw potato was added to the hydrogen peroxide.

d What does this graph tell you about the effect of temperature on an enzyme-catalysed reaction?

e Why does temperature have this effect on the enzyme-catalysed reaction but not on the reaction catalysed by manganese(IV) oxide?

5 Students investigated two protease enzymes, A and B. They added samples of enzymes A and B to test tubes containing solutions at a range of pH values. After 20 minutes they tested the activity of the enzyme in each test tube.

The following table shows their results.

pH of solution in test tube	2	4	6	8	10	12
Activity of enzyme A in arbitrary units	0	0	12	32	24	8
Activity of enzyme B In arbitrary units	26	20	6	0	0	0

a Name **two** variables that the students should have controlled in this investigation.

b Give **one** way in which the students could have made this investigation more reliable.

c What conclusions can the students make from these results about the enzymes A and B?

d The students are told that the two enzymes are pepsin (produced in the stomach) and trypsin (produced by the pancreas).

Suggest which letter represents pepsin and which letter represents trypsin. Give reasons for your answer.

Practice questions

1 Figure 1 is a diagram of the human digestive system.

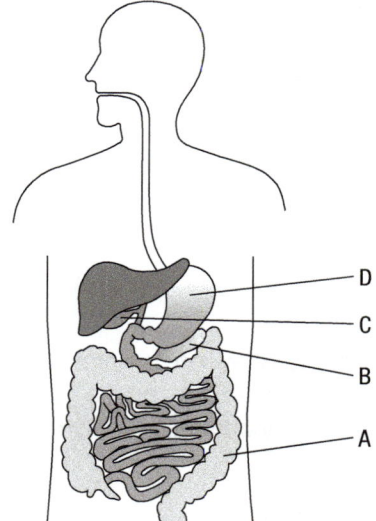

Figure 1

a i Name the parts labelled A, B, C, and D. (4)
ii Copy and complete the following sentences using the letters A–D from the Figure 1:
Part .... churns the food by muscular action.
Fats are emulsified by bile from part ....
Protease enzymes are produced by parts .... and
....
Water is reabsorbed in part .... (5)

b *In this question you will be assessed on using good English, organising information clearly, and using specialist terms where appropriate.*

Describe the role of digestive enzymes in breaking down large, insoluble food molecules into small, soluble molecules that can be absorbed. (6)

2 Figure 2 represents large molecules found in our body.

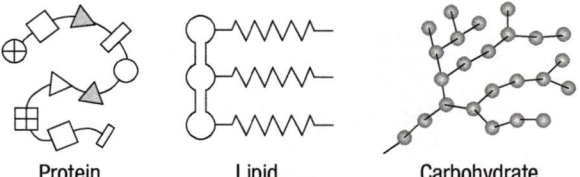

| Protein | Lipid | Carbohydrate |

Figure 2

a Name the small molecules represented by these symbols:

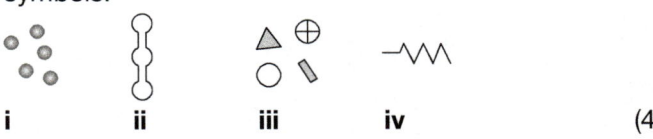

i   **ii**   **iii**   **iv** (4)

b Enzymes are proteins.
i What is an enzyme? (1)
ii Explain how the specific shape of an enzyme is essential for it to work. (3)
iii Give the term used to describe an enzyme that has lost its specific shape. (1)
iv Name **two** factors that can cause an enzyme to lose its shape. (2)

c Name **three** other types of proteins found in the body. Describe the function of each type of protein. (6)

3 Enzymes have many uses in the home and in industry.

a Which type of organisms are used to produce these enzymes? (1)

b Babies may have difficulty digesting proteins in their food. Manufacturers of baby food use enzymes to pre-digest the protein in the food to overcome this difficulty.

Copy and complete the following sentences:
i Proteins are pre-digested using enzymes called .... (1)
ii This pre-digestion produces .... (1)

c A manufacturer of baby food uses enzyme V to pre-digest protein. He tries four new enzymes – W, X, Y, and Z – to see if he can reduce the time taken to pre-digest the protein.

Figure 3 shows the time taken for the enzymes to completely pre-digest the protein. The manufacturer uses the same concentration of enzyme and the same mass of protein in each experiment.

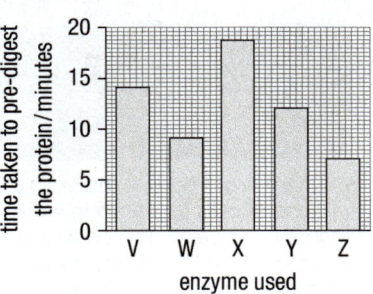

Figure 3

i How long did it take enzyme V to pre-digest the protein? (1)
ii Which enzyme would you advise the baby food manufacturer to use? Give a reason for your answer. (2)
iii State **two** other factors that should be controlled in the manufacturer's investigations. (2)

6.1

Responding to change

Figure 1 *Your body is made up of millions of cells that have to work together. Whatever you do with your body, whether it's walking to school or playing on the computer, your movements need to be coordinated*

You need to know what is going on in the world around you. Your **nervous system** makes this possible. It enables you to react to your surroundings and coordinate your behaviour.

Your nervous system carries electrical signals (**impulses**) that travel fast – between 1 metre and 120 metres per second. This means you can react to changes in your surroundings very quickly.

The nervous system

Like all living things, you need to avoid danger, find food, and, eventually, find a mate! This is where your nervous system comes into its own. Your body is particularly sensitive to changes in the world around you. Any changes (known as **stimuli**) are picked up by cells called **receptors**.

Receptor cells, such as the light receptor cells in your eyes, are similar to most animal cells. They have a nucleus, cytoplasm, and a cell membrane. These receptors are usually found clustered together in special **sense organs**, such as your eyes and your skin. Humans and other animals have many different types of sensory receptor (Figure 2).

??? Did you know ... ?

Some male moths have receptors which are so sensitive they can detect the scent of a female several kilometres away and follow the scent trail to find her!

Figure 2 *This fennec fox relies on its sensory receptors to detect changes in the environment*

How your nervous system works

Once a sensory receptor detects a stimulus, the information (sent as an electrical impulse) passes along special cells called **neurones**. These are usually found in bundles that contain hundreds or even thousands of neurones, which are known as **nerves**.

The impulse travels along the neurone until it reaches the **central nervous system**, or CNS. The CNS is made up of the brain and the spinal cord. The cells that carry impulses from your sense organs to your central nervous system are called **sensory neurones**.

Your brain gets huge amounts of information from all the sensory receptors in your body. It coordinates the response to the information, and sends impulses out along special cells. These cells, called **motor neurones**, carry information from the CNS to the rest of your body. They carry impulses to make the right bits of your body – the **effector organs** – respond (Figure 3).

Effector organs are muscles or glands. Your muscles respond to the arrival of impulses by contracting. Your glands respond by releasing (secreting) chemical substances, for example, your salivary glands produce and release extra saliva when you smell food cooking.

The way your nervous system works can be summed up as:

receptor → sensory neurone → coordinator (CNS) → motor neurone → effector

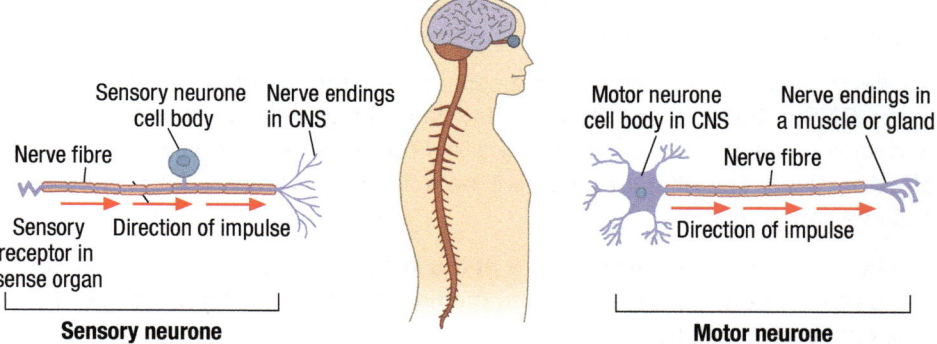

Sensory neurones carry impulses to the CNS. The information is processed and impulses are sent out along motor neurones to produce an action.

Figure 3 *The rapid responses of our nervous system allow us to respond to our surroundings quickly – and in the right way!*

Study tip

Be careful to use the terms neurone and nerve correctly.

Talk about **impulses** (*not* messages) travelling along a neurone.

Key points

- The nervous system uses electrical impulses to enable you to react quickly to your surroundings and coordinate what you do.

- Cells called receptors detect stimuli (changes in the environment).

- Impulses from receptors pass along sensory neurones to the brain or spinal cord (CNS). The brain coordinates the response, and impulses are sent along motor neurones from the brain (CNS) to the effector organs.

- Effectors include muscles and glands.

Summary questions

1 **a** What is the main function of the nervous system?
   b What is the difference between a neurone and a nerve?
   c What is the difference between a sensory neurone and a motor neurone?

2 Make a table to show the different types of sense receptor. For each one, give an example of the sort of things it responds to, for instance, touch receptors respond to an insect crawling on your skin.

3 Explain what happens in your nervous system when you see a piece of fruit, pick it up, and eat it.

6.2 Reflex actions

Your nervous system lets you take in information from your surroundings and respond in the right way. However, some of your responses are so fast that they happen without giving you time to think.

When you touch something hot, or sharp, you pull your hand back before you feel the pain. If something comes near your face, you blink. Automatic responses like these are known as **reflexes**.

What are reflexes for?

Reflexes are very important both for humans and for other animals. They help you to avoid danger or harm because they happen so fast. There are also lots of reflexes that take care of your basic bodily functions. These functions include breathing and moving food through your gut.

Reflexes do not involve the conscious areas of your brain. It would make life very difficult if you had to think consciously about those things all the time – and it would be fatal if you forgot to breathe!

How do reflexes work?

Reflex actions often involve just three types of neurone. These are:
- sensory neurones
- motor neurones
- **relay neurones** – these connect a sensory neurone and a motor neurone, and are found in the CNS.

An electrical impulse passes from a sensory receptor along a sensory neurone to the CNS. It then passes along a relay neurone (usually in the spinal cord) and straight back along a motor neurone. From there, the impulse arrives at an effector organ. The effector organ will be a muscle or a gland. This pathway is called a **reflex arc**.

The key point about a reflex arc is that the impulse bypasses the conscious areas of your brain. The result is that the time between the stimulus and the reflex action is as short as possible.

How synapses work

Your neurones are not joined up directly to each other. There are junctions between them called **synapses**, which form physical gaps between the neurones. The electrical impulses travelling along your neurones have to cross these synapses. They cannot leap the gap. Look at Figure 1 to see what happens next.

The reflex arc in detail

Look at Figure 2. It shows what would happen if you touched a hot object.
- When you touch the object, a receptor in your skin is stimulated. An electrical impulse from a receptor passes along a sensory neurone to the central nervous system – in this case, the spinal cord.

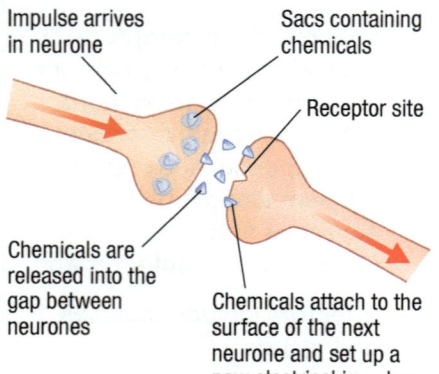

Impulse arrives in neurone

Sacs containing chemicals

Receptor site

Chemicals are released into the gap between neurones

Chemicals attach to the surface of the next neurone and set up a new electrical impulse

Figure 1 *When an impulse arrives at the junction between two neurones, chemicals called neurotransmitters are released that cross the synapse and arrive at receptor sites on the next neurone. This starts up a new electrical impulse in the next neurone*

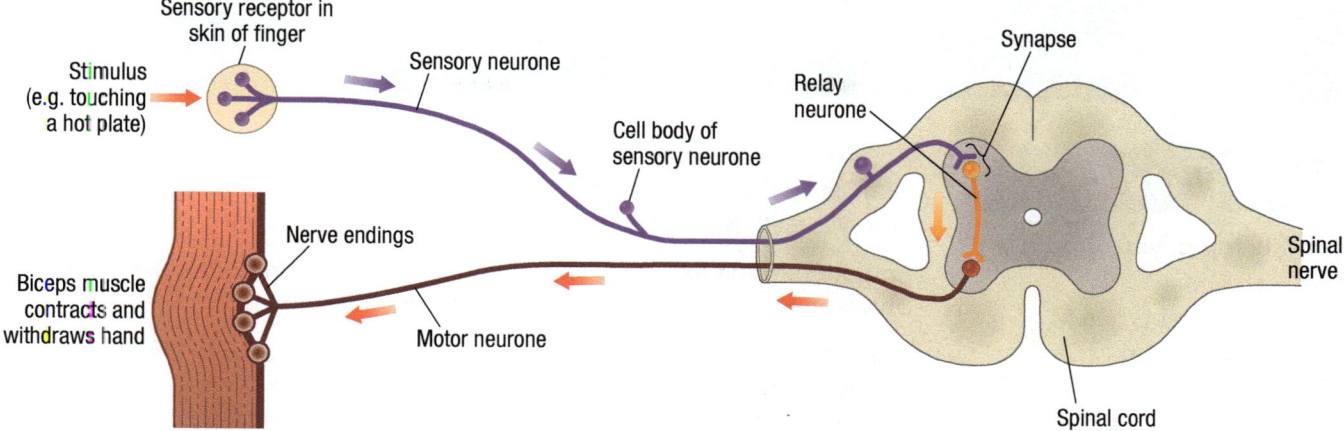

Figure 2 *The reflex action that moves your hand away from something hot can save you from being burnt. Reflex actions are quick and automatic – you do not think about them*

- When an impulse from the sensory neurone arrives at the synapse with a relay neurone, a neurotransmitter is released which acts as a chemical messenger. This chemical crosses the synapse to the relay neurone, where it sets off a new electrical impulse that travels along the relay neurone. The diffusion of the neurotransmitter across the synapse is slower than the electrical impulse in the neurones, but it makes it possible for the impulse to cross the gap between them.

- When the impulse reaches the synapse between the relay neurone and a motor neurone returning to the arm, another neurotransmitter is released. Again, the chemical crosses the synapse and starts a new electrical impulse travelling down the motor neurone to an effector. When the impulse reaches the effector organ, it is stimulated to respond. In this example, the impulses arrive in the muscles of the arm, causing them to contract. This action moves the hand rapidly away from the source of pain. If the effector organ is a gland, it will respond by releasing (secreting) chemical substances.

The reflex pathway is not very different from a normal conscious action. However, in a reflex action the coordinator is a relay neurone either in the spinal cord or in the unconscious areas of the brain. The whole reflex is very fast indeed.

An impulse also travels up the spinal cord to the conscious areas of your brain. You know about the reflex action, but only after it has happened.

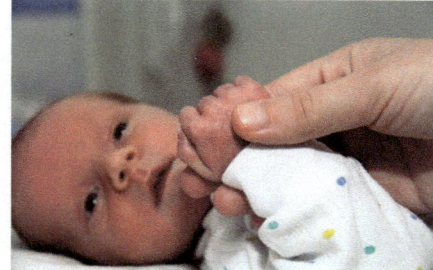

Figure 3 *Newborn babies have a number of special reflexes that disappear as they grow. This grasp reflex is one of them*

Summary questions

1 **a** Why are reflexes important?
   b Why is it important that reflexes don't go to the conscious areas of your brain?

2 Explain why some actions, such as breathing and swallowing, are reflex actions, whilst others, such as speaking and eating, are under your conscious control.

3 Make a flow chart to explain what happens when you step on a pin. Make sure you include an explanation of how a synapse works.

Key points

- Some responses to stimuli are automatic and rapid, and are called 'reflex actions'. They involve sensory, relay, and motor neurones.

- Reflex actions control everyday bodily functions, such as breathing and digestion, and help you to avoid danger.

- There are gaps between neurones called synapses. The release of chemicals into the synapse allows the impulse to cross from one neurone to another.

6.3 Animal behaviour

You see animal behaviour going on around you all the time in pets and farm animals, in the wildlife in your home, garden, and the countryside – and most of all in yourself. Biologists define behaviour as:

an action made in response to a stimulus, which modifies the relationship between the organism and the environment.

This means that animals pick up changes in the world around them and respond in a way that changes the situation in some way. Some of the behaviour that rules the lives of animals is **innate** or instinctive and some of it is **learnt**.

Different types of behaviour

- Innate behaviour is found in all members of a species. It isn't learnt by the animal. It is the result of specific nerve pathways laid down as the embryo develops. Some innate behaviour involves very basic responses. For example, invertebrates such as woodlice move randomly until they find themselves somewhere dark and damp – then they stop moving. This allows them to stay safe from drying out and hidden from predators. But some very complex behaviours, from competing for a mate and building a web or a nest to travelling thousands of miles in migration, are also innate behaviours (see 6.4 Animal communications).

Figure 1 *This tiny, beautiful nest is made by the hummingbird from spider webs, moss, and down in a complex piece of innate behaviour. No one teaches her how to do it – and she must get it right first time or her offspring will not survive*

- **Habituation** happens when a stimulus is repeated time after time and nothing happens, good or bad. Eventually the animal stops responding to that stimulus, and the response does not return when it meets the stimulus again. Habituation is particularly important in the development of young animals as they learn not to react to the neutral features of the natural world. For example, many different types of animals have to learn to ignore the movement and noise of the wind or the ocean, or their nervous systems would be constantly firing off false alarms.

Figure 2 *Imprinting makes sure that these tiny ducklings stay close to their mother and so stay as safe as possible*

- **Imprinting** is a very specialised form of learning behaviour that is seen only in very young animals. At a receptive stage of its early life, the young animal identifies with and attaches itself emotionally to another large organism. This is usually the parent. Once an animal has imprinted it will follow the parent whilst it is young, and relate to other similar animals throughout its life. Imprinting enables the animal to recognise other animals of the same species. If the parent isn't available, the baby animal will imprint on any available object – including humans!

- **Classic conditioning** takes place when animals learn to associate an existing unconditioned reflex with a new stimulus. The most famous example of this is the work of the Russian scientist Pavlov, who carried out research into the behaviour of dogs. When dogs see and smell their food, they begin to salivate before they start eating. Ringing a bell does not affect the production of saliva. Pavlov tried ringing a bell before presenting his experimental dogs with their food. In time, the dogs began salivating when they heard the bell, even when there was no food to see, smell, or eat. The normal food response had become conditioned. The dogs learnt that the bell was a signal for food arriving and so their bodies responded.

- **Operant conditioning** is also known as trial-and-error learning. This takes place when a piece of trial behaviour by an animal is either rewarded (something good happens, such as it gets food) or punished (something bad happens, such as the animal gets hurt). An animal will usually repeat the behaviour several times. If something good happens every time, the behaviour is likely to be repeated and used in the long term. If the behaviour always results in punishment, it is likely that the animal will stop trying it. Operant learning plays an important part in the way animals, including people, learn in everyday life. It is also widely used by people who want to train animals to carry out particular behaviours.

Figure 3 *When scientists raise rare birds, such as the Californian condor, from eggs, they use puppets of the parents and stay hidden to prevent the young birds imprinting on people rather than their own kind*

⌘ links

You will learn more about animal behaviours in 6.4 Animal communications and 6.5 Reproductive behaviours.

Figure 4 *Cats are not naturally gifted with technology, but if they once jump onto a desk near a computer, they quickly learn that it is a warm place to settle down for a nap – but only when the computer is switched on*

Summary questions

1 Define animal behaviour.

2 Make a table to compare innate behaviour, imprinting, habituation, classic conditioning, and operant conditioning.

3 Find out more about two of the following types of behaviour: innate behaviour, imprinting, habituation, or classic conditioning. Write a brief report on each, including at least one example of a scientific investigation into the behaviour and several examples from the natural world.

Key points

- The different behaviours displayed by animals include:
  - innate behaviour
  - imprinting
  - habituation
  - classic conditioning
  - operant conditioning.

6.4 Animal communications

A lot of animal behaviour is based on communication between both members of the same species, and members of different species. Animals use a wide variety of types of signals to communicate with each other. Three of the most common forms of communication involve sound signals, visual signals, and chemical signals.

Sound signals

To communicate using sound, animals need to be able to make sounds, hear sounds, and interpret sounds. Humans shape sounds into speech to communicate everything from simple needs to sophisticated ideas. People can also use speech to communicate with other species of animals, such as cats, dogs, cattle, and sheep, even if they do not understand what we say to them.

Social primates such as chimpanzees use a wide range of sounds to communicate complex emotions, from friendship to rage. Other mammals also use sound to communicate. This ranges from the howling of wolves that maintains contact between the pack members and defends their territory, to the gentle sounds made by a mother sheep as she bonds with her newborn lamb.

It is not only mammals that use sound to communicate. Birds sing to mark their territories, find possible mates, and warn off rivals. Frogs and toads do a similar thing, inflating their throats to amplify the sound so it travels as far as possible (Figure 1). Male crickets and cicadas make sounds by rubbing parts of their bodies together. These sounds are to attract a mate, and they can be very loud. Different animals can hear sounds in different ranges. For example, bats make sounds that are too high for humans to hear. These sounds allow the bats to use echolocation to fly in the dark, and they also act as communication between the bats so they don't collide in mid-air. Humpbacked whales make very low sounds that can travel for miles underwater. Scientists think these sounds act as a signal between males and females during mating behaviour.

Figure 1 *The communication sounds made by these tiny tropical frogs are used to mark territories and attract females, but they are so loud that they are easily heard by other species too*

Visual signals

Animals do not only use sound to communicate. Animals such as humans that communicate using sound often rely heavily on visual communication as well. Raised eyebrows, a clenched fist, tension in the lips, dilated pupils – visual cues like these can tell us whether someone is happy or sad, attracted to us, or angry, without them saying a word. In many different species of animals, and between species of animals, visual cues indicate whether an animal is hungry or sleepy, playful or aggressive. Many of these visual signals can be read by more than one species (Figure 2).

Many forms of visual communication in the animal world are linked to selecting a mate, reproduction, and raising offspring. Examples include the magnificent feathers of the peacock and birds of paradise; the red, swollen body parts that signal that a female primate is ready to mate; the gaping beaks of baby birds that stimulate their parents to feed them; or the large eyes of many young mammals that trigger a protective response in adults.

Figure 2 *Communication across species – the playful body language of a sled dog communicates to a polar bear. Instead of the bear killing and eating the dog, the animals ended up playing together*

Other visual signals warn of danger. The black-and-yellow colours of wasps, bees, and hornets warn that they sting. The coral snake uses bright colours to warn that it is poisonous (Figure 3). Many beetles, such as the common garden ladybug, use strong reds and yellows to warn potential predators that they don't taste good.

Chemical signals

Chemical signals are key to communication in many animal species – they even play an important role for humans. Chemical signals include both scents that you are aware of and chemicals known as pheromones that you cannot consciously smell but that affect the way you behave. Chemical signals are widely used by animals to identify both members of the same species and individuals within the species. Pheromones are often involved in sexual attraction, and they are used to communicate who is dominant in a social group. In some species, including horses, cattle, sheep, and cats, the males curl back their top lip to capture the chemical signals from fertile females, or the faeces and urine of rival males, using a specialised sense organ.

Chemical signals are very important in defence for many animals, especially insects and other invertebrates. Squirting unpleasant smelling and/or tasting substances protects many insects from birds and other predators. In social insects such as bees and ants, chemicals are used to communicate everything from where to find food to giving warning of attack by another colony.

Figure 3 *Lying snakes. The coral snake at the top of this image is deadly poisonous. The milk snake below is harmless, but mimics a coral snake well enough to communicate the same warning - milk snakes are usually left alone!*

∞ **links**

You will learn more about reproductive behaviour in 6.5 Reproductive behaviours.

Figure 4 *The female oak eggar moth produces a chemical signal that is picked up by the male moth's feathery antennae. By following her chemical signals, the male can find her and mate*

Summary questions

1 Describe the three main ways in which animals communicate.

2 Give examples of how sound might be used:

  a to communicate territory boundaries

  b to show aggression towards another animal

  c to help find a mate.

3 *Communication between species is almost as important as communication between members of the same species. Discuss this statement, and give examples of different types of communication to support your arguments.*

Key points

- Animals use a variety of types of signal to communicate with each other.

- These signals include sound signals, visual signals, and chemical signals.

6.5

Reproductive behaviours

In terms of survival of the species, the most important thing that any animal can do is reproduce. For many animals this involves finding a mate and then trying to make sure that as many offspring as possible survive to adulthood. Here are some of the many different ways of reproducing and parenting that have evolved over time.

Finding a mate

Animals that use sexual reproduction need to find and select a mate. To give their offspring the best possible chance of survival, they need to select the best quality mate who will pass on the best alleles to the offspring. Choosing a mate usually involves a process of courtship. This is a set of behaviours that help the animals to select a mate and then build a bond between the pair. In most species the females decide which males they will mate with, so courtship behaviour has evolved to advertise the quality of each male as a potential mate. Sometimes this involves direct fighting between the males (Figure 1), and the females stay with the winner (this happens in most deer species, wild horses, and sheep). Often the competition is less direct – it may involve singing (many birds, frogs, and insects), or displaying dramatic feathers or markings such as those shown in Figure 2 (peacocks, birds of paradise, macaws), bringing gifts of food (many spiders), or building a nest (many birds, sticklebacks). Displaying males do not risk physical damage in the competition, and the females get to choose the mate who seems strongest and fittest. Once the female has chosen her mate, they may form a strong pair bond through a variety of different behaviours and displays – usually including mating.

Animals have a number of different mating strategies. These include:

- Finding a mate for life, for example, swans, albatrosses, and black vultures. This is rare in the animal kingdom.
- Having several different mates over a lifetime, for example, in a pride of lions the dominant male, who breeds with all the females, may change several times over the breeding life of any one female.
- Having one mate for a single breeding season, for example, many types of penguins and garden birds.
- Having several mates over a single breeding season, for example, kittens or puppies in a litter may have several fathers.

Figure 1 *When males compete by fighting they show off their strength but risk injury and even death*

Parental care

Some animals produce many offspring and leave them to look after themselves. Most of the offspring die or are eaten, but a few usually survive. This strategy is common in invertebrates, fish, amphibians, and reptiles. Other species have developed special behaviours for rearing their young. These behaviours often involve the parents in the care of their offspring. Species where one or both of the parents takes care of the offspring usually produce relatively few offspring, but more of them survive. Parenting behaviour can be risky for the parents. It uses up time and resources and makes them vulnerable to starvation and to predators. For some animals, parenting can even result in death. But parental care can be a very successful evolutionary strategy. It increases the chances of the offspring surviving. It also increases the chances of those offspring reproducing and passing the parental genes on to the next generation.

Figure 2 *The dramatic display of the male peacock has evolved to convince females that he is a high-quality mate, but without risking his life*

Examples of parental behaviour include:

- Mouth brooding – some fish and frogs keep their eggs and even their young in their mouths until the eggs hatch or the young are big enough to look after themselves.

- Egg-laying and incubation – many reptiles and birds lay eggs that contain food for the developing offspring, so they can grow to a stage at which they are more likely to survive before they hatch. Birds also incubate their eggs, keeping them warm as the young develop inside and then caring for the offspring after hatching. Some young birds can feed themselves immediately, but they need to be taught where to find food and how to avoid danger. Some species hatch featherless and helpless. They need dedicated parental care, including feeding, until their feathers grow and they are strong enough to fly.

- Young mammals develop inside their mother, and a small number of offspring are produced at one time. Once the babies are born the mother usually displays strong parenting behaviour. She produces milk to feed her young. Male mammals are less likely to be involved in rearing their offspring. Some young mammals, such as foals, camels, and lambs, are born able to walk and keep up with the adults. They learn from their mothers how to feed themselves and how to behave. Some mammals, such as mice, kittens, and chimpanzees, are born very helpless and need continual care until they can see and walk. Often the parental care continues for months and even years. Primate babies take longest to grow and mature. The female orangutan looks after her baby for around 8 years before having another. For chimpanzees, growing up takes 3–5 years. Parental behaviour involves many things, from feeding and grooming to teaching the infant how to find food and how to behave with other animals. Human babies are the slowest-growing primates.

Figure 3 *Providing food and protection to a large brood of babies is very exhausting for the parents*

Figure 4 *The behaviour needed to take care of a baby for many years can put a mother at risk but it greatly increases the chances of her offspring surviving to adulthood*

Key points

- Sexual reproduction requires the finding and selection of a suitable mate, and can involve courtship behaviours that advertise the individual's quality.

- Animals have different mating strategies, including selecting a mate for life, several mates over a lifetime, a mate for a single breeding season, or several mates over one breeding season.

- Some animals have developed special behaviours for rearing their young.

- Parental care can be a successful evolutionary strategy, giving an increased chance of survival for the offspring and an increased chance of parental genes being passed on by the offspring, but it may involve risks to the parents.

Summary questions

1 What are the main mating strategies seen in the animal kingdom?

2 Animals select their mates in different ways.

   a Males usually show some form of courtship display. Why do they do this?

   b Describe three different types of courtship behaviour and suggest the advantages and disadvantages to the animals in the examples you have chosen.

3 a There are two basic strategies for the production of offspring. Give both, and explain the advantages and disadvantages of each one.

   b *Parental care may involve risks to the parents, but it is worth it for the evolutionary benefits*. Discuss this statement.

6.6 Human use of animal behaviours

Learning objectives

After this topic, you should know:

- ways in which people use conditioning when training captive animals for specific purposes.

Figure 1 *When human babies learn to eat solid food they get the double reward of food and praise from their carers*

links

You learnt about different types of behaviours in 6.3 Animal behaviour, 6.4 Animal communications, and 6.5 Reproductive behaviours.

Long before scientists began to study behaviour, people knew how to change the behaviour of animals, often using what is now called operant conditioning. The same ideas are also used when parents bring up their children, teaching them how to behave in the home and in society.

Training animals

For much of human history people have shared their lives with animals, from the dogs and cats in their homes to the horses they ride and the cows, sheep, and goats they use to give them food, milk, and leather. Whenever people use animals, they train them and change their behaviour to suit their purposes. The main way in which they do this is to use operant conditioning.

In natural operant conditioning, a random piece of behaviour by an animal either gives it a benefit or harms it. People have adapted this natural learning process to train animals for specific purposes. Desirable behaviour is rewarded by a treat, or by fuss and attention. Undesirable behaviour is 'punished' by withholding treats or attention. In modern society, the way we use animals has become increasingly sophisticated, but the basics of training remain the same.

Training sniffer dogs

People today can fly all over the world. There are huge gatherings of people at sporting, musical, and political events. Most of the people who travel, and who attend big events, do so for entirely innocent reasons. Unfortunately, some people are not so innocent. There are two major problems for police and security forces:

- The movement of illegal drugs from the countries where they are grown to the countries where they will be used. Illegal drugs are worth millions of pounds to the criminals who sell them, but they cause immeasurable harm to the people who use them. Most countries try to prevent the movement of illegal drugs.
- The use of terrorist tactics to try and change the way people think. Terrorists frequently use explosives and guns to kill innocent people and destroy airliners. Countries around the world want to protect their citizens from the actions of terrorists.

Sniffer dogs are an important part of the fight against both illegal drugs and terrorists. Dogs have a highly developed sense of smell that is up to 10 000 times better than that of humans. They can be trained to detect substances such as drugs or explosives, and to indicate their find to a handler. The training of sniffer dogs involves operant conditioning.

In the early stages of training, the dog is exposed to the smell it has to detect. When it approaches the substance and responds, it is rewarded using treats, praise, or a toy. If it doesn't sniff the substance it is ignored. Once the dog always responds to the scent of the substance, the task gets harder – the substance is hidden. When the dog finds the substance and responds, it is again rewarded. Finally, the dog is only rewarded when it detects the desired object and responds in a particular way, for example, when it sits and stares or barks.

Figure 2 *Sniffer dogs can be trained through operant conditioning to indicate the presence of substances, from drugs or explosives to body parts*

These highly trained dogs and their handlers are used at airports and large gatherings of people around the globe, and play an important role in keeping us safe.

Training police horses

Police forces around the world work to prevent crime and to capture criminals. At large public gatherings, including religious events, sporting occasions, and political rallies, the police also maintain public order. To help them, they often use horses. Police horses allow the police to see what is happening, and to move through a crowd – people will give way to horses more easily than to someone on foot. The police also use horses when searching for people in difficult, inaccessible terrain.

Like all riding horses, police horses must be obedient and responsive, able to walk forward, backwards or sideways on command, and must be safe in traffic.

Police horses also have to cope calmly with large crowds, lots of noise, sudden bangs and explosions, musical instruments, fire, and even rioting people. Their natural behaviour would be to panic and run away. This is where operant conditioning is very important. Potential police horses are gradually exposed to different difficult situations. Every time they respond calmly, they are rewarded by praise and sometimes food. They develop a strong bond of trust with their riders and will eventually move calmly through the loudest crowd and the most violent demonstration, helping to maintain law and order and keep people safe.

Figure 3 *The calm behaviour of police horses is the result of careful conditioning as they are trained*

Summary questions

1 Describe how operant conditioning is used to change behaviour.

2 Dogs are often used to help to detect people trapped in rubble after an earthquake or other natural disaster. Explain carefully the difficulties that have to be overcome, and how these dogs might be trained.

3 Look into three different examples of how people have trained captive animals for different purposes over time. Discuss the different purposes, how the behaviour of the animal has to be changed, and the different ways in which this has been done.

Key points

- Humans can make use of conditioning when training captive animals for specific purposes, including:
  - sniffer dogs
  - police horses.

Summary questions

1 This question is about animal responses. Match up the beginnings and ends of the sentences:

a	Many processes in the body …	**A**	… effector organs.
b	The nervous system allows you …	**B**	… secreted by glands.
c	The cells that are sensitive to light …	**C**	… to react to your surroundings and coordinate your behaviour.
d	Hormones are chemical substances …	**D**	… are found in the eyes.
e	Muscles and glands are known as …	**E**	… are known as nerves.
f	Bundles of neurones …	**F**	… are controlled by hormones.

2 a What is the job of your nervous system?

   b Where in your body would you find nervous receptors that respond to:

     i light **ii** sound **iii** heat **iv** touch?

   c Draw and label a simple diagram of a reflex arc. Explain carefully how a reflex arc works, and why it allows you to respond quickly to danger.

3 a Describe habituation and explain why it is such important behaviour in the development of young animals.

   b Konrad Lorenz hatched geese from eggs in incubators and found that the goslings would follow him around. More recently scientists hatched rare birds under the wings of microlight aircraft. When the birds grew up they followed the aircraft on the migration routes that they would normally learn from their parents.

     i What type of behaviour is described in this text?

     ii What is the value of this behaviour in normal animal development?

     iii From the text, explain how scientists are using this behaviour to help save rare species of birds.

4 A Skinner box is a piece of apparatus for testing how animals learn. The box contains a lever. If the lever is pushed or pressed, a pellet of food is delivered. Rats or pigeons placed in a Skinner box will push the lever accidentally as they move and explore the box. They very quickly learn that pressing the lever gets them a reward, and then spend all their time in the box pressing the lever and eating the food.

   a What type of behaviour is this?

   b How does this behaviour differ from the behaviour seen in Pavlov's dogs?

   c Discuss how humans make use of the behaviour demonstrated in a Skinner box to train animals they want to use for specific purposes.

5 a What are the **three** main ways in which animals communicate?

   b Discuss the main advantages and disadvantages of the three methods described in **a**, giving examples.

6 a What is courtship?

   Look at Figure 1.

Figure 1

   Explain how these two very different male animals advertise their qualities to the females of their species.

   b Discuss the advantages and disadvantages of each courtship strategy.

   c Some animals produce large numbers of offspring but provide no parental care. Explain how this works, and what its limitations are.

   d Describe three different types of parental care, and explain how each benefits the offspring and involves risks as well as benefits for the parents.

Practice questions

1 The body is coordinated by detecting stimuli in the environment, processing this information, and responding to it.

a Copy and complete the sequence to show the parts involved:

.... → coordination centre → .... (2)

b A dog responds to stimuli.

Figure 1

Write down the name of the organ which detects:
i chemical substances
ii sound and body movements
iii light
iv pressure. (4)

c Describe the **two** ways in which coordination centres can send information to initiate a response. (2)

2 A man touches a hot saucepan and immediately his hand pulls away as a reflex action.

a Copy and complete the pathway below showing a reflex action:

stimulus → .... → sensory neurone → .... → .... → effector → .... (4)

b Name the stimulus and effector in this reflex action:
i stimulus **ii** effector. (2)

c The impulse is transferred from one neurone to the next across a synapse.
i Describe how the electrical impulse in a neurone is passed across a synapse. (2)
ii Where in the nervous system are synapses located? (1)

d The nerve pathway from the hand to the spinal cord and back to the effector is 1.2 metres long. The time it takes for the impulse to reach the effector is 0.02 seconds. Calculate the speed of the impulse. (2)

3 A group of students were investigating the knee-jerk reflex. They wanted to find out how the speed of the hammer affected the distance the lower leg moved.

Figure 2 shows how the experiment was set up.

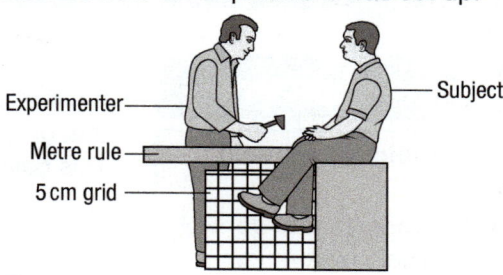

Experimenter — Subject
Metre rule
5 cm grid

Figure 2

Each trial was recorded on a video. A frame was taken every 33 milliseconds. The video was then played using single-frame advance. The number of frames for the hammer to move to the knee was found. The faster the speed, the smaller was the number of frames. The video was also used to find the distance moved by the subject's toe.

In each trial, the experimenter held the hammer 20 cm from the subject's knee and then hit the subject's tendon. The experimenter used the hammer at a different speed in each trial.

The following table shows some of the results.

Trial number	1	2	3	4	5	6	7	8	9	10
Distance hammer moved to knee (cm)	20	20	20	20	20	20	20	20	20	20
Number of frames it took the hammer to move to the knee	15	14	12	10	9	8	7	6	2	2
Distance moved by toe (cm)	0	0	5	5	4	10	10	10	10	10

a From the table, identify the independent variable, the dependent variable, and a control variable. (3)

b Give **two** advantages of using a video to make the measurements. (2)

c Suggest how the accuracy of this experiment could have been improved. (1)

d Draw a conclusion from the results of the experiment. (2)

7.1 Principles of homeostasis

The conditions inside your body are known as its **internal environment**. Your organs cannot work properly if this keeps changing. Many of the processes that go on inside your body aim to keep everything as constant as possible. This balancing act is called **homeostasis**.

Homeostasis involves your nervous system, your hormone system, and many of your body organs. Internal conditions that are controlled include:

- temperature
- the water content of the body
- the ion content of the body
- blood glucose levels.

◯◯ links

For more information on the nervous system, including control systems and receptors, look back to 6.1 Responding to change.

How hormones work

Hormones are chemical substances that coordinate many body processes. Special endocrine glands make and secrete (release) these hormones into your blood. The hormones are then carried around your body in the bloodstream to their target organs. Hormones regulate the functions of many organs and cells. They can act very quickly, but often their effects are quite slow and long-lasting. A number of hormones are important in the processes of homeostasis.

> ### Study tip
> Always say that hormones are 'secreted' by glands – do not use the word 'excreted' instead. 'Excreted' refers to waste production, and hormones are not waste products.

Figure 1 *Everything you do, from eating a meal to running a marathon, affects your internal environment*

Controlling water and ions

Water moves in and out of your body cells by osmosis. How much it moves depends on the concentration of mineral ions (such as those in salt) and the amount of water in your body. If too much water moves into or out of your cells, they can be damaged or destroyed.

◯◯ links

You can find out more about osmosis in 1.7 Osmosis.

You take water and ions into your body as you eat and drink. You lose water as you breathe out, and in your sweat. You lose ions in your sweat as well. You also lose water and ions in your urine, which is made in your **kidneys**.

Your kidneys can change the amount of ions and water lost in your urine, depending on your body conditions. They help to control the balance of water and mineral ions in your body. The concentration of the urine produced by your kidneys is controlled by both nerves and hormones.

For example, imagine drinking a lot of water all in one go. Your kidneys will remove the extra water from your blood and you will produce lots of very pale urine.

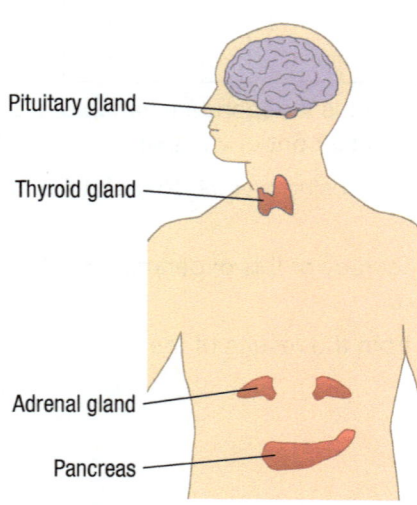

Pituitary gland

Thyroid gland

Adrenal gland

Pancreas

Figure 2 *Hormones act as chemical messages. They are made in glands in one part of the body, but have an effect somewhere else*

Controlling temperature

It is vital that your deep **core body temperature** is kept at 37 °C. Your enzymes work best at this temperature. At only a few degrees above or below normal body temperature, the reactions in your cells no longer take place at the ideal speed and you may die.

Your body controls your temperature in several ways. For example, you can sweat to cool down and shiver to warm up. Sweating causes the body to cool down because energy is transferred from the skin surface as the water in the sweat evaporates. The skin cools down as this happens, so the blood flowing through the skin will also be cooled down. Your nervous system is essential in coordinating the way your body responds to changes in temperature.

Once your body temperature drops below 35 °C, you are at risk of dying from **hypothermia**. For example, in the UK, several hundred old people die from the effects of cold each year, as do a number of younger people who get lost on mountains or try to walk home in the snow after a night out.

If your body temperature goes above about 40–42 °C, your enzymes and cells don't work properly. This means that you may die of heat stroke or heat exhaustion.

Controlling blood sugar

When you digest a meal, lots of glucose (simple sugar) passes into your blood. Left alone, your blood glucose levels would keep changing. The levels would be very high straight after a meal, but very low again a few hours later. This would cause chaos in your body.

However, the concentration of glucose in your blood is kept constant by hormones made in your pancreas. The pancreas acts as a coordination centre. This means your body cells are provided with the constant supply of energy that they need.

Figure 3 *You can change your behaviour to help control your temperature, for example, by adding extra clothing or turning up the heating when it's really cold, or wearing less or lighter clothing when it is hot*

Study tip

Sweating affects both temperature *and* water content of the body.
It cools the body by transferring energy from the skin to evaporate the water.

Key points

- Homeostasis is the process by which automatic control systems, including your nervous system, your hormones, and your body organs, maintain almost constant internal conditions.

- Homeostasis is important because body cells need almost constant conditions to work properly.

- Humans need to maintain a constant internal environment, controlling levels of water, ions, and blood glucose as well as temperature.

Summary questions

1 a Define a hormone.
  b How does coordination and control by hormones differ from coordination and control by the nervous system?

2 Why is it important to control:
  a water levels in the body
  b body temperature
  c glucose (sugar) levels in the blood?

3 a Look at the marathon runners in Figure 1. List the ways in which running is affecting their:
    i water balance
    ii ion balance
    iii temperature.
  b It is much harder to run a marathon in a costume than in running clothes. Explain why this is.

7.2 Controlling body temperature

After this topic, you should know:

- how your body monitors its temperature
- how your body maintains a relatively constant core temperature regardless of the external conditions.

Study tip

Learn how to spell thermoregulatory centre, and remember that it is in the brain.

Figure 1 *People in different parts of the world live in conditions of extreme heat and extreme cold and still maintain a constant internal body temperature*

Wherever you go and whatever you do, your body temperature needs to stay at around 37 °C. This is the temperature at which your enzymes work best. Your skin temperature can vary enormously without problems. It is the temperature deep inside your body, known as the core body temperature, which must be kept as stable as possible.

At only a few degrees above or below normal body temperature, your enzymes don't function properly. Many things can affect your internal body temperature, including:

- energy transferred in your muscles during exercise
- fevers caused by disease
- the external temperature rising or falling.

Basic temperature control

You can change your clothing, light a fire, and turn on the heating or air-conditioning to help control your body temperature. However, it is your internal control mechanisms that are most important. Body temperature is monitored and controlled by the **thermoregulatory centre** in your brain. This centre contains receptors that are sensitive to temperature changes in the blood flowing through the brain itself.

Extra information comes from the temperature receptors in the skin. These send impulses to the thermoregulatory centre, giving information about the skin temperature. The receptors are so sensitive they can detect a difference in temperature as small as 0.5 °C. Impulses from the skin prepare the thermoregulatory centre so it reacts rapidly to any changes in blood temperature that may follow.

If your temperature starts to go up, your sweat glands release more sweat, which cools the body down. As you lose more water through sweating when it is hot or you are exercising hard, it is important to take in more fluid through your drink and/or food to balance this loss. Your skin also looks redder as more blood flows through it, cooling you down. If your temperature starts to go down, you will look pale as less blood flows through your skin, transferring less energy to the surroundings.

Cooling the body down

Ext

If you get too hot, your enzymes denature and can no longer catalyse the reactions in your cells. When your core body temperature begins to rise, impulses are sent from the thermoregulatory centre to the body so more energy is transferred to the environment to cool you down:

- The blood vessels that supply your surface skin capillaries dilate (open wider). This is called vasodilation and it lets more blood flow through the capillaries. Your skin flushes, so you transfer more energy to the surroundings by radiation.
- Your rate of sweating goes up so your sweat glands are producing more sweat. This extra sweat cools your body down as the water evaporates from your skin, transferring energy from the skin to the environment. In humid weather, when the water in sweat does not evaporate, it is much harder to keep cool.

Ext Keeping warm

It is just as dangerous for your core temperature to drop as it is for it to rise. If you get very cold, the rate of the enzyme-controlled reactions in your cells falls too low. When this happens, you don't transfer enough energy for the reactions of your metabolism and your cells begin to die. If your core body temperature starts to fall, impulses are sent from your thermoregulatory centre to the body to reduce energy transfers and so minimise cooling.

- The blood vessels that supply your skin capillaries constrict (close up) to reduce the flow of blood through the capillaries. This vasoconstriction reduces the energy transferred to the environment through the surface of the skin.
- Your muscles contract and relax rapidly, causing you to shiver. These muscle contractions need lots of respiration, transferring energy to the tissues and raising your body temperature. As you warm up, the shivering stops.

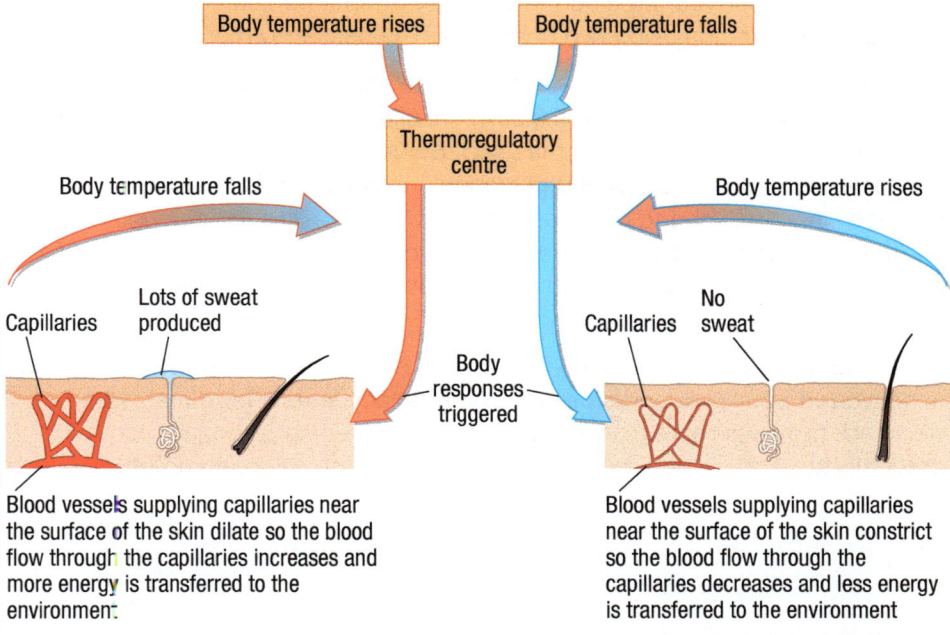

Figure 2 *Changes in your core body temperature set off automatic responses to oppose the changes and maintain a steady internal temperature*

links

For more on enzyme reactions, see 5.3 Factors affecting enzyme action.

Study tip

Never say that capillaries dilate or constrict. They are not able to do this as they have no muscle cells! Instead, it is the blood vessels supplying the capillaries that dilate or constrict.

Also, blood vessels *never* move! Either more blood flows in vessels near the skin surface, or more blood flows in the vessels lower down.

Key points

- Your body temperature is monitored and controlled by the thermoregulatory centre in your brain.
- Your body responds to cool you down or warm you up if your core body temperature changes, so it is maintained at around 37 °C.
- The blood vessels that supply the capillaries in the skin dilate and constrict to control the blood flow to the surface, controlling the transfer of energy to the environment.
- Energy is transferred through the evaporation of water in sweat from the surface of the skin to cool the body down.
- Shivering involves contraction of the muscles, transferring energy from respiration to warm the body.

Summary questions

1 a Why is it so important to maintain a body temperature of about 37 °C?
 b Explain why it is so important that the core body temperature does not rise above around 40 °C or fall much below 35 °C.

2 Explain the role of:
 a the thermoregulatory centre in the brain, and
 b the temperature sensors in the skin in maintaining a constant core body temperature.

3 Explain how the body responds to both an increase and a decrease in core temperature to return the core temperature to normal levels.

Controlling blood glucose

It is essential that your cells have a constant supply of the glucose they need for respiration. To achieve this, one of your body systems controls your blood sugar levels to within very narrow limits.

Insulin and the control of blood glucose levels

When you digest a meal, large amounts of glucose pass into your blood. Without a control mechanism, your blood glucose levels would vary significantly. They would range from very high after a meal to very low several hours later – so low that cells would not have enough glucose to respire.

This situation is prevented by your pancreas. The pancreas is a small pink organ found under your stomach. It constantly monitors and controls your blood glucose concentration using two hormones. The best known of these is **insulin**.

When your blood glucose concentration rises after you have eaten a meal, the pancreas produces and releases insulin. Insulin allows glucose to move from the blood into your cells where it is used. Soluble glucose is also converted to an insoluble carbohydrate called glycogen. Insulin controls the storage of glycogen in your liver and muscles. This is where most of the glucose is stored, and the glycogen can be converted back into glucose when it is needed. As a result, your blood glucose concentration stays stable within a narrow range of concentrations.

When the glycogen stores in the liver and muscles are full, any excess glucose is converted into lipids and stored. If you regularly take in food that, when digested, provides more glucose than can be stored as glycogen, you will gradually store more and more lipids and may eventually become **obese**.

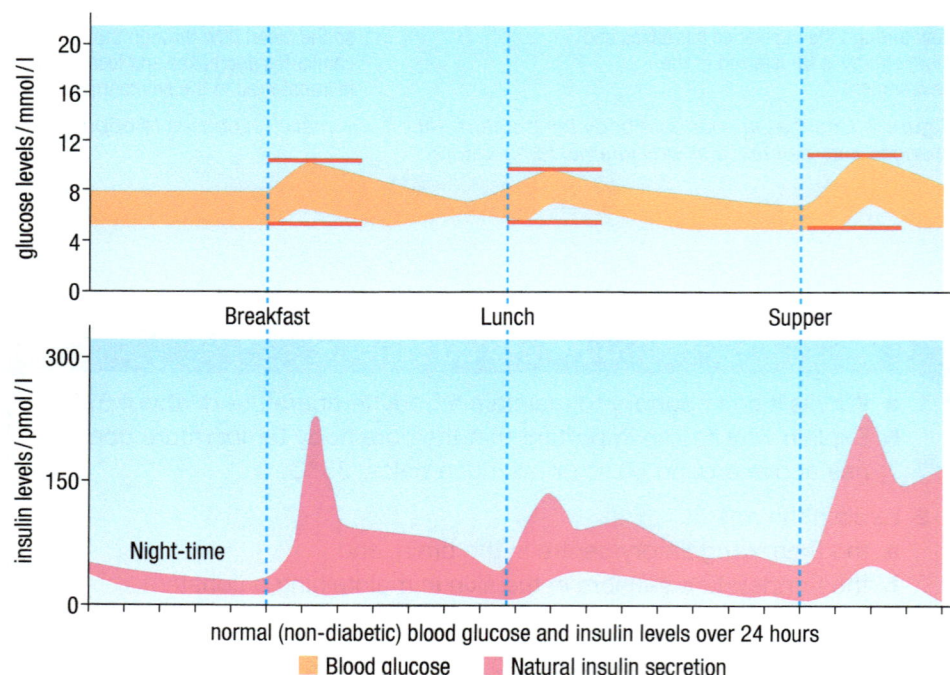

Figure 1 *Insulin is secreted from the pancreas after meals to keep your blood glucose concentration stable within narrow limits*

Glucagon and control of blood glucose levels

The control of your blood sugar doesn't just involve insulin. When your blood glucose concentration falls below the ideal range, the pancreas produces and secretes a second hormone, **glucagon**. Glucagon triggers the conversion of glycogen to glucose in the liver. In this way, the stored glucose is released back into the blood.

By using two hormones and the glycogen store in your liver, your pancreas keeps your blood glucose concentration fairly constant. It does this using feedback control, which involves switching between the two hormones.

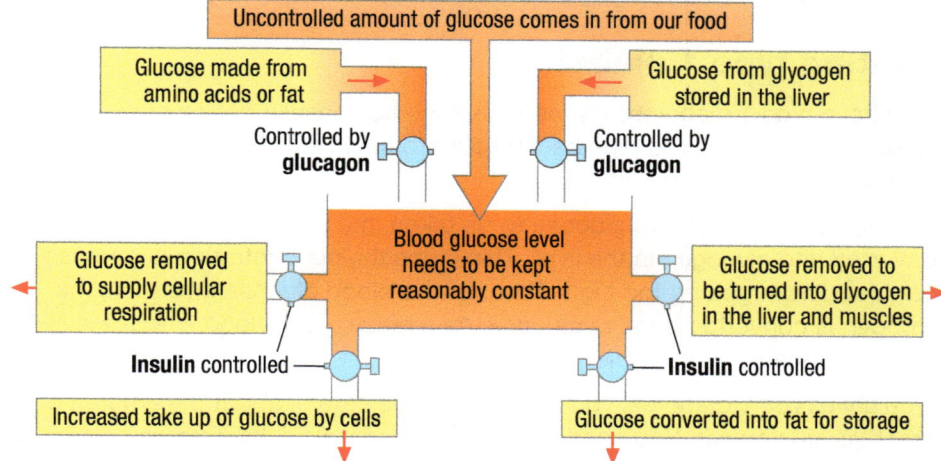

Figure 2 *This model of your blood glucose control system shows the blood glucose as a tank. It has both controlled and uncontrolled inlets and outlets. Control is given by the hormones insulin and glucagon*

What causes diabetes?

If your pancreas does not make enough (or any) insulin, your blood glucose concentration is not controlled. You have **type 1 diabetes**.

Without insulin your blood glucose levels get very high after you eat. Eventually your kidneys excrete glucose in your urine. You produce lots of urine and feel thirsty all the time. Without insulin, glucose cannot get into the cells of your body, so you lack energy and feel tired. You break down fat and protein to use as fuel instead, so you lose weight. Type 1 diabetes usually starts in young children and teenagers, and there seems to be a genetic element to the development of the disease.

Type 2 diabetes is another, very common type of diabetes. It gets more common as people get older and is often linked to obesity, lack of exercise or both. There is also a strong genetic tendency to develop type 2 diabetes. In type 2 diabetes, the pancreas still makes insulin, although it may make less than your body needs. Most importantly, your body cells stop responding properly to the insulin you make. In countries such as the UK and the USA, levels of type 2 diabetes are rising rapidly as the number of obese people in the population increases.

Summary questions

1 Define the following terms:
   a hormone   **b** insulin   **c** diabetes   **d** glycogen.

2 **a** Explain how your pancreas keeps your blood glucose level constant.
   b Why is it so important to control the level of glucose in your blood?

3 Explain the difference between type 1 and type 2 diabetes.

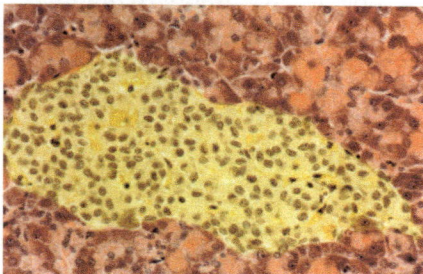

Figure 3 *Part of the pancreas. The tissue stained red makes digestive enzymes, and the central yellow area contains the cells that make insulin*

Key points

- Your blood glucose concentration is monitored and controlled by your pancreas.

- If blood glucose levels are too high, the pancreas produces the hormone insulin, which allows glucose to move from the blood into the cells and to be stored as glycogen in the liver and muscles.

- When blood glucose levels fall, the pancreas produces a second hormone, glucagon, which causes glycogen to be converted back into glucose and released into the blood.

- In type 1 diabetes, the blood glucose may rise to fatally high levels because the pancreas does not produce enough insulin.

- In type 2 diabetes, the body stops responding to its own insulin. Obesity is a significant factor in the development of type 2 diabetes.

7.4

Treating diabetes

Before there was any treatment for diabetes, people would waste away. Eventually they would fall into a coma and die.

The treatment of diabetes has developed over the years and continues to improve today. There are now some very effective ways of treating people with diabetes, although over the long term, even well-managed diabetes may cause problems with the circulatory system, the kidneys or the eyesight.

Treating type 1 diabetes

If you have type 1 diabetes, you need replacement insulin before meals. Insulin is a protein that would be digested in your stomach, so it is usually given as an injection to get it into your blood.

This injected insulin allows glucose to be taken into your body cells and converted into glycogen in the liver. This stops the concentration of glucose in your blood from getting too high. Then, as the blood glucose levels fall, the glycogen is converted back to glucose. As a result, your blood glucose levels are kept as stable as possible.

If you have type 1 diabetes, you also need to be careful about the amount of carbohydrate you eat. You need to have regular meals. Like everyone else, you need to exercise to keep your heart and blood vessels healthy. However, taking exercise needs careful planning to keep your blood sugar levels steady and your cells supplied with glucose as your cells respire more rapidly to produce the energy needed for your muscles to work.

Insulin injections treat diabetes successfully but they do not cure it. Until a cure is developed, someone with type 1 diabetes has to inject insulin every day of their life.

Using insulin from other organisms

In the early 1920s, Frederick Banting and Charles Best made some dogs diabetic by removing their pancreases. Then they gave them extracts of pancreas taken from other dogs. People now know these extracts contained insulin. Banting and Best realised that extracts of animal pancreas could keep people with diabetes alive. Many dogs died in the search for a successful treatment. However, the lives of millions of people have been saved over the years since.

For years, insulin from animals was used to treat people with diabetes, although there were problems. Animal insulin is not identical to human insulin and the supply depended on how many animals were killed for meat. So sometimes there was not enough insulin to go around.

In recent years, genetic engineering has been used to develop bacteria that can produce pure human insulin. This is genetically identical to natural human insulin and the supply is constant. This is now used by most people with type 1 diabetes. However, some people do not think this type of interference with genetic material is ethical.

Ways of delivering the insulin are changing too. More and more people with diabetes wear a pump that delivers insulin automatically all the time, although the individual still has to carry out lots of blood tests to make sure their glucose levels are correct.

Figure 1 *The treatment of Type 1 diabetes involves regular blood glucose tests and insulin injections to keep blood glucose levels constant*

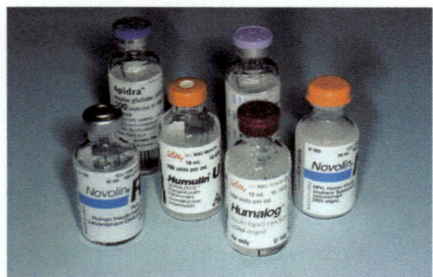

Figure 2 *Treatments such as human insulin allow a person to manage type 1 diabetes and live with it, but they do not cure the condition*

Curing type 1 diabetes

Scientists and doctors want to find a treatment that means people with diabetes never have to take insulin again. However, so far no treatment is widely available.

- Doctors can transplant a pancreas successfully. However, the operations are quite difficult and rather risky. These transplants are still only carried out on a few hundred people each year in the UK. There are 250 000 people in the UK with type 1 diabetes and there are simply not enough donors available. What's more, the patient exchanges one sort of medicine (insulin) for another (immunosuppressants).
- Transplanting the pancreatic cells that make insulin from both dead and living donors has been tried, with very limited success so far.

In 2005, scientists produced insulin-secreting cells from embryonic stem cells and used them to cure diabetes in mice. In 2008, UK scientists discovered a completely new technique. Using genetic engineering, they turned mouse pancreas cells that normally make enzymes into insulin-producing cells. Other groups are using adult stem cells from diabetic patients to try the same idea.

Scientists hope that eventually they will be able to genetically engineer faulty human pancreatic cells so they work properly. Then they will be able to return them to the patient with no rejection issues. It still seems likely that the best long-term cure will be to use stem cells from human embryos that have been specially created for the process. However, for some people, this is not ethically acceptable.

Although much more research is needed, scientists hope that before too long, type 1 diabetes will be an illness they can cure rather than simply treat and manage.

Treating type 2 diabetes

If you develop type 2 diabetes, which is linked to obesity, lack of exercise, and old age, you can often deal with it without needing to inject insulin. Many people can restore their normal blood glucose balance by taking three simple steps:
- eating a balanced diet with carefully controlled amounts of carbohydrates
- losing weight
- doing regular exercise.

If this doesn't work there are drugs that:
- help insulin work on the body cells more effectively
- help your pancreas make more insulin
- reduce the amount of glucose you absorb from your gut.

Only if all of these treatments do not work will you end up having insulin injections.

Type 2 diabetes usually affects older people. However, it is becoming more and more common in young people who are very **overweight**.

⚭ links

For information on embryonic stem cells, look back to 2.3 Stem cells.

Figure 3 *Losing weight and taking exercise seem simple ways to overcome type 2 diabetes. However, some people object to being given this advice and ignore it until they need medication to control the diabetes*

Summary questions

1 It is a common misconception that diabetes is treated only by using insulin injections.
   a Explain why this is not always true for people with type 1 diabetes.
   b Explain why treatment with insulin injections is relatively uncommon for people with type 2 diabetes.
2 a Compare modern insulin treatment with the original insulin used to treat diabetics and evaluate the two treatments.
   b Transplanting a pancreas to replace natural insulin production seems to be the ideal treatment for type 1 diabetes. Compare this treatment with insulin injections and explain why it is not more widely used.

Key points

- Type 1 diabetes may be controlled by injecting insulin, careful diet control, and exercise.
- Type 2 diabetes can be controlled by careful diet, exercise, and by drugs that help the cells respond to insulin.

Summary questions

1 a What is homeostasis?

b Write a paragraph explaining why control of the conditions inside your body is so important.

2 a Negative feedback is involved in many different aspects of homeostasis. Draw a clear diagram showing how feedback systems are important in the control of:
  i water balance in the body
  ii temperature control in the body
  iii control of the blood sugar in the body.

b Draw and label a diagram to summarise the general principles of feedback control in the body.

3 In August 2003 a heatwave hit Europe. Figure 1 shows the effect it had on the number of deaths in Paris.

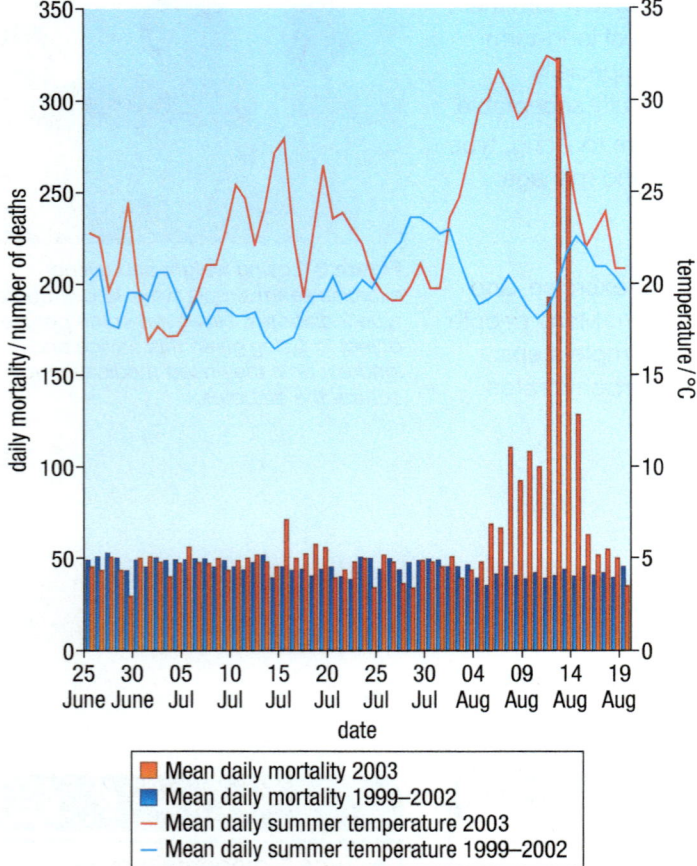

Figure 1

a What effect did the Paris heatwave have on deaths in the city?

b From the data, what temperature begins to have an effect on the death rate?

c Explain why more people die when conditions are very hot.

4 Figure 2 shows the blood glucose levels of a non-diabetic person and someone with type 1 diabetes managed with regular insulin injections. They both eat at the same times.

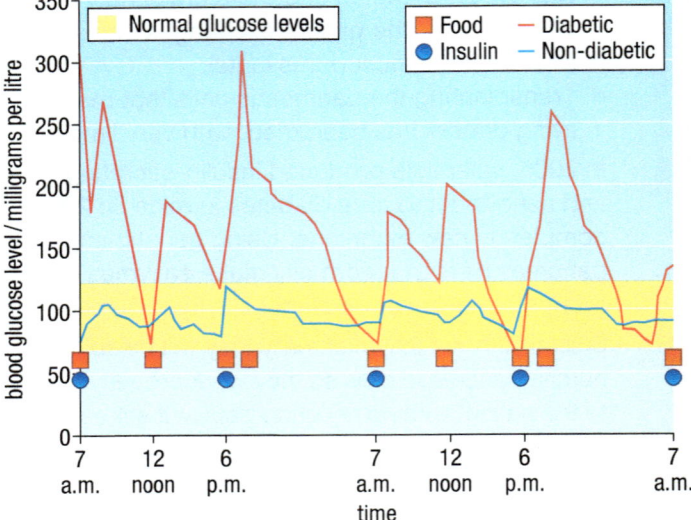

Figure 2

Use Figure 2 to help you to answer the questions below:

a What happens to the blood glucose levels in both individuals after eating?

b What is the range of blood glucose concentration of the non-diabetic person?

c What is the range of blood glucose concentration of the person with diabetes?

d Figure 2 shows the effect of regular insulin injections on the blood glucose level of someone with diabetes. Why are the insulin injections so important to their health and well-being? What does this data suggest are the limitations of insulin injections?

e People with diabetes have to monitor the amount of carbohydrate in their diet. Explain why.

Practice questions

1 Automatic control systems in the body keep conditions inside the body relatively constant.

a Name this process. (1)

b Give **three** examples of internal conditions that are controlled automatically. (3)

c Name the organ of the body that contains:
  i   the thermoregulatory centre
  ii receptors for blood glucose. (2)

2 A walker falls through thin ice into very cold water.

The walker's core body temperature falls. He may die of hypothermia (when core body temperature falls too low).

a i   Which part of the brain monitors the fall in core body temperature? (1)
  ii How does this part of the brain detect the fall in core body temperature? (2)

b Whilst in the water the walker begins to shiver. Shivering helps to stop the core body temperature falling too quickly. Explain how. (2)

c The walker had been drinking alcohol. Alcohol causes changes to the blood vessels supplying the skin capillaries, making the skin look red.
  i   Describe the change to the blood vessels. (1)
  ii The walker is much more likely to die of hypothermia than someone who has not been drinking alcohol. Explain why. (2)

3 The kidney is the main organ involved in regulating the amount of water in the body.

a Describe how the blood is filtered in the kidney so that waste molecules are excreted in the urine, but useful molecules are not. (4)

b The volume of urine excreted depends on the concentration of water in the blood.
  i   Where is the water content of the blood monitored? (1)
  ii Name the hormone that acts on the kidney to control how much water is excreted. (1)
  iii Name the gland that releases this hormone. (1)

c *In this question you will be assessed on using good English, organising information clearly, and using specialist terms where appropriate.*

On a very hot day, a builder starts work. After one hour he becomes very thirsty. He visits the toilet and passes a small volume of concentrated urine. He drinks two large bottles of water and feels much better. Half an hour later he visits the toilet again and passes a large volume of dilute urine.

Use the words named in part **b** to explain how the builder's body has responded to the water content of his blood since he started work. (6)

8.1 Pathogens and disease

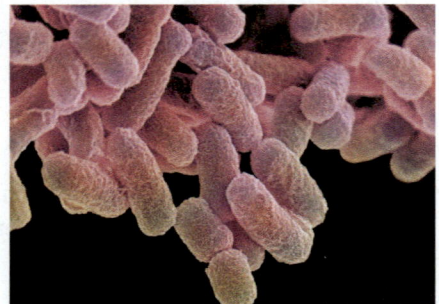

Figure 1 *Many bacteria are very useful to humans but some, such as these E. coli, are pathogens and cause disease*

∞ **links**

Find out more about the structure of bacteria by looking back to 1.2 Eukaryotes and prokaryotes.

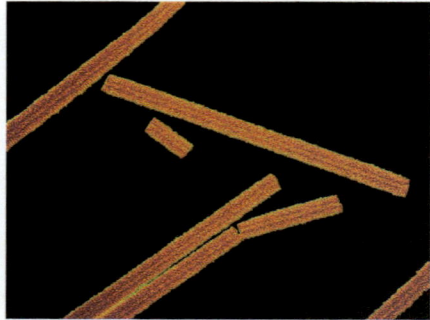

Figure 2 *These tobacco mosaic viruses cause disease in plants*

∞ **links**

For more information on bacteria that are resistant to antibiotics, see 8.5 Changing pathogens.

Infectious diseases are found all over the world, in every country. Some infectious diseases are fairly mild, such as the common cold and tonsillitis. Others are known killers, such as influenza and HIV/Aids.

An infectious disease is caused by a **microorganism** entering and attacking your body. People can pass these microorganisms from one person to another. This is what is meant by **infectious**.

Microorganisms that cause disease are called **pathogens**. Common pathogens are bacteria and viruses. Pathogens can attack all sorts of different organisms, including plants.

The differences between bacteria and viruses

Bacteria are single-celled living organisms that are much smaller than animal and plant cells.

Although some bacteria cause disease, many are harmless and some are really useful to us. People use them to make food like yoghurt and cheese, to treat sewage, and to make medicines. Bacteria are also important in the environment, as decomposers, and in your body.

Pathogenic bacteria are the minority – but they are significant because of the major effects they can have on individuals and society.

 Did you know … ?

Scientists estimate that most people have between 1 kg and 2 kg of bacteria in their guts, made up of around 500 different species.

Viruses are even smaller than bacteria. They usually have regular shapes. Viruses cause diseases in every type of living organism, from people to bacteria.

How pathogens cause disease

Once bacteria and viruses are inside your body, they reproduce rapidly. This is how they make you ill.

- Bacteria divide rapidly by splitting in two (called binary fission). They often produce toxins (poisons) which affect your body. Sometimes they directly damage your cells.
- Viruses take over the cells of your body as they reproduce, damaging and destroying the cells. They very rarely produce toxins.

Common disease symptoms are a high temperature, headaches, and rashes. These are caused by the damage and toxins produced by the pathogens. The symptoms also appear as a result of the way your body responds to the damage and toxins.

You catch an infectious disease when you pick up a pathogen from someone else who is infected with the disease.

Understanding infections

Pathogens have caused infectious diseases in human beings for many thousands of years. People have been writing about diseases such as tuberculosis, smallpox, and plague throughout recorded history. However, although people have recognised the symptoms of infectious diseases for many centuries, it is only in the past 150–200 years that they have really understood the causes of these diseases and how they are spread.

It has been the work of people such as Ignaz Semmelweis and Louis Pasteur that has helped people reach the understanding of pathogens that they have today.

The work of Ignaz Semmelweis

Semmelweis was a doctor in the mid-1850s. At the time, many women in hospital died from childbed fever a few days after giving birth. However, no one knew what caused it.

Semmelweis noticed that his medical students went straight from dissecting a dead body to delivering a baby without washing their hands. The women delivered by medical students and doctors rather than midwives were much more likely to die. Semmelweis wondered if they were carrying the cause of disease from the corpses to their patients.

When another doctor died from symptoms identical to childbed fever after cutting himself whilst working on a body, Semmelweis became convinced that the fever was caused by some kind of infectious agent. He therefore insisted that his medical students wash their hands before delivering babies. Immediately, fewer mothers died from the fever. However, other doctors were very resistant to Semmelweis's ideas.

Other discoveries

Also in the mid- to late 19th century:

- Louis Pasteur showed that microorganisms caused disease. He also developed vaccines to prevent the spread of diseases such as anthrax and rabies.
- Joseph Lister started to use antiseptic chemicals to destroy pathogens before they caused infection in operating theatres.
- As microscopes improved, it became possible to see pathogens ever more clearly too. This helped to convince people that they were really there!

Understanding how infectious diseases are spread from one person to another makes it possible for us to take a number of measures to prevent their spread.

Figure 3 *Scientists and doctors are still discovering more about the role of pathogens in disease. Barry Marshall won a Nobel Prize in 2005 after a long battle to convince other doctors and scientists that most stomach ulcers are actually caused by a bacterium called Helicobacter pylori*

Study tip

Remember the ways to stop the spread of infections in hospitals:
- more hand washing
- greater use of disinfectants
- better cleaning.

Summary questions

1 a What causes infectious diseases?
  b How do viruses differ from bacteria in the way they cause disease?
  c How do pathogens make you ill?

2 Give five examples of things that people now know they can do to reduce the spread of pathogens and so lower the risk of disease, for example, hand-washing in hospitals.

3 Explain carefully why you think it took so long for people to recognise the causes of the infectious diseases that have caused illness and death in human populations for many thousands of years.

Key points

- Infectious diseases are caused by microorganisms, such as bacteria and viruses, called pathogens.

- Bacteria and viruses reproduce rapidly inside your body. Bacteria can produce toxins that make you feel ill.

- Viruses live and reproduce inside cells, causing damage.

Defence mechanisms

Learning objectives

After this topic, you should know:

- how your white blood cells protect you from disease.

There are a number of ways in which pathogens spread from one person to another. The more pathogens that get into your body, the more likely it is that you will get an infectious disease.

- **Droplet infection** – When you are ill, you expel tiny droplets full of pathogens from your breathing system when you cough, sneeze, or talk. Other people breathe in the droplets, along with the pathogens they contain, so they pick up the infection, for example flu (influenza), tuberculosis, or the common cold.

Figure 1 *Droplets carrying millions of pathogens fly out of your mouth and nose at up to 100 miles an hour when you sneeze*

- **Direct contact** – Some diseases, such as impetigo and sexually transmitted diseases such as genital herpes, are spread by direct contact of the skin.
- **Contaminated food and drink** – Eating raw, undercooked, or contaminated food, or drinking water containing sewage, can spread disease such as diarrhoea, cholera, or salmonellosis. You get these diseases by taking large numbers of microorganisms straight into your gut.
- **Through a break in your skin** – Pathogens such as HIV/Aids or hepatitis can enter your body through cuts, scratches, and needle punctures.

When people live in crowded conditions with no sewage treatment, infectious diseases can spread very rapidly.

Preventing microorganisms getting into your body

Each day, you come across millions of disease-causing microorganisms. Fortunately, your body has several ways of stopping these pathogens getting inside:

- Your skin covers your body and acts as a barrier. It prevents bacteria and viruses from reaching the tissues beneath that can be infected.
- If you damage or cut your skin, you bleed. Your blood quickly forms a clot, which dries into a scab. The scab forms a seal over the cut, stopping pathogens getting in through the wound.
- Your breathing system could be a weak link in your body defences. Every time you breathe you draw air, which is full of pathogens, into the airways of the lungs. However, your breathing system produces sticky liquid, called mucus. This mucus covers the lining of your lungs and tubes, such as the bronchi and bronchioles. It traps the pathogens. The mucus is then moved up and out of your body or swallowed down into your gut where the acid destroys the microorganisms.
- In the same way, the stomach acid destroys most of the pathogens you take in through your mouth.

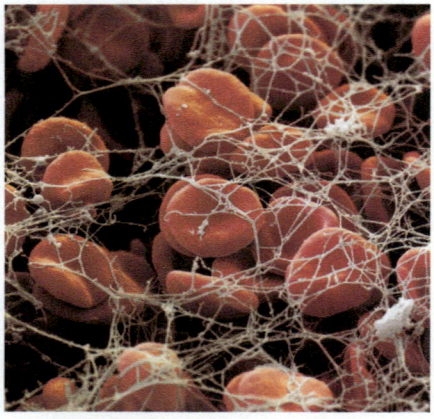

Figure 2 *When you get a cut, the platelets in your blood set up a chain of events to form a clot that dries into a scab. This stops pathogens from getting into your body. It also stops you from bleeding to death*

⬭⬭ links

To find out more about your blood and clotting, see 4.3 Transport in the blood.

How white blood cells protect you from disease

In spite of your body's defence mechanisms, some pathogens still get inside your body. Once there, they will meet your second line of defence – the white blood cells of your immune system.

The white blood cells help to defend your body against pathogens in several ways, summarised in Table 1.

Table 1 *Ways in which your white blood cells destroy pathogens and protect you against disease*

Role of white blood cell	How it protects you against disease
Ingesting microorganisms 	Some white blood cells ingest (take in) pathogens, digesting and destroying them so they can't make you ill. This process is called **phagocytosis**.
Producing antibodies Antibody Antigen Bacterium White blood cell Antibody attached to antigen	Some white blood cells produce special chemicals called antibodies. These target particular bacteria or viruses and destroy them. You need a unique antibody for each type of pathogen. When your white blood cells have met and produced antibodies against a particular pathogen once, the right antibodies can be made very quickly if that pathogen gets into the body again.
Producing antitoxins Antitoxin molecule Toxin and antitoxin joined together Toxin molecule Bacterium	Some white blood cells produce antitoxins. These counteract (cancel out) the toxins (poisons) released by pathogens.

??? Did you know ... ?

Mucus produced from your nose turns green when you have a cold. This happens because some white blood cells contain green-coloured enzymes. When you have a cold these white blood cells destroy the cold viruses and any bacteria in the mucus of your nose. The dead white blood cells, along with the dead bacteria and viruses, are removed in the mucus, making it look green.

Summary questions

1 **a** What are the four main ways in which diseases are spread?
   b For each method of spread, explain how the pathogens are passed from one person to another.

2 Certain diseases mean that you cannot fight infections very well. Explain why the following symptoms would make you less able to cope with pathogens:
   a Your blood won't clot properly.
   b The number of white cells in your blood falls.

3 Here are three common habits. Explain carefully how each one helps to prevent the spread of disease:
   a Washing your hands before preparing a salad.
   b Throwing away tissues after you have blown your nose.
   c Regularly wiping kitchen surfaces with disinfectant.

4 Explain in detail how the white blood cells in your body work.

Key points

- Your white blood cells help to defend you against pathogens by ingesting them (phagocytosis), making antibodies, and making antitoxins.

8.3 Immunity

Every cell has unique proteins called antigens on its surface. The antigens on the microorganisms that get into your body are different to the ones on your own cells. Your immune system recognises they are different. Your white blood cells then make specific antibodies, which attach to the antigens and destroy that particular pathogen. The first time you meet a new pathogen you get ill because there is a delay whilst your body sorts out the antibody that is needed.

Some of your white blood cells (the memory cells) will then 'remember' the right antibody needed to destroy a particular pathogen. If you meet that pathogen again, these memory cells can make the same antibody very quickly. This time you completely destroy the invaders before they have time to make you feel unwell. You are immune to that disease.

Vaccination

Some pathogens, such as the bacteria that cause meningitis, can make you seriously ill very quickly. In fact, you can die before your body manages to make the right antibodies. Fortunately, you can be protected against many of these serious diseases by **immunisation** (also known as **vaccination**).

Immunisation involves giving you a **vaccine**. A vaccine is usually made of a dead or inactivated form of the disease-causing microorganism. It works by stimulating your body's natural immune response to invading pathogens (Figure 2).

A small amount of dead or inactive pathogen is introduced into your body. This gives your white blood cells the chance to develop the right antibodies against the pathogen without you getting ill.

Then, if you meet the live pathogens, your white blood cells can respond rapidly. They can make the right antibodies just as if you had already had the disease, so that you are protected against it.

Doctors use vaccines to protect us against both bacterial diseases (such as tetanus and diphtheria) and viral diseases (such as polio, measles, and mumps). For example, the MMR vaccine protects against measles, mumps, and rubella. Vaccines have saved millions of lives around the world. One disease – smallpox – has been completely wiped out by vaccinations. Doctors hope that polio will also disappear in the next few years.

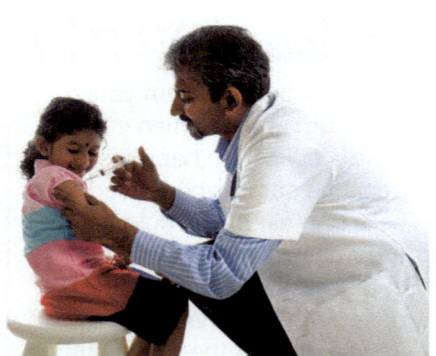

Figure 1 *No one likes having a vaccination very much – but they save millions of lives around the world every year!*

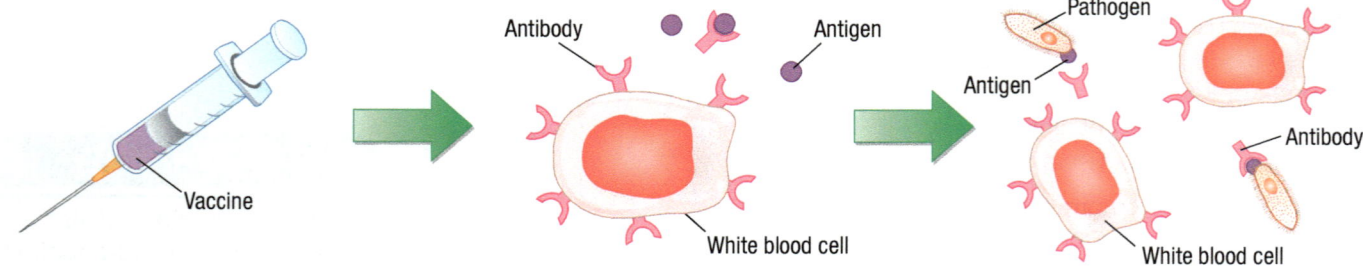

Small amounts of dead or inactive pathogen are put into your body, often by injection.

The antigens in the vaccine stimulate your white blood cells to make antibodies. The antibodies destroy the antigens without any risk of you getting the disease.

You are immune to future infections by the pathogen. That's because your body can respond rapidly and make the correct antibody as if you had already had the disease.

Figure 2 *This is how vaccines protect you against dangerous infectious diseases*

Herd immunity

If a large proportion of the population is immune to a disease, the spread of the pathogen is very much reduced. This is known as **herd immunity**. If, for any reason, the number of people taking up a vaccine falls, the disease can reappear. This is what happened in the UK in the 1970s when there was a scare about the safety of whooping cough vaccine. Vaccination rates fell from over 80% to around 30%. In the following years, thousands of children got whooping cough again and a substantial number died. Yet the vaccine was as safe as any medicine – and when people eventually realised this and enough children were vaccinated for herd immunity to be effective again, the number of cases of the disease and associated deaths quickly fell once more.

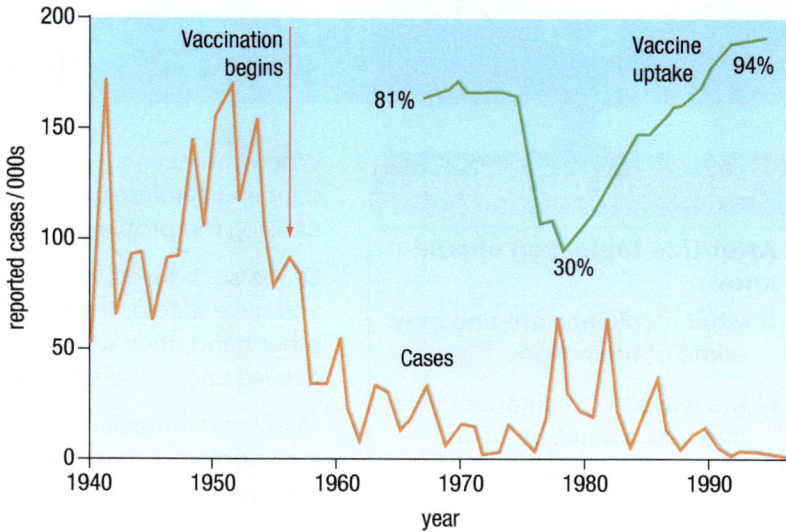

Figure 3 *Graph showing the effect of the whooping cough scare on both uptake of the vaccine and the number of cases of the disease (Source: Open University)*

No medicine is completely risk-free. Very rarely, a child will react badly to a vaccine with tragic results. Making the decision to have your baby immunised can be difficult. Yet because vaccines are so successful, you rarely see the terrible diseases that they protect you against. A hundred years ago, nearly 50% of all deaths of children and young people were caused by infectious diseases. The development of antibiotics and vaccines means that now only 0.5% of all deaths in the same age group are due to infectious disease. Before vaccination, many children were also left permanently damaged by serious infections. Parents today are often aware of the very small risks associated with vaccination, but sometimes forget about the terrible dangers of the diseases children are being vaccinated against.

Society needs as many people as possible to be immunised against as many diseases as possible to keep the pool of infection in the population very low. On the other hand, there is a remote chance that something may go wrong with a vaccination. For the great majority, vaccination is the best option both for the child and for society.

Study tip

High levels of antibodies do not stay in your blood forever – immunity is the ability of white blood cells called memory cells to produce the right antibodies quickly if you are re-infected by a pathogen.

Key points

- You can be immunised against a disease by introducing small amounts of dead or inactive pathogens into your body (vaccination).

- Vaccines stimulate white blood cells to produce antibodies to destroy the pathogen. This produces immunity to that pathogen because the body can respond rapidly, making the correct antibody as if you had had the disease.

- If a large proportion of the population is immune to a pathogen the spread of disease is greatly reduced.

- The MMR vaccine protects children against measles, mumps, and rubella.

Summary questions

1 a What is an antigen?
  b What is an antibody?
  c Give an example of **one** bacterial and **one** viral disease that you can be immunised against.

2 Explain carefully, using diagrams if they help you:
  a how the immune system of your body works
  b how vaccines use your natural immune system to protect you against serious diseases.

3 Explain why vaccines can be used against both bacterial and viral diseases.

8.4

Using drugs to treat disease

When you have an infectious disease, you generally take medicines that contain useful **drugs**. Often the medicine doesn't affect the pathogen that is causing the problems – it just eases the symptoms and makes you feel better.

Drugs such as aspirin and paracetamol are very useful as painkillers. When you have a cold, they will help relieve your headache and sore throat. On the other hand, they will have no effect on the viruses that have entered your tissues and made you feel ill.

Many of the medicines you can buy at a chemists or supermarket relieve your symptoms but do not kill the pathogens, so they do not cure you any faster. You have to wait for your immune system to overcome the pathogens.

Antibiotics

Drugs that make us feel better are useful, but what people really need are drugs that can cure infectious diseases. You can use antiseptics and disinfectants to kill bacteria outside the body and so reduce the risk of infections spreading, but these are generally far too poisonous to use inside your body – they would kill you and your pathogens at the same time!

The drugs that have really changed the way in which doctors treat infectious diseases are antibiotics. These are medicines that can work inside your body to kill the bacteria that cause diseases.

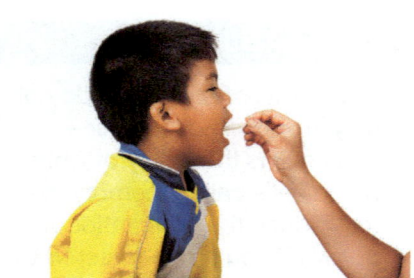

Figure 1 *Taking a painkiller will make this child feel better, but he will not actually get well any faster as a result*

Discovering penicillin

In the early 20th century, doctors and scientists were on the lookout for chemicals that might kill bacteria and so cure some of the terrible infectious diseases of the day. In 1928, Alexander Fleming was growing lots of bacteria on agar plates to investigate them. Fleming was rather careless, and his lab was quite untidy. He often left the lids off his plates for a long time and forgot about experiments he had set up!

After one holiday, Fleming saw that lots of his culture plates had mould growing on them. He noticed a clear ring in the jelly around some of the spots of mould. Something had killed the bacteria covering the jelly.

Fleming saw how important this was. He called the mould penicillin. He worked hard to extract a juice from the mould, but he couldn't get much penicillin and he couldn't make it survive, even in a fridge. So Fleming couldn't prove that an extract of the mould would actually kill bacteria and make people better. By 1934, he gave up on penicillin and went on to do different work.

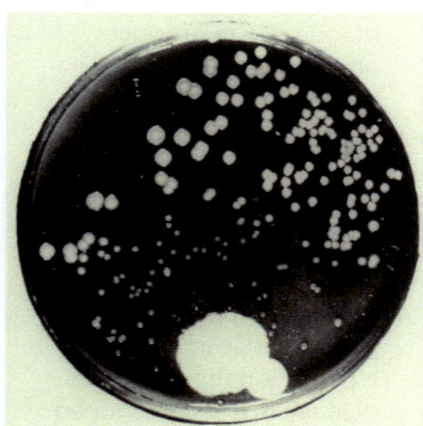

Figure 2 *Alexander Fleming was on the lookout for something that would kill bacteria. Because he noticed the effect of this mould on his cultures, millions of lives have been saved around the world*

About 10 years after penicillin was first discovered, Ernst Chain and Howard Florey set about trying to find better ways of extracting penicillin from the mould so that they could use it on people. They gave some penicillin they extracted to a man dying of a blood infection. The effect was amazing and he recovered – until the penicillin ran out. Florey and Chain even tried to collect unused penicillin from the patient's urine, but in spite of this he died.

Chain and Florey kept working and eventually they managed to make penicillin on an industrial scale, producing enough antibiotic to supply the demands of the Second World War. Doctors have used it as a medicine ever since.

How antibiotics work

Antibiotics such as penicillin work by killing the infective bacteria that cause disease whilst they are inside your body. They damage the bacterial cells without harming your own cells. They have had an enormous effect on our society. Doctors can now cure bacterial diseases that killed millions of people in the past.

If you need antibiotics, you usually take a pill or syrup, but if you are very ill antibiotics may be fed straight into your bloodstream so that they reach the pathogens in your cells as quickly as possible. Some antibiotics kill a wide range of bacteria. Others are very specific and only work against very specific bacteria. It is very important that the right antibiotic is chosen and used.

Unfortunately, antibiotics are not the complete answer to the problem of infectious diseases. They have no effect on diseases caused by viruses. The problem with viral pathogens is that they reproduce inside the cells of your body. It is extremely difficult to develop drugs that kill the viruses without damaging the cells and tissues of your body at the same time.

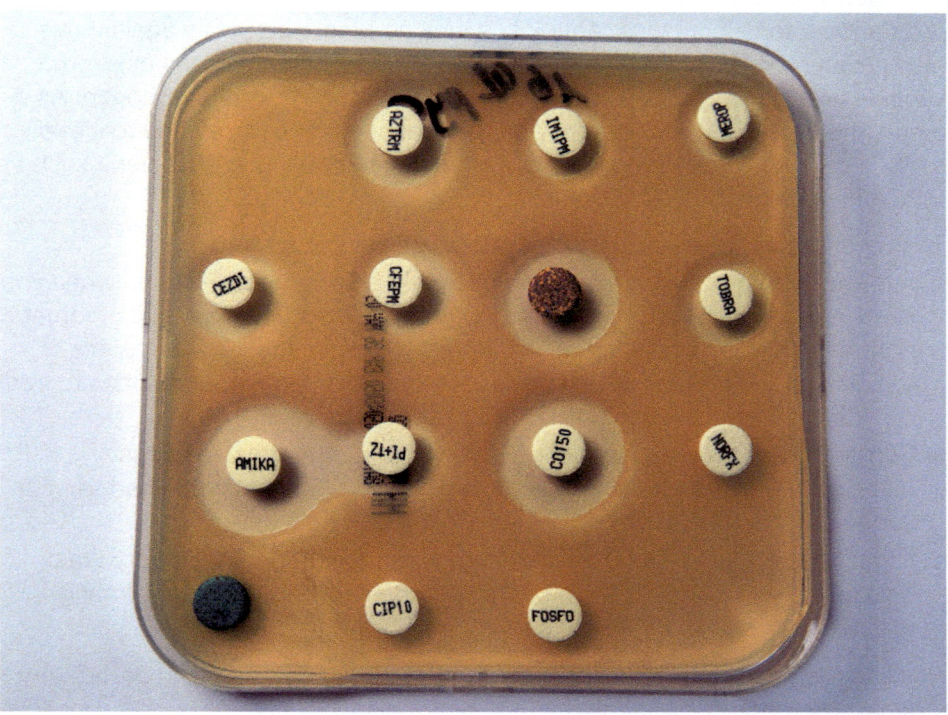

Figure 3 *Penicillin was the first antibiotic. Now there are many different ones that kill different types of bacterium. Here, several different antibiotics are being tested*

Study tip

Don't confuse antiseptic, antibodies, and antibiotics:
- Antiseptic is a liquid that kills microorganisms in the environment.
- Antibiotics are drugs that kill bacteria (*not* viruses) in the body.
- Antibodies are proteins made by white blood cells to kill pathogens (both bacteria and viruses).

Key points

- Some medicines relieve the symptoms of disease but do not kill the pathogens which cause it.
- Antibiotics cure bacterial diseases by killing the bacteria inside your body.
- Antibiotics do not destroy viruses because viruses reproduce inside body cells. It is difficult to develop drugs that can destroy viruses without damaging your body cells.

Summary questions

1. What is the main difference between drugs such as paracetamol and drugs such as penicillin?
2. a How did Alexander Fleming discover penicillin?
   b Why was it so difficult to make a medicine out of penicillin?
   c Who developed the industrial process that made it possible to mass-produce penicillin?
3. Explain why it is so much more difficult to develop medicines against viruses than it has been to develop antibacterial drugs.

8.5

Changing pathogens

Study tip

Remember that mutations occur by chance only. They are not caused by the antibiotic.

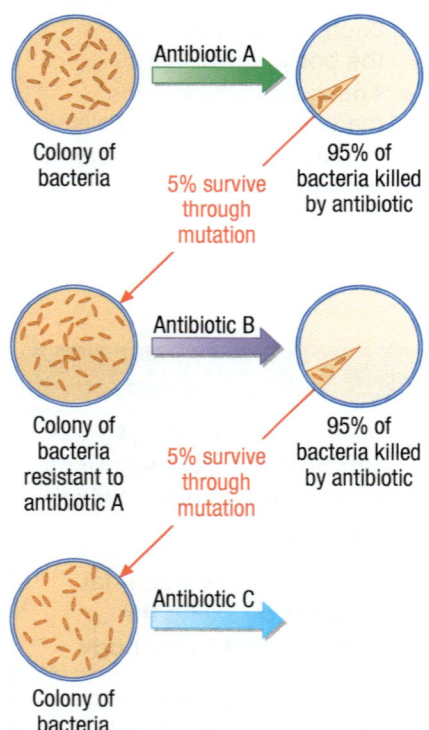

Figure 1 *Bacteria can develop resistance to many different antibiotics in a process of natural selection, as this simple model shows*

If you are given an antibiotic and use it properly, the bacteria that have made you ill are killed off. However, some bacteria develop resistance to antibiotics. They have a natural **mutation** (change in their genetic material) that means they are not affected by the antibiotic. These mutations happen by chance and they produce new strains of bacteria by **natural selection**.

More types of bacteria are becoming resistant to more antibiotics, so bacterial diseases are becoming more difficult to treat. Over the years, antibiotics have been used inappropriately and when they are not really needed. This has increased the rate at which antibiotic-resistant strains have developed.

Antibiotic-resistant bacteria

Normally, an antibiotic kills the bacteria of a non-resistant strain. However, individual resistant bacteria survive and reproduce, so the population of resistant bacteria increases. Antibiotics may no longer be active against this new resistant strain of the pathogen. What's more, in some cases existing vaccines are no longer effective against the mutated, resistant pathogen. As a result, the new strain will spread rapidly because no one is immune to it and there is no effective treatment. This is what has happened with bacteria such as MRSA (see below).

Ext

To prevent more resistant strains of bacteria appearing:

Ext

- It is important not to overuse antibiotics. It's best to use them only when you really need them. For this reason, doctors no longer use antibiotics to treat non-serious infections such as mild throat or ear infections. Also, since antibiotics don't affect viruses, people should not demand antibiotics to treat an illness which their doctor thinks is caused by a virus.
- Some antibiotics treat very specific bacteria, whilst others treat many different types of bacteria. The right type of antibiotic must be used to treat each bacterial infection.
- It is also important that people finish their course of medicine every time. This is to make sure that even bacteria in the early stages of developing resistance are killed by the antibiotic.

Hopefully, steps like these will slow down the rate of development of resistant strains.

The MRSA story

Hospitals use a lot of antibiotics to treat infections. As a result of natural selection, some of the bacteria in hospitals are resistant to many antibiotics. This is what has happened with the bacterium known as **MRSA**, which stands for methicillin-resistant *Staphylococcus aureus*.

As doctors and nurses move from patient to patient, these antibiotic-resistant bacteria are spread easily.

MRSA alone now causes or contributes to over 1000 deaths every year in UK hospitals and care homes, yet a number of simple measures can reduce the spread of microorganisms such as MRSA.

- Antibiotics should only be used when they are really needed.
- Specific bacteria should be treated with specific antibiotics.
- Medical staff should wash their hands with soap and water or alcohol gel between patients, and wear disposable clothing or clothing that is regularly sterilised.
- Hospitals should have high standards of hygiene so that they are really clean.
- Patients who become infected with antibiotic-resistant bacteria should be looked after in isolation from other patients.
- Visitors to hospitals and care homes should wash their hands as they enter and leave.

Simple common-sense measures such as these can have a considerable effect in reducing deaths from antibiotic-resistant bacteria.

Medicines for the future

In recent years, doctors have found strains of bacteria that are resistant to even the strongest antibiotics. In these cases, there is nothing more that antibiotics can do for a patient and he or she may well die. Scientists are constantly looking for new antibiotics. However, it isn't easy to find chemicals that kill bacteria without damaging human cells.

Penicillin and several other antibiotics are made by moulds. Scientists are collecting soil samples from all over the world to try to find another mould to produce a new antibiotic against antibiotic-resistant bacteria such as MRSA. They are also spreading the search much wider than moulds. For example, crocodiles have teeth full of rotting meat, they live in dirty water, and fight a lot, but the terrible bites they inflict on each other do not become infected. Scientists have extracted peptides from crocodile and alligator blood which seem to act as antibiotics, and which they hope to turn into human medicines.

Similarly:

- Scientists are analysing the protective slime that covers fish. They have isolated proteins in the slime that have antibiotic properties, which may be useful.
- Scientists in Germany and Australia have found that certain types of honey (used since the time of the Ancient Egyptians to help heal wounds) have antibiotic properties that kill many bacteria, including MRSA. Doctors are using manuka honey dressings to treat infected wounds.

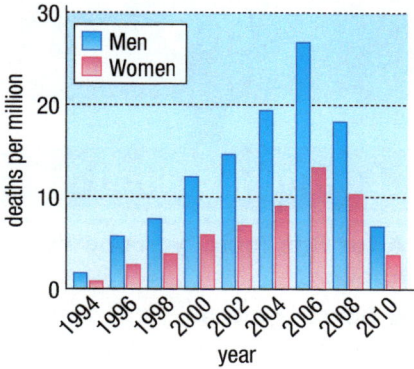

Figure 2 *The number of deaths in which MRSA played a part, 1993–2010 (Source: Office for National Statistics)*

Figure 3 *Scientists are looking throughout the natural world for new antibiotics – including undertaking investigations into crocodile blood*

Key points

- Mutations of pathogens produce new strains.
- Antibiotics kill the non-resistant strain of bacteria, but individual resistant pathogens survive and reproduce, so the population increases by natural selection.
- The new resistant strain can spread rapidly as people are not immune to it and antibiotics and vaccination may not be effective against it.
- To prevent antibiotic resistance getting worse, it is important not to overuse antibiotics, and to use them correctly when they are needed.
- The development of antibiotic resistance means new antibiotics must be developed.

Summary questions

1 Make a flow chart to show how bacteria develop resistance to antibiotics.

2 **a** Is MRSA a bacterium or a virus?
  b How does MRSA illustrate the importance of not using antibiotics too frequently or when they are not really necessary?

3 Use Figure 2 to help you answer the following questions:
  a How could you explain the increase in deaths linked to MRSA?
  b Suggest reasons for the difference in deaths from MRSA between men and women.
  c Deaths from MRSA and other hospital-acquired infections are falling. How could you explain this fall, which is still continuing in most places?

8.6 Growing and investigating bacteria

Learning objectives

After this topic, you should know:

- how to grow an uncontaminated culture of bacteria in the lab

- how uncontaminated cultures are used

- why bacteria are cultured at lower temperatures in schools than in industry.

Did you know ...?

You are surrounded by disease-causing bacteria all the time. If you cultured bacteria at 37 °C (human body temperature) there would be a very high risk of growing some dangerous pathogens.

To find out more about microorganisms, you need to culture them. This means you grow very large numbers of them so that you can see all of the bacteria (the colony) as a whole. Many microorganisms can be grown in the laboratory. This helps scientists learn more about them. Scientists can find out what nutrients microorganisms need for growth and can investigate which chemicals are best at killing them. Bacteria are the most commonly cultured microorganisms.

Growing microorganisms in the lab

To culture (grow) microorganisms, you must provide them with everything they need. This means giving them a liquid or gel containing nutrients – a **culture medium**. This contains carbohydrate as an energy source, various minerals, a nitrogen source so they can make proteins, and sometimes other chemicals. Most microorganisms also need warmth and oxygen to grow.

You usually provide the nutrients in **agar** jelly. Hot agar containing all the nutrients your bacteria will need is poured into a Petri dish. It is then left to cool and set before you add the microorganisms.

You must take great care when you are culturing microorganisms. The bacteria you want to grow may be harmless. However, there is always the risk that a **mutation** (a change in the DNA) will take place and produce a new and dangerous pathogen.

You also want to keep the pure strains of bacteria you are culturing free from any other microorganisms. Such contamination might come from your skin, the air, the soil, or the water around you. Investigations need uncontaminated cultures of microorganisms. Whenever you are culturing microorganisms, you must carry out strict health and safety procedures to protect yourself and others (Figure 1).

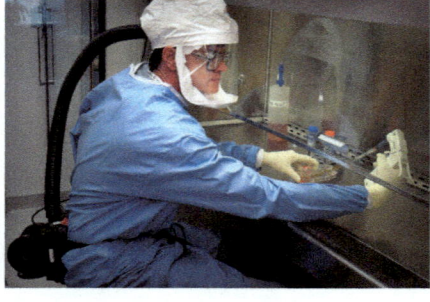

Figure 1 *When working with the most dangerous pathogens, scientists need to be very careful. Sensible safety precautions are needed when working with microorganisms*

Growing useful organisms

You can prepare an uncontaminated culture of microorganisms in the laboratory by following a number of steps:

Step 1
The Petri dishes on which you will grow your microorganisms must be sterilised before using them. The nutrient agar, which will provide food for the microorganisms, must also be sterilised. This kills off any unwanted microorganisms. You can use heat to sterilise glass dishes. A special oven called an autoclave is often used. It sterilises by using steam at high pressure. Plastic Petri dishes are often bought ready-sterilised. UV light or gamma radiation is used to kill the bacteria.

Study tip

Make sure you understand that you sterilise solutions and equipment to kill all the bacteria already on them. Otherwise the bacteria would grow and contaminate the culture you are studying.

Step 2

The next step is to **inoculate** the sterile agar with the microorganisms you want to grow.

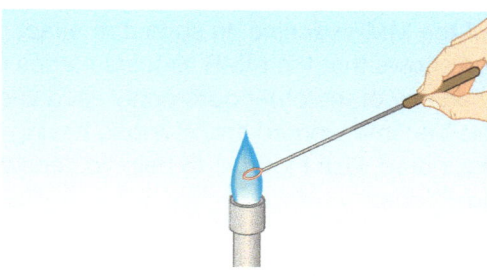

Sterilise the inoculating loop used to transfer microorganisms to the agar by heating it in the flame of a Bunsen burner until it is red hot and then letting it cool. Do not put the loop down or blow on it as it cools.

Dip the sterilised loop in a suspension of the bacteria you want to grow and use it to make zigzag streaks across the surface of the agar. Replace the lid on the dish as quickly as possible to avoid contamination.

Fix the lid of the Petri dish with adhesive tape to prevent microorganisms from the air contaminating the culture – or microbes from the culture escaping. Do not seal all the way around the edge, so that oxygen can get into the dish and prevent harmful anaerobic bacteria from growing.

The Petri dish should be labelled and stored upside down to stop condensation falling onto the agar surface.

Figure 2 *Culturing microorganisms safely in the laboratory*

Step 3

Once you have inoculated your plates, the secured Petri dishes need to be incubated (kept warm) for several days so the microorganisms can grow (Figure 3). In school and college laboratories, the maximum temperature at which cultures are incubated is 25 °C. This greatly reduces the likelihood that you will grow pathogens that might be harmful to people. In industrial conditions, bacterial cultures, (e.g., genetically modified insulin-producing bacteria), are often grown at higher temperatures to enable the microorganisms to grow more rapidly. A hospital lab would also incubate human pathogens at 37 °C, so they grow as fast as possible to be identified.

Summary questions

1 a Why do scientists culture microorganisms in the laboratory?
  b What is agar jelly, and why is it so important in setting up bacterial cultures?

2 When you set up a culture of bacteria in a Petri dish (Figure 2), you give the bacteria everything they need to grow as fast as possible. However, these ideal conditions do not last forever. What might limit the growth of the bacteria in a culture on a Petri dish?

3 a Why do you grow bacteria at 25 °C or below in the school lab when this is not their optimum temperature for growth?
  b Why are bacteria cultured at much higher temperatures in industrial plants?

Required practical

You can investigate the effect of disinfectants and antibiotics on the growth of bacteria by adding circles of filter paper soaked in different types or concentrations of disinfectant or antibiotic when you set up your culture plate. An area of clear jelly indicates that the bacteria have been killed or cannot grow.

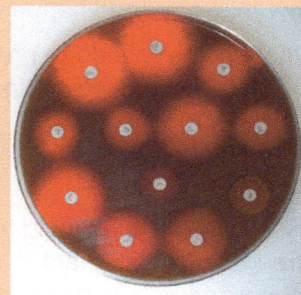

Figure 3 *Culturing microorganisms makes it possible for us to observe how different chemicals affect them*

Key points

- An uncontaminated culture of microorganisms can be grown using sterilised Petri dishes and agar. You sterilise the inoculating loop before use and fix the lid of the Petri dish to prevent unwanted microorganisms getting in. The culture is left upside down at about 25 °C for a few days.

- Uncontaminated cultures are needed so you can investigate the effect of chemicals such as disinfectants and antibiotics on microorganisms.

- Cultures should be incubated at a maximum temperature of 25 °C in schools and colleges to reduce the likelihood of harmful pathogens growing, although in industry they are cultured at higher temperatures.

Summary questions

1 Bacteria and viruses are very small.

  a How do they get into the body?

  b How do bacteria cause the symptoms of disease?

  c How do viruses make you ill?

2 **a** Vancomycin is an antibiotic that doctors used for patients infected with MRSA and other antibiotic-resistant bacteria. Now they are finding that some infections are resistant to vancomycin. Explain how this may have happened.

  b What can people do to prevent the problem of antibiotic resistance getting worse?

3 **a** How would you set up a culture of bacteria in a school lab?

  b Describe how you would test to find out the right strength of disinfectant to use to wash the school floors.

4 The body has a number of defences against pathogens. Some of them are very general and work against any pathogen that enters your body. Others are very specific to a particular pathogen.

Describe all of the ways in which your body defends you against pathogens causing disease.

5 Vaccination uses your body's natural defence system to protect you against disease.

  a Explain how vaccination works.

  b Why are you vaccinated against some diseases and not others?

6 Measles is a disease that affects millions of people around the world. Approximately every three minutes, someone, somewhere dies of measles – a total of 164 000 people each year. Others are left blind or brain-damaged by the disease.

Mumps is another common infectious disease, which is usually relatively mild but can cause sterility in men, deafness, and meningitis.

Rubella is a mild disease in adults but can cause serious problems in the developing fetus if a pregnant woman becomes infected.

The MMR (measles, mumps, and rubella) vaccine protects children against all of these viral diseases.

However, a doctor who has since been discredited and struck off the medical register started a scare story about the safety of the MMR vaccine. In spite of the fact that all the evidence shows that the MMR vaccine carries no more risk than any of the other commonly used vaccines, many people became worried and stopped having their children vaccinated. Use Figure 1 to help you answer the following questions:

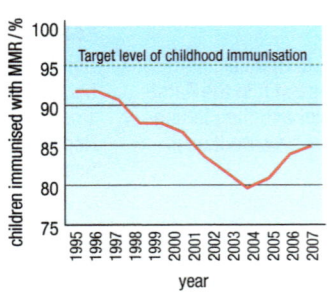
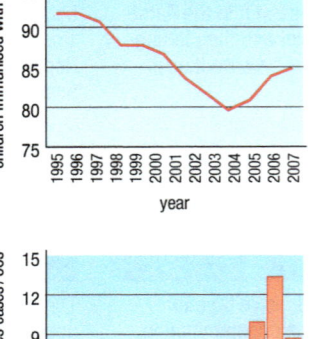
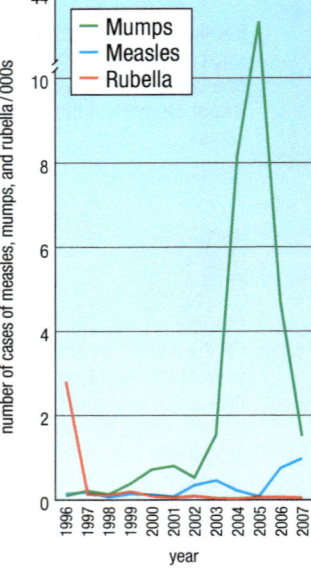

Figure 1

a Based on this evidence, when do you think the MMR scare story was published?

b What happened to the number of children infected with measles after the scare?

c What happened to the number of cases of mumps in 2005?

d Why did the number of cases of measles and mumps remain stable for several years after the numbers of children being vaccinated started to fall?

e In 2011, levels of uptake of the MMR vaccine were over 90% for children aged 5 years and under. The aim is to get levels up to 95% of the population as soon as possible.

  i Why is it important to get vaccination levels so high?

  ii Describe the pattern you would expect to see in the number of cases of measles and mumps over the next few years, and explain your answer.

Practice questions

1 Copy and complete the following sentences:

   a Microorganisms that cause infection are called .... (1)

   b These microorganisms may be bacteria or .... (1)

   c Bacteria may produce .... that make us feel ill. (1)

2 Polio is a disease caused by a virus. In the UK, children are given polio vaccine to protect them against the disease.

   a Copy and complete the following sentences:

   i It is difficult to kill the polio virus inside the body because .... (1)

   ii The vaccine contains an .... form of the polio virus. (1)

   iii The vaccine stimulates the white blood cells to produce ...., which destroy the virus. (1)

   b Figure 1 shows the number of cases of polio in the UK between 1948 and 1968.

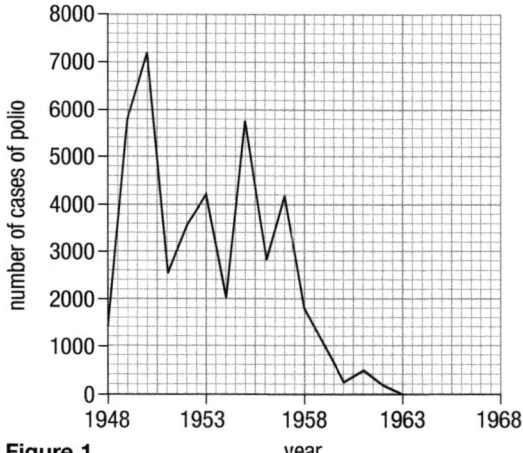

Figure 1

   i In which year was the number of cases of polio highest? (1)

   ii Polio vaccination was first used in the UK in 1955. How many years did it take for the number of cases of polio to fall to zero? (1)

   iii There have been no cases of polio in the UK for many years. However, children are still vaccinated against the disease.
   Suggest one reason for this. (1)

3 The blood system supplies body tissues with essential materials.

   a Blood contains red blood cells, white blood cells, and platelets.

   i Give the function of white blood cells. (1)

   ii Give the function of platelets. (1)

   iii Figure 2 shows a magnified red blood cell.

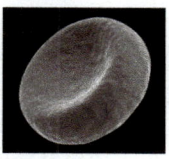

Figure 2

   The average diameter of a real red blood cell is 0.008 millimetres.
   In Figure 2, the diameter of the red blood cell is 10 millimetres.
   Use the formula below to calculate the magnification of the photograph:
   diameter on photograph
      = real diameter × magnification (2)

   iv Some blood capillaries have an internal diameter of approximately 0.01 millimetres.
   Use the information given in part to explain why only one red blood cell at a time can pass through a capillary. (2)

   v Red blood cells transport oxygen.
   Describe how oxygen is moved from the lungs to the tissues. (3)

   b Two students did the same step-up exercise for three minutes.

   One of the students was fit. The other student was unfit.

   Figure 3 shows how the students' heart rates changed during and after the exercise.

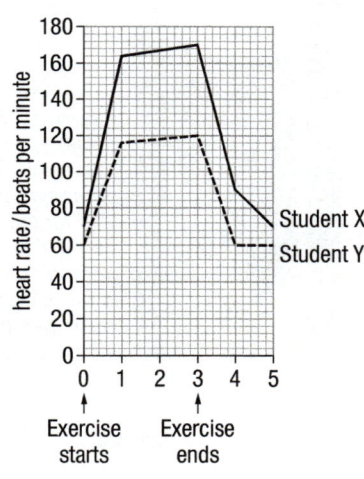

Figure 3

   i Use the information in the Figure 3 to suggest which student was the fittest.
   Explain your choice. (3)

   ii Explain the advantage to the students of the change in heart rate during exercise. (4)

9.1 Photosynthesis

Learning objectives

After this topic, you should know:

- the raw materials for photosynthesis

- the energy source for photosynthesis

- how plants absorb light.

Study tip

Learn the word and chemical equations for photosynthesis. Remember that the process needs light, which normally comes from the Sun.

Figure 1 *The oxygen produced as a by-product of photosynthesis is vital for life on Earth. You can demonstrate that it is produced using water plants such as this* Cabomba

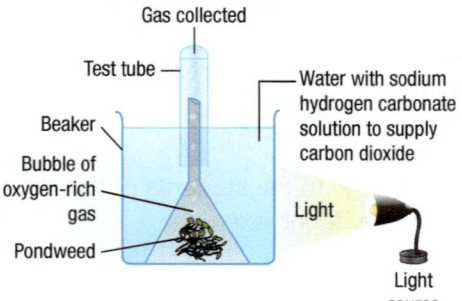

Gas collected

Test tube

Beaker

Bubble of oxygen-rich gas

Pondweed

Water with sodium hydrogen carbonate solution to supply carbon dioxide

Light

Light source

Figure 2 *Diagram to show simple apparatus for measuring the rate of photosynthesis*

Like all living organisms, plants and algae need food to provide them with the chemicals needed for growth and respiration. However, plants don't need to eat – they can make their own food by photosynthesis. This takes place in the green parts of plants (especially the leaves) when it is light. Algae can also carry out photosynthesis.

The process of photosynthesis

The cells in algae and plant leaves are full of small green parts called chloroplasts, which contain a green substance called chlorophyll. During photosynthesis, light is absorbed by the chlorophyll in the chloroplasts. This energy is then used to convert carbon dioxide from the air plus water from the soil into a simple sugar called **glucose**. This chemical reaction also produces oxygen gas as a by-product. The oxygen is released into the air, which you can then use when you breathe it in.

Photosynthesis can be represented by the equation:

$$\text{carbon dioxide + water} \xrightarrow{\text{light}} \text{glucose + oxygen}$$

$$6CO_2 + 6H_2O \xrightarrow{\text{light}} C_6H_{12}O_6 + 6O_2$$

Some of the glucose produced during photosynthesis is used immediately by the cells of the plant for respiration. However, a lot of the glucose is converted into insoluble starch or other large molecules.

Practical

Demonstrating photosynthesis in a water plant

You will need this experimental method to carry out the **Required practical** in 9.2 *Limiting factors*

It isn't easy to measure the rate of photosynthesis of a land plant in a school laboratory. However, using water plants such as *Cabomba* or *Elodea*, you can show that a plant is photosynthesising by observing the oxygen it gives off (Figure 1). As a water plant photosynthesises it produces bubbles of colourless gas which you can see. You can count the number of bubbles produced in a given time, or you can collect the gas and measure the volume produced in a given time (Figure 2). Both methods give you a way of measuring the rate of photosynthesis. The gas produced by the plant will relight a glowing splint, showing that it is rich in oxygen.

Leaf adaptations

For photosynthesis to be successful, a plant needs plenty of carbon dioxide, light, and water. The leaves of plants are perfectly adapted as organs of photosynthesis because:

- most leaves are broad, giving them a large surface area for light to fall on
- they contain chlorophyll in the chloroplasts to absorb light energy
- they have air spaces that allow carbon dioxide to get to the cells, and oxygen to leave them by diffusion
- they have veins, which bring plenty of water in the xylem to the cells of the leaves and remove the products of photosynthesis in the phloem.

These adaptations mean the plant can photosynthesise as much as possible whenever there is light available.

Algae are aquatic so they are adapted to photosynthesising in water. They have a large surface area and absorb carbon dioxide dissolved in the water around them. The oxygen they produce also dissolves in the water around them as it is released.

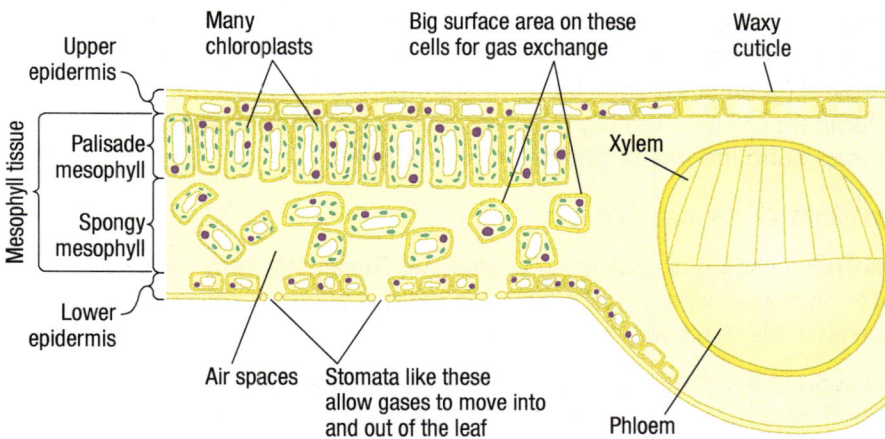

Figure 3 *A section (slice) through a leaf, showing the different tissues and how they are adapted for photosynthesis*

 Did you know ...?

Every year, plants produce about 368 000 000 000 tonnes of oxygen, so there is plenty to go around!

Summary questions

1 a Where does a plant get the carbon dioxide and water that it needs for photosynthesis, and how does it get the light it needs?
  b Where do algae get carbon dioxide and water from?

2 Describe the path taken by a carbon atom as it moves from being part of the carbon dioxide in the air to being part of a starch molecule in a plant.

3 Explain carefully why a water plant kept in the light for 24 hours will produce gas that will relight a glowing splint, whereas a water plant kept in the dark for 24 hours produces a gas which can put out a glowing splint.

⊂⊃ **links**

For more on the structure and function of plant cells, see 1.1 Animal and plant cells.

Study tip

Practise labelling the parts and cells in the cross-section of a leaf, and be sure that you know the function of each one.

Key points

- During photosynthesis, light energy is absorbed by chlorophyll in the chloroplasts of the green parts of the plant. It is used to convert carbon dioxide and water into sugar (glucose). Oxygen is released as a by-product.

- Photosynthesis can be represented by the following word and balanced symbol equation:

carbon dioxide $\xrightarrow[\text{energy}]{\text{light}}$ glucose
+ + +
water oxygen

$$6CO_2 + 6H_2O \xrightarrow[\text{energy}]{\text{light}} C_6H_{12}O_6 + 6O_2$$

- Leaves are well adapted to maximise the amount of photosynthesis which can take place.

9.2 Limiting factors

You may have noticed that plants grow quickly in the summer, yet they hardly grow at all in the winter. Plants need light, warmth, and carbon dioxide if they are going to photosynthesise and grow as fast as they can. Sometimes any one or more of these things can be in short supply, limiting the amount of photosynthesis a plant can manage. This is why they are known as **limiting factors**.

Light

The most obvious factor affecting the rate of photosynthesis is light intensity. If there is plenty of light, lots of photosynthesis can take place. In low light, photosynthesis will stop whatever the other conditions are around the plant. For most plants, the brighter the light, the faster the rate of photosynthesis.

Temperature

Temperature affects all chemical reactions, including photosynthesis. As the temperature rises, the rate of photosynthesis increases as the reaction speeds up. However, photosynthesis is controlled by enzymes. Most enzymes are destroyed (denatured) once the temperature rises to around 40–45 °C. So, if the temperature gets too high, the enzymes controlling photosynthesis are denatured and the rate of photosynthesis will fall.

Carbon dioxide levels

Plants need carbon dioxide to make glucose. The atmosphere is only about 0.04% carbon dioxide, which often limits the rate of photosynthesis. Increasing the carbon dioxide levels will increase the rate at which photosynthesis takes place.

On a sunny day, carbon dioxide levels are the most common limiting factor for plants. The carbon dioxide levels around a plant tend to rise at night, when the plant respires but doesn't photosynthesise. As light and temperature levels increase in the morning, the carbon dioxide around the plant is used up.

In a garden, woodland, or field (rather than a lab or greenhouse where conditions can be controlled), light, temperature, and carbon dioxide levels interact, and any one of them might be the factor that limits photosynthesis.

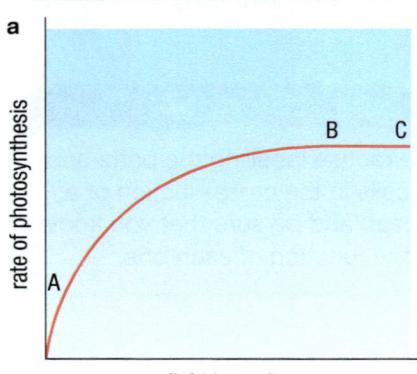

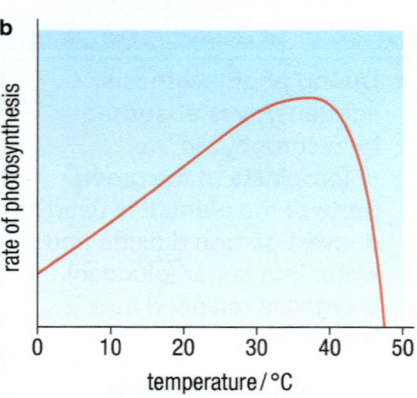

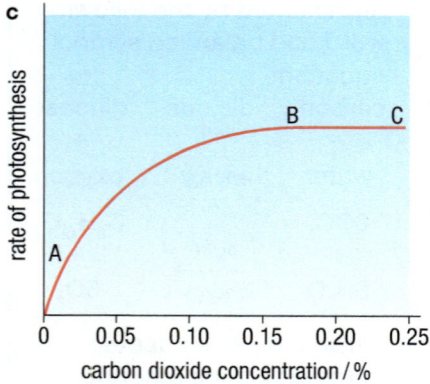

Figure 1 *Light, temperature, and carbon dioxide levels all affect the rate of photosynthesis in a plant*

Required practical

Investigating the effect of variables on the rate of photosynthesis

Using water plants and the apparatus described on page 114, you can investigate:

- the effect of light intensity on the rate of photosynthesis by changing the distance of a light source from the plant
- the effect of temperature on the rate of photosynthesis by varying the temperature of the water surrounding the plant
- the effect of carbon dioxide levels on the rate of photosynthesis by changing the concentration of sodium hydrogen carbonate in the surrounding water.

Remember to keep all other variables constant each time.

Making the most of photosynthesis

The more a plant photosynthesises, the faster it grows. Farmers want their plants to grow as fast and as big as possible to make the best profit. Out in the fields it is almost impossible to influence the growing conditions, but by using greenhouses farmers can artificially control the environment of their plants. Most importantly, in greenhouses the atmosphere is warmer inside than out. This speeds up the rate of photosynthesis so plants grow faster, flower and fruit earlier, and produce higher yields. You can also use greenhouses to protect delicate fruits from harsh weather conditions.

Controlling a crop's environment

Companies use big commercial greenhouses to control the temperature and the levels of light and carbon dioxide. The levels are varied to get the fastest possible rates of photosynthesis. For example:

- Carbon dioxide levels are increased during the day when light levels are at their highest, so they do not act as a limiting factor.
- The optimum temperature for enzyme activity is maintained all the time.
- Artificial lighting is used to prolong the hours of photosynthesis and to increase the light intensity.

As a result, the plants photosynthesise for as long as possible and grow increasingly quickly. Artificial lighting can be used to give year-round production by plants, and to grow plants out of their normal season or out of their normal habitat.

The greenhouses are huge and conditions are controlled using computer software. This all costs a lot of money, but controlling the environment has many benefits. Turnover is fast, which means profits can be high. The crops are clean and unspoilt. There is no ploughing or preparing the land, and crops can be grown where the land is poor.

It takes a lot of energy to keep conditions in the greenhouses just right, but fewer staff are needed. Monitoring systems and alarms are vital in case things go wrong, but for plants grown in controlled greenhouse conditions, limiting factors no longer limit their rate of photosynthesis and growth.

Summary questions

1 What are the **three** main limiting factors that affect the rate of photosynthesis in a plant?

2 **a** In each of these situations, **one** factor in particular is most likely to be limiting photosynthesis. In each case listed below, suggest which factor this is, and explain why the rate of photosynthesis is limited.
   i A wheat field first thing in the morning.
   ii The same field later on in the day.
   iii Plants growing on a woodland floor in a cold winter.
   iv Plants growing on a woodland floor in a hot summer.
   b Why is it impossible to be certain which factor is involved in each of these cases?

3 Look at graph **a** in Figure 1.
   a Explain what is happening between points A and B on the graph.
   b Explain what is happening between points B and C on the graph.
   c Now look at graph **b** in Figure 1. Explain why it is a different shape to the other two graphs shown in Figure 1.

?? Did you know ... ?

The first recorded greenhouse was built in about 30 AD for Tiberius Caesar, a Roman emperor who wanted to eat cucumbers out of season.

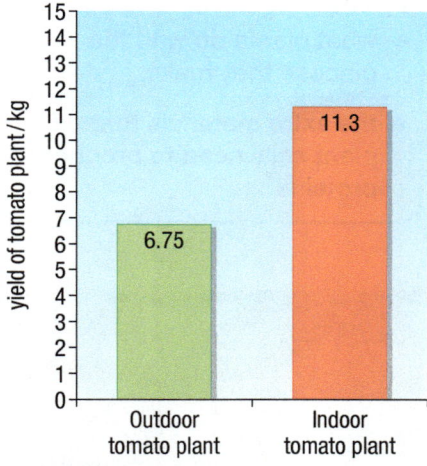

Figure 2 *One piece of American research showed that the crop yield inside a greenhouse was almost double that of crops grown outdoors*

Figure 3 *Controlling the temperature, light, and carbon dioxide level in a greenhouse removes limiting factors and enables farmers to achieve the biggest yields possible in the shortest time*

Key points

- The rate of photosynthesis may be limited by shortage of light, low temperature, and shortage of carbon dioxide.

- Farmers can artificially control the light intensity, temperature, and carbon dioxide concentration when growing crops in greenhouses to increase the rate of photosynthesis and so increase the yield of the crops.

9.3 How plants use glucose

Learning objectives

After this topic, you should know:

- what plants do with the glucose they make
- the extra materials that plant cells need to produce proteins

Figure 1 *Worldwide, photosynthesis in algae produces more oxygen and biomass than photosynthesis in plants does, but people often forget all about them*

??? Did you know …?

Some algal cells are very rich in oils. They are even being considered as a possible source of biofuels for the future.

⬭⬭ links

For more information on transport in plants, see 9.4 Transport systems in plants. For more on osmosis in plants, see 1.7 Osmosis.

Plants and algae make glucose when they photosynthesise. This glucose is vital for their survival. Some of the glucose produced during photosynthesis is used immediately by the plant and algal cells. They use it for respiration, to provide energy for cell functions such as growth and reproduction.

Using glucose

Plant cells and algal cells, like any other living cells, respire all the time. They use some of the glucose produced during photosynthesis as they respire. The glucose is broken down using oxygen to provide energy for the cells. Carbon dioxide and water are the waste products of the reaction. Chemically, respiration is the reverse of photosynthesis.

The energy transferred in respiration is used to build up smaller molecules into bigger molecules. Some of the glucose made in photosynthesis is changed into insoluble starch for storage (see below). Plants and algae also build up glucose into more complex carbohydrates such as cellulose. They use this to strengthen their cell walls.

Plants use some of the glucose from photosynthesis to make amino acids. They do this by combining sugars with **nitrate ions** and other **mineral ions** from the soil. These amino acids are then built up into proteins to be used in their cells (see later). This needs energy from respiration.

Algae also make amino acids. They do this by taking the nitrate ions and other materials they need from the water they live in.

Plants and algae also use glucose from photosynthesis and energy from respiration to build up fats and oils. These may be used in the cells as an energy store. In addition, plants often use fats or oils as an energy store in their seeds. Seeds provide lots of energy for the new plant as it germinates.

Starch for storage

Plants make food by photosynthesis in their leaves and other green parts, but the food is needed all over the plant. It is moved around the plant in the phloem.

Plants convert some of the glucose produced in photosynthesis into starch to be stored. Glucose is soluble in water. If it were stored in plant cells, it could affect the way water moves into and out of the cells by osmosis. Lots of glucose stored in plant cells could affect the water balance of the whole plant.

Starch is insoluble in water, so it will have no effect on the water balance of the plant. This means that plants can store large amounts of starch in their cells.

Starch is the main energy store in plants and it is found all over a plant:

- Starch is stored in the cells of the leaves. It provides an energy store for when it is dark or when light levels are low.
- Starch is also kept in special storage areas of a plant. For example, many plants produce **tubers** and bulbs which are full of stored starch, to help them survive through the winter. You often take advantage of these starch stores, found in vegetables such as potatoes, by eating them yourself.

Practical

Making starch

The presence of starch in a leaf is evidence that photosynthesis has taken place. You can test for starch using the iodine solution test. Take a leaf from a plant kept in the light and a leaf from plant kept in the dark for at least 24 hours. Leaves have to be specially prepared so the iodine solution can reach the cells. Just adding iodine solution to a leaf is not enough, because the waterproof cuticle keeps the iodine out so it can't react with the starch. Also, the green chlorophyll would mask any colour changes if the iodine did react with the starch.

You therefore need to treat the leaves by boiling them in ethanol, to destroy the waxy cuticle and then to remove the colour. The leaves are then rinsed in hot water to soften them. After treating the leaves, add iodine solution to them both. Iodine solution turns blue-black in the presence of starch.

Figure 2 *The leaf on the right has been kept in the dark. Its starch stores have been used for respiration or moved to other parts of the plant. The leaf on the left has been in the light and been able to photosynthesise. The glucose has been converted to starch, which is clearly visible when it reacts with iodine and turns blue-black*

Safety: Take care when using ethanol. It is volatile, highly flammable, and harmful. Always wear eye protection. No naked flames – use a hot water bath to heat ethanol.

Study tip

Two important points to remember:
- Plants **respire** 24 hours a day to release energy.
- Glucose is soluble in water, but starch is insoluble.

Mineral ions, proteins, and plants

Plants make amino acids and build them up into proteins. However, to make amino acids from the carbohydrates produced by photosynthesis, plants must take in nitrates from the soil. Plants take up nitrate ions from the soil through their roots. The nitrate ions are moved into the root hair cells against a concentration gradient by active transport.

Some carnivorous plants, such as the Venus flytrap and sundews, are especially adapted to live in nutrient-poor soil. They can survive because they obtain most of their nutrients from the animals, such as insects, that they catch. Special enzymes then digest the insects. The carnivorous plants use the nutrients from the digested bodies of their victims in place of the nutrients that they cannot get from the poor soil in which they grow.

Figure 3 *The Venus flytrap – an insect-eating plant*

Key points

- Plant and algal cells use the soluble glucose they produce during photosynthesis for respiration; to convert into insoluble starch for storage; to produce fats or oils for storage; and to produce fats, proteins, or cellulose for use in the cells and cell walls.

- Plant and algal cells also need other materials, including nitrate ions, to make the amino acids which make up proteins.

Summary questions

1 List as many ways as possible in which a plant uses the glucose produced by photosynthesis.

2 **a** Why is some of the glucose made by photosynthesis converted to starch to be stored in the plant?

  b Where might you find starch in a plant?

  c How could you show that a potato is a store of starch?

3 Explain why relatively few plants grow successfully in bogs, yet carnivorous plants are often found growing there.

9.4 Transport systems in plants

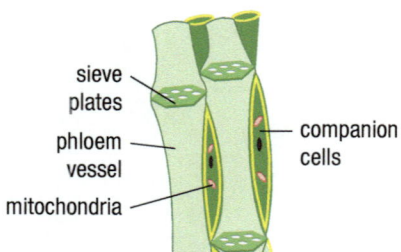

Figure 1 *The adaptations of the phloem cells mean that they are well adapted to their function of moving dissolved food around the plant*

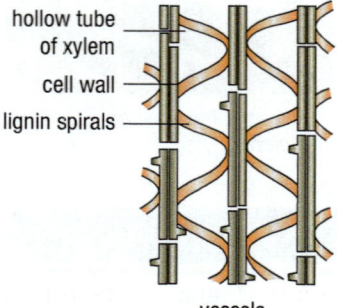

Figure 2 *The adaptations of the xylem cells mean they are well adapted to their function of moving water from the roots to the leaves of a plant and providing strength and support*

Plants make glucose (a simple sugar) by photosynthesis in the leaves and other green parts. This glucose is needed all over the plant. Similarly, water and mineral ions move into the plant from the soil through the roots, but they are needed by every cell of the plant. Water moves through the plant in the transpiration stream. Plants have two separate transport systems to move substances around their bodies. The phloem and the xylem contain different types of specialised cells, each adapted to their function.

Phloem – moving food

The phloem tissue transports the sugars made by photosynthesis from the leaves to the rest of the plant. This includes the growing areas of the stems and roots where the dissolved sugars are needed for making new plant cells. Food is also transported to the storage organs where it is needed to provide an energy store for the winter.

Phloem is a living tissue – the phloem cells are alive. The movement of dissolved sugars from the leaves to the rest of the plant is called **translocation**.

Phloem cells form long tubes that reach from the roots to the leaves. The cell walls between the cells break down to form specialised sieve plates. These allow water carrying dissolved food to move freely up and down the tubes to where it is needed. Phloem cells lose their internal structures but they are supported by companion cells that help to keep them alive. The mitochondria of the companion cells provide the energy needed to move dissolved food up and down the plant in the phloem.

Xylem – moving water and mineral ions

The xylem tissue is the other transport tissue in plants. It carries water and mineral ions from the soil around the plant to the stem and the leaves. Xylem is also important in supporting the plant. The structure of the xylem cells is adapted to their functions in two main ways:

- The cells are alive when they first form, but a special woody chemical called lignin builds up in spirals in the cell walls. The cells die and form long, hollow tubes that allow water and mineral ions to move freely through them from one end of the plant to the other.

- The spirals and rings of woody material in the xylem cells make them very strong and help them withstand the pressure of water moving up the plant. They also help support the plant stem.

Practical

Evidence for movement through xylem

You can demonstrate the movement of water up the xylem by placing celery stalks in water containing a coloured dye. After a few hours, slice the stem in several places – you will see the coloured circles where the water and dye have been moved through the xylem.

In woody plants like trees, the xylem makes up the bulk of the wood and the phloem is found in a ring just underneath the bark. This makes young trees in particular very vulnerable to damage by animals – if a complete ring of bark is eaten, transport in the phloem stops and the tree will die.

Figure 3 *Without protective collars on the trunks, deer would destroy the transport tissue of young trees like these and kill them before they could become established in the woodland*

Why is transport so important?

It is vital to move the food made by photosynthesis around the plant – all the cells need sugars for respiration as well as for providing materials for growth. The movement of water and dissolved mineral ions from the roots is equally important – the mineral ions are needed for the production of proteins and other molecules within the cells.

The plant needs water for photosynthesis, when carbon dioxide and water combine to make glucose (plus oxygen). It also needs water to hold the plant upright. When a cell has plenty of water inside it, the vacuole presses the cytoplasm against the cell walls. This pressure of the cytoplasm against the cell walls gives support for young plants and for the structure of the leaves. For young plants and soft-stemmed plants – although not trees – this is the main method of support, combined with the strength of the xylem vessels.

Summary questions

1 a Why does a plant need a transport system?
  b Explain why a constant supply of sugar and water is so important to the cells of a plant.

2 Make a table to compare the xylem and phloem in a plant.

3 A local woodland trust has set up a scheme to put protective plastic covers around the trunks of young trees. Some local residents are objecting to this, saying it spoils the look of the woodland. Explain exactly why this protection is necessary, and what impact not protecting the trees would have on the wood.

Summary questions

1 **a** Give the word and balanced symbol equations for photosynthesis.

b Much of the glucose made in photosynthesis is turned into an insoluble storage compound. What is this compound?

The values in the following table show the mean growth of two sets of oak seedlings. One set was grown in 85% full sunlight, the other set in only 35% full sunlight.

c

Year	Mean height of seedlings grown in 85% full sunlight (cm)	Mean height of seedlings grown in 35% full sunlight (cm)
2005	12	10
2006	16	12.5
2007	18	14
2008	21	17
2009	28	20
2010	35	21
2011	36	23

i Plot a graph to show the growth of both sets of oak seedlings.

ii Using what you know about photosynthesis and limiting factors, explain the difference in the growth of the two sets of seedlings.

2 Palm oil is made from the fruit of oil palms. Large areas of tropical rainforests have been destroyed to make space to plant these oil palms, which grow rapidly.

a Why do you think that oil palms grow rapidly in the conditions that support a tropical rainforest?

b Where does the oil in the oil palm fruit come from?

c What is it used for in the plant?

d How else is glucose used in the plant?

3 Compare the adaptations of plant leaves for the exchange of carbon dioxide, oxygen, and water vapour with the adaptations of the roots for the absorption of water and mineral ions.

4 Figure 1 shows the two main transport tissues in plants.

a Name tissue A and tissue B.

b Explain, including labelled diagrams, how the structure of tissue A is adapted for its function in the plant.

c Explain, including labelled diagrams, how the structure of tissue B is adapted for its function in the plant.

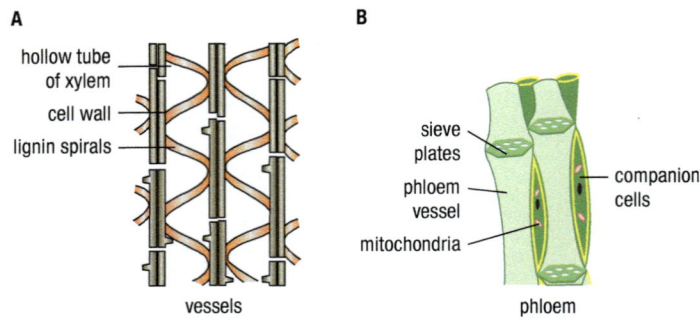

A
hollow tube of xylem
cell wall
lignin spirals
vessels

B
sieve plates
phloem vessel
mitochondria
companion cells
phloem

Figure 1

5 Farmers who grow their crops in large greenhouses control many different factors. Use the graph below to explain why it is so important for them to control.

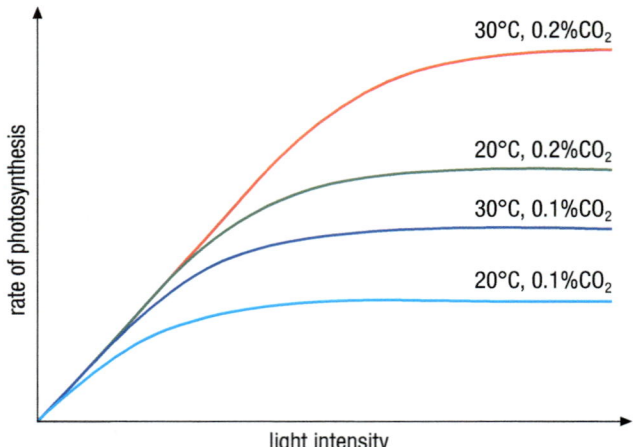

rate of photosynthesis
light intensity
30°C, 0.2%CO₂
20°C, 0.2%CO₂
30°C, 0.1%CO₂
20°C, 0.1%CO₂

Figure 2

a Light levels

b Carbon dioxide levels

c Temperature

Practice questions

1 a Copy and complete the word equation for photosynthesis:

.... + water $\xrightarrow{\text{light energy}}$ glucose + ... (2)

b Geraniums are green plants that grow in gardens.

Where does the light energy for photosynthesis in the geranium come from? How does the geranium absorb this light energy? (3)

c On a frosty morning in December, the rate of photosynthesis in the geranium plant is very slow.

Suggest which factors may be limiting and why. (4)

d Some of the glucose produced by the geranium plant is used for respiration. Give **three** other ways in which the plant uses the glucose produced in photosynthesis. (3)

2 a Water and food molecules need to be transported around the plant. **List A** contains words about these processes. **List B** contains explanations. Match the words in list A to the correct explanation in list B:

List A

Translocation

Xylem

Stomata

Phloem

Transpiration stream

List B

The using up of glucose in respiration

The transport of water from roots to leaves.

The evaporation of water.

Openings in the lower surface of the leaf for gas exchange.

The cells that transport sugars around the plant.

The cells that transport water around the plant.

The movement of sugars from the leaves to other tissues. (5)

b Water is needed for photosynthesis. Give **two** other essential roles for water within the plant. (2)

c Explain why plants take up larger amounts of water from the soil in hot, dry, and windy conditions. (4)

d Describe the process by which root hairs take up mineral ions from the soil against a concentration gradient. (2)

e Why do plants need nitrate ions from the soil? (1)

3 A farmer has decided to grow strawberry plants in polytunnels.

The tunnels are enclosed spaces with walls made of plastic sheeting. The farmer decides to set up several small polytunnels, as models, so that he can work out the best conditions for the strawberry plants to grow. He needs help from a plant biologist, who provides some data.

The data is shown in Figure 2.

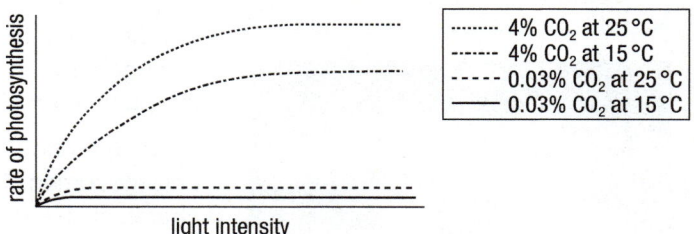

Figure 2

a *In this question you will be assessed on using good English, organising information clearly, and using specialist terms where appropriate.*

You are advising the farmer. Using all the information given, describe the factors the farmer should consider when building his model tunnels so that he can calculate the optimal conditions for growing strawberry plants. (6)

b Biologists often use models in their research. Suggest **one** reason why. (1)

4 Explain carefully why a water plant kept in the light for 24 hours will produce a gas that will relight a glowing splint, whereas a water plant kept in the dark for 24 hours will produce a gas that will put out a glowing splint. (4)

10.1 Inheritance

Learning objectives

After this topic, you should know:

- how parents pass on genetic information to their offspring

- why you resemble your parents but are not identical to them

- where the genetic information is found in a cell

- how both genes and environmental causes influence the characteristics of individuals.

Figure 1 *This mother cat and her kittens are not identical, but they are obviously related*

∞ **links**

For more on how sexual reproduction gives rise to genetic variety, see 2.2 Cell division in sexual reproduction and 10.2 Types of reproduction.

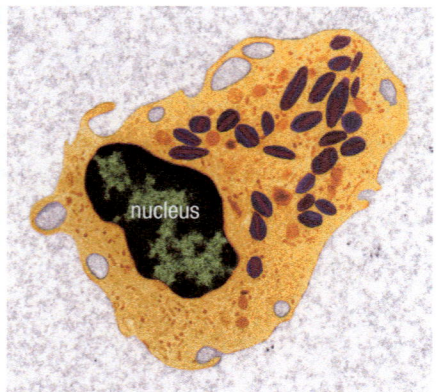

Figure 2 *The DNA that carries the genetic information is found in the nucleus of a cell*

Young animals and plants resemble their parents. For example, horses have foals, people have babies, and chestnut trees produce conkers that grow into little chestnut trees. Many of the smallest organisms that live in the world around you are actually identical to their parents, so what makes humans the way they are?

Why do you resemble your parents?

Most families have characteristics that you can see are clearly passed from generation to generation. Characteristics such as nose shape, eye colour, and dimples are inherited. They are passed on to you from your parents.

Your resemblance to your parents is the result of information carried by genes. These are passed on to you in the sex cells (gametes) from which you developed. This genetic information determines what you will be like.

Chromosomes and genes

The genetic information is carried in the nucleus of your cells and is passed from generation to generation during reproduction. The nucleus contains all the plans for making and organising a new cell and a whole new organism.

Inside the nuclei of all your cells, there are thread-like structures called chromosomes. The chromosomes are made up of large molecules of a special chemical called DNA (deoxyribonucleic acid). This is where the genetic information – the coded information that determines inherited characteristics – is actually stored.

DNA is a long molecule made up of two strands that are twisted together to make a spiral. This is known as a double helix – imagine a ladder that has been twisted round.

∞ **links**

Find out more about DNA in 10.4 From Mendel to modern genetics.

Each different type of organism has a different number of chromosomes in its body cells. For example, humans have 46 chromosomes, potatoes have 48, and chickens have 78. Chromosomes always come in pairs. You have 23 pairs of chromosomes in all your normal body cells. You inherit half your chromosomes from your mother and half from your father.

Each of your chromosomes contains thousands of genes joined together. These are the units of inheritance.

Each gene is a small section of the long DNA molecule. Genes control what an organism is like. They determine its size, its shape, and its colour. Genes work at the level of the molecules in your body to control the development of all the different characteristics you can see. They do this by controlling all the different proteins made in your body. This means that genes determine which enzymes and other proteins are made in your body.

Your chromosomes are organised so that both of the chromosomes in a pair carry genes controlling the same things. This means your genes also come in pairs – one from your father and one from your mother.

Some of your characteristics are decided by a single pair of genes. For example, there is one pair of genes which decides whether or not you will have dimples when you smile. However, most of your characteristics are the result of several different genes working together. For example, your hair and eye colour are both the result of several different genes.

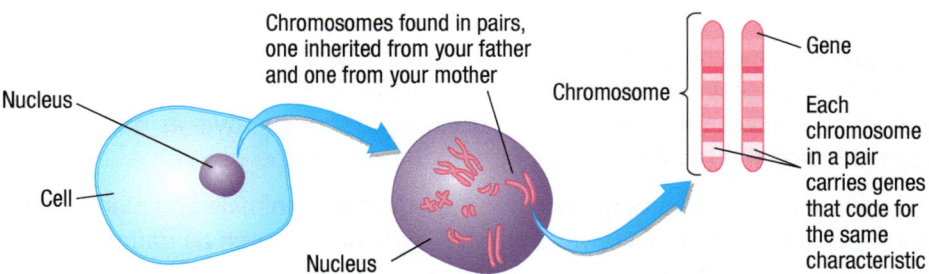

Figure 3 *The relationship between a cell, the nucleus, the chromosomes, and the genes*

Similarities and differences

Although individuals of the same kind of organism have characteristics in common, they do not all look identical. This is partly due to the genes they have inherited (**genetic causes**). However, the conditions in which individuals develop also affect their characteristics. For example, genetically identical plants would grow differently if they were planted in a shady spot compared with bright sunshine. These external influences are known as **environmental causes**.

The differences in characteristics between most organisms of the same species are due to a combination of genetic and environmental causes.

Study tip

Make sure you know the difference between chromosomes, genes, and DNA:

one chromosome → many genes → lots of DNA.

??? Did you know …?

Although scientists have analysed the entire human genome (the total amount of genetic information in the chromosomes), they are still not sure exactly how many genes humans have. At present, they think the human genome is made up of between 20 000 and 25 000 genes.

∞ links

To find out more about the effect of genetic and environmental factors on characteristics, see 10.3 Causes of variation.

Key points

- Parents pass on genetic information to their offspring in the sex cells (gametes).

- The genetic information is found in the nucleus of your cells. The nucleus contains chromosomes, and chromosomes carry the genes that control the characteristics of your body.

- Chromosomes are normally found in pairs.

- Differences in the characteristics of individuals may be due to genetic causes, environmental causes, or both.

Summary questions

1 a What is the basic unit of inheritance?
   b Offspring inherit information from their parents, but do not look identical to them. Why not?

2 a Why do chromosomes come in pairs?
   b Why do genes come in pairs?
   c How many genes do scientists think humans have?

3 a Most organisms from the same species would not look the same even if they were given identical growing conditions. Explain why.
   b If genetically identical organisms have the same growing conditions, they should look identical. Explain why.
   c In reality, even genetically identical organisms do not have exactly the same characteristics. Why is this?

10.2

Types of reproduction

Reproduction is essential to living things. It is during reproduction that genetic information is passed on from parents to their offspring. There are two very different ways of reproducing – **asexual reproduction** and **sexual reproduction**.

Asexual reproduction

Asexual reproduction involves only one parent. There is no joining of special sex cells (gametes) and there is no genetic variation in the offspring. Asexual reproduction gives rise to genetically identical offspring known as **clones**. Their genetic material is identical to that of both the parent and each other.

Asexual reproduction is very common in the smallest animals and plants, and in fungi and bacteria too. However, many larger plants, such as daffodils, strawberries, and brambles, also reproduce asexually.

The cells of your body reproduce asexually all the time. They divide into two identical cells for growth and to replace worn-out tissues.

Figure 1 *A mass of daffodils such as this can contain hundreds of identical flowers because they come from bulbs that reproduce asexually. They also reproduce sexually using their flowers*

Sexual reproduction

Sexual reproduction involves a male sex cell and a female sex cell from two parents. These two special sex cells (gametes) join together to form a zygote which goes on to develop into a new individual.

The offspring that result from sexual reproduction inherit genetic information from both parents. This means that you will have some characteristics from both of your parents, but won't be identical to either of them. This introduces variation. The offspring of sexual reproduction show much more variation than the offspring from asexual reproduction.

- In plants, the gametes involved in sexual reproduction are found within ovules and pollen.
- In animals, they are called ova (eggs) and sperm.

Sexual reproduction is risky because it relies on the sex cells from two individuals meeting. However, it also introduces variation. That's why you find sexual reproduction in organisms ranging from single-celled organisms to people.

Variation

Why is sexual reproduction so important? The variation it produces is a great advantage in making sure that a species survives. Variation makes it more likely that at least a few of the offspring will have the ability to survive difficult conditions.

If you take a closer look at how sexual reproduction works, you can see how variation appears in the offspring.

Different genes control the development of different characteristics. Most things about you, such as your hair and eye colour, are controlled by several different pairs of genes. Each gene will have different forms, or **alleles**. Each allele will result in a different protein. A few of your characteristics are controlled by one single pair of genes, with just two possible alleles.

For example, there are genes that decide whether:

- your earlobes are attached closely to the side of your head or hang freely
- your thumb is straight or curved
- you have dimples when you smile
- you have hair on the second segment of your ring finger.

You can use these genes to help you understand how inheritance works.

Figure 2 *Although there are some likenesses in this family group, the variety caused by the mixing of genetic information in the generations is clear*

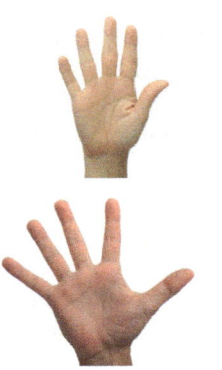
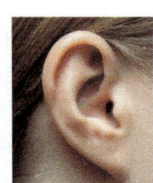
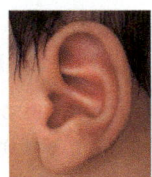

Figure 3 *These are all human characteristics that are controlled by a single pair of genes. They can help us to understand how sexual reproduction introduces variation and how inheritance works*

You will get a random mixture of genetic information from your parents, which is why you don't look exactly like either of them!

Key points

- In asexual reproduction, there is no joining of gametes and only one parent. There is no genetic variation in the offspring.

- The genetically identical offspring of asexual reproduction are known as clones.

- In sexual reproduction, male and female gametes join together. The mixture of genetic information from two parents leads to genetic variation in the offspring.

- Different genes control the development of different characteristics of an organism. Some characteristics are controlled by a single gene. Each gene may have different forms called alleles.

Summary questions

1 Define the following terms:
   - **a** asexual reproduction
   - **b** sexual reproduction
   - **c** gamete
   - **d** variation.

2 Compare the advantages and disadvantages of sexual reproduction with those of asexual reproduction.

3 Some animals such as *Hydra* and some plants such as daffodils reproduce both asexually and sexually.
   - **a** How do daffodils reproduce **i** asexually and **ii** sexually?
   - **b** How does this help to make *Hydra* and daffodils very successful organisms?
   - **c** Explain the genetic differences between a *Hydra*'s sexually and asexually produced offspring.

10.3

Causes of variation

Learning objectives

After this topic, you should know:

- some variation in a population is the result of genetic differences

- some variation in a population is affected by environmental factors.

Figure 1 *However much this Shetland ponyeats, it will never be as tall as the thoroughbred. It just isn't in the genes!*

Figure 2 *The differences in these cows are partly genetic and partly down to their environment, from the milk they drank as calves to the quality of the grass they eat each day*

Have a look at the ends of your fingers and notice the pattern of your fingerprints. No one else in the world will have exactly the same fingerprints as you. Even identical twins have different fingerprints. Animals and plants of the same species, even individuals with the same parents, show variation. What factors make each organism so different from the others around it?

Nature – genetic variation

The genes you inherit determine a lot about you. An apple seed will never grow into an oak tree. Environmental factors, such as the weather or soil conditions, do not matter. The basic characteristics of every species are determined by the genes they inherit.

Certain characteristics are clearly inherited, such as the species you belong to. The shapes of basic characteristics, from leaves to earlobes, as well as whether an organism is male or female, are also the result of genetic information inherited from the parent organisms.

Variation between individuals arises as a result of sexual reproduction, through which a new combination of genes appears every time a female and a male gamete fuse together. Sometimes one individual will show a new variation when a different characteristic results from a mutation. But your genes are only part of the story.

Nurture – environmental variation

Some differences between the organisms of a species are entirely due to the environment they live in. For example, you may have a scar as a result of an accident or an operation. People in countries where there is a lot of cheap food are often overweight. People from countries where there are food shortages as a result of droughts or floods are often very thin. The variation in the populations is a result of their environments.

Combined causes of variation

Genes play a major part in deciding how an organism will look, but the conditions in which it develops are important, too. Genetically identical plants can be grown under different conditions of light or soil nutrients. The resulting plants do not look identical. Plants deprived of light, carbon dioxide, or nutrients do not make as much food as plants with plenty of everything. They will be smaller and weaker. They have not been able to fulfil their genetic potential.

Many of the differences between individuals of the same species are the result of both their genes and their environment. For example, you inherit your hair colour and skin colour from your parents. However, whatever your inherited skin colour, it will be darker if you live in a sunny environment. If your hair is brown or blonde, it will be lighter if you live in a sunny country. You may have a genetic tendency to be overweight, but if you never have enough to eat you will be very thin.

Investigating variation

It is quite easy to produce genetically identical plants to investigate variation. You can then put them in different situations to see how the environment affects their appearance. Scientists also use groups of animals which are genetically very similar to investigate variation. You cannot easily do this in a school laboratory.

The only genetically identical humans are identical twins who come from the same fertilised egg. Scientists are very interested in identical twins, to find out how similar they are as adults. It would be unethical to take identical twins away from their parents and have them brought up differently to investigate environmental effect, but there are cases of identical twins who were adopted by different families. Some scientists have researched these separated identical twins and compared them with identical twins brought up together.

Often they look and act in a remarkably similar way. Scientists measure features such as height, weight, and IQ (a measure of intelligence). The evidence shows that some of the differences are mainly due to genetics and some are largely due to our environment. Height appears to be largely genetic, for example, but the environment has a relatively large effect on body mass.

Continuous and discontinuous variation

Look around at the people in your classroom, in the local market, or in a crowd watching sport. There are many different comparable features that vary across all the individuals. There will be some very short people, some very tall people, and many different heights in between. The same is true for foot size, weight, and intelligence. These characteristics show **continuous variation**, which means there is a gradual transition between the two extremes. Features which show continuous variation are usually determined by a number of different genes and are also affected by the environment (e.g., the availability of food, the impact of disease).

However not all features exist in a wide variety of forms. Some characteristics are either present or they are not. For example, you are female or male and you have blood group A, B, AB, or O. These features show **discontinuous variation**. Characteristics that show discontinuous variation are often determined by a single gene (or chromosome, in the case of sex), and there is little or no environmental impact on the features.

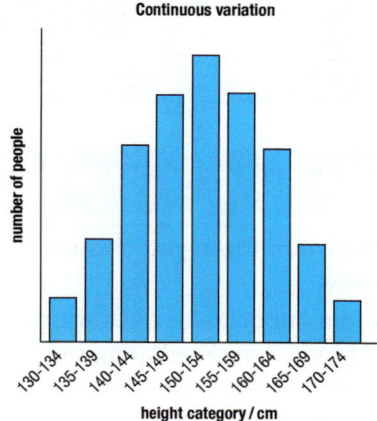

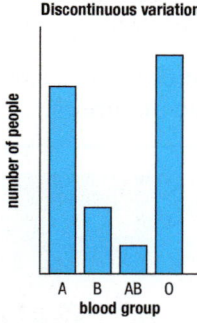

Figure 3 *The difference between characteristics showing continuous and discontinuous variation is very clear when you measure how they are distributed in a population*

Study tip

- Genes control the development of characteristics.
- Characteristics may be changed by the environment.

Key points

The causes of variation between organisms include:

- Genetic variation – different characteristics arise as a result of sexual reproduction or mutation (inherited characteristics).
- Environmental variation – different characteristics are caused by an organism's environment (acquired characteristics).
- A combination of genetic and environmental causes.

Summary questions

1 Give three causes for the variation seen between individual members of the same species of animals or plants.

2 Look at the graphs in Figure 3. Use them to help you to explain the difference between continuous and discontinuous variation in characteristics.

3 You are given 20 pots containing identical cloned seedlings all the same height. Explain how you would investigate the effects of temperature on the growth of these seedlings, compared with the impact of their genes.

10.4

From Mendel to modern genetics

Figure 1 *Gregor Mendel was the father of modern genetics. His work with peas was not recognised in his own lifetime, but now people know just how right he was!*

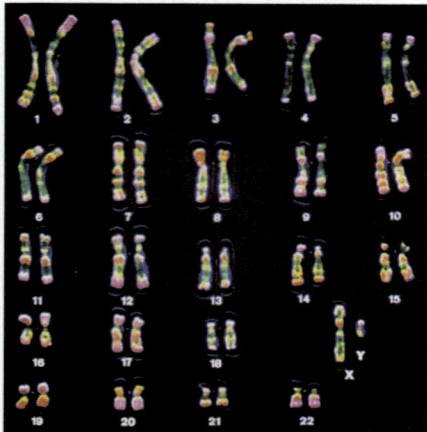

Figure 2 *This special photo, called a karyotype, shows the 23 pairs of human chromosomes. You can see the XY chromosomes (bottom right) which tell you they are from a male*

Until about 150 years ago people had no idea how information was passed from one generation to the next. Today you can predict the outcome of a single genetic cross.

Mendel's discoveries

The man who first worked out how characteristics are passed from one generation to another was Gregor Mendel, an Austrian monk born in 1822. He carried out breeding experiments using peas. In one set of experiments Mendel cross-bred green peas with yellow peas and counted the different offspring carefully. He found that the colours were inherited in clear, predictable patterns.

Mendel explained his results by suggesting that there are separate units of inherited material. He realised some characteristics were dominant over others and that they never mixed together. Mendel kept records of everything he did, and analysed his results. This was almost unheard of in those days. Eventually in 1866 Mendel published his findings. He explained some of the basic laws of genetics using mathematical models in ways that are still used today.

Mendel was ahead of his time. As no one knew about genes or chromosomes, people simply didn't understand his theories. He died 20 years later with his ideas still ignored – but convinced that he was right. Sixteen years after Mendel's death, his work was finally recognised. By 1900, people had seen chromosomes through a microscope. Other scientists discovered Mendel's papers and repeated his experiments. When they published their results, they gave Mendel the credit for what they observed.

From then on ideas about genetics developed rapidly. It was suggested that Mendel's units of inheritance might be carried on the chromosomes seen under the microscope. The science of genetics as you know it today was born.

Understanding genetics

Scientists have built on the work of Gregor Mendel. They now understand how genetic information is passed from parent to offspring. For example, they know that humans have 23 pairs of chromosomes. In 22 cases, each chromosome in the pair is a similar shape. Each one has genes carrying information about the same things. One pair of chromosomes is different – these are the **sex chromosomes**. Two X chromosomes mean you are female. One X chromosome and a much smaller one, known as the Y chromosome, mean you are male. Every egg contains an X chromosome. A sperm has a 50/50 chance of containing an X chromosome or a Y chromosome. Every time an egg and sperm combine, it is chance whether the egg will fuse with a sperm carrying an X chromosome or one carrying a Y chromosome. There is a 50% chance that a fertilised egg will be female and a 50% chance that it will be male.

Punnet squares

You can use diagrams and models to help you understand genetics. The Punnett square is a very useful diagram. You can use it to help you predict the possible offspring from any genetic cross. Punnett squares can be drawn for whole chromosomes or individual genes.

When you use a Punnett square, always show the **genotype** (the genetic makeup) of the parents for the characteristic in question as well as their **phenotype** (physical characteristics). Work out the possible gametes that might be formed You can then use these gametes to work out the possible genotypes of the offspring. From this you can work out the possible phenotypes of the offspring as well. Figure 3 shows a Punnett square for the inheritance of the sex chromosomes.

Always give the genotypes and phenotypes of the parents.

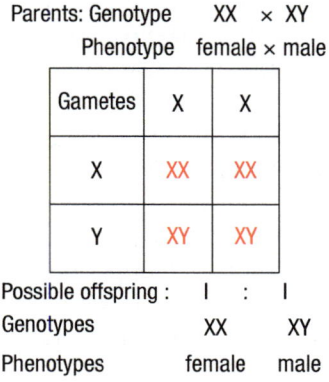

Always give the possible genotypes and phenotypes of the offspring and the probability that they will be produced.

Figure 3 *A Punnett square for the inheritance of the sex chromosomes*

Summary questions

1 Copy and complete this paragraph using the following words:

male   sex chromosomes   23   X   22   XX   Y

Humans have .... pairs of chromosomes. In .... pairs the chromosomes are always the same. The final pair are known as .... If you inherit .... you will be female, whilst an .... and a .... chromosome make you ....

2 a How did Mendel's experiments with peas convince him that there were distinct units of inheritance that were not blended together in the offspring?

  b Why didn't people accept Mendel's ideas at first?

  c The development of the microscope played an important part in helping convince people that Mendel was right. How?

3 A couple have two sons and are expecting another baby. Some people tell them that they are almost certain to have a girl this time. Explain why this is not true.

Key points

- Gregor Mendel was the first person to suggest separately inherited factors, which are now called genes.

- In human body cells the sex chromosomes determine whether you are female (XX) or male (XY).

- You can interpret genetic diagrams of monohybrid inheritance and sex inheritance.

- You can construct genetic diagrams to predict characteristics.

Ext

10.5 Inheritance in action

Learning objectives

After this topic you should know:

- that different genes control different characteristics of an organism
- that some characteristics are dominant and some are recessive
- what the terms homozygous and heterozygous mean.

Study tip

Make sure that you use the terms gene and allele correctly. A gene is the overall term for a section of DNA which codes for one characteristic. An allele is a particular form of one gene.

Most of your characteristics, such as your eye colour and nose shape, are controlled by a number of genes. However, some characteristics, such as dimples or having attached earlobes, are controlled by a single gene. Often, there are only two possible alleles for a particular feature. However, sometimes you can inherit one of a number of different possibilities. You can make biological models that help you predict the outcome of any genetic cross.

How inheritance works

The chromosomes you inherit carry your genetic information in the form of genes. Many of these genes have different forms, or alleles. The genes operate at a molecular level to control your characteristics, because they code for the proteins that are made. Each allele codes for a different protein. The combination of alleles you inherit will determine your characteristics. Picture a gene as a position on a chromosome. An allele is the particular form of information in that position on an individual chromosome. For example, the gene for dimples may have the dimple (D) or the no-dimple (d) allele in place. Because you inherit one allele from each parent, you will have two alleles controlling whether or not you have dimples. When you chose the letters to represent alleles, try to choose letters that look different in upper case and lower case, such as T and t, A and a, or D and d. If the upper and lower case letters look similar it can be very confusing when you are using a Punnet square.

Genetic terms

Some words are useful when you are working with biological models such as Punnett squares or family trees:

- Genotype – this describes the genetic makeup of an individual regarding a particular characteristic, for example, **Dd** or **dd**.

  - Phenotype – this describes the physical appearance of an individual regarding a particular characteristic, for example, dimples or no dimples.

  - **Homozygous** – an individual with two identical alleles for a characteristic, for example, **DD** or **dd**.

  - **Heterozygous** – an individual with different alleles for a characteristic, for example, **Dd**.

  - **Dominant** alleles – alleles that control the development of a characteristic even when they are only present on one of your chromosomes, for example, the allele for dimples. You use an upper case letter to represent dominant alleles, for example, **D** for dimples. So, if you inherit **DD** or **Dd** from your parents, you will have dimples.

  - **Recessive** alleles – alleles that only control the development of a characteristic if they are present on both chromosomes (i.e., when no dominant allele is present), for example, the allele for no dimples. You use a lower case letter to represent recessive alleles, for example, **d** for no dimples. You will only show the characteristic of no dimples if you have inherited two recessive alleles, **dd**.

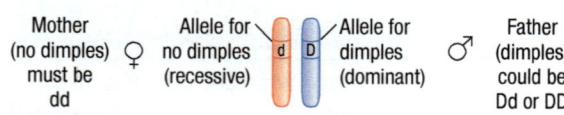

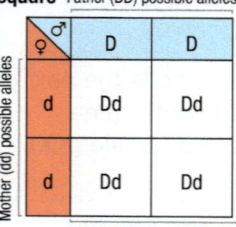 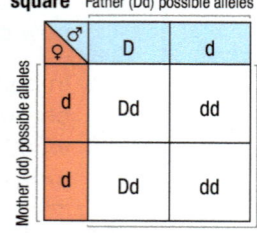

Figure 1 *The different forms of genes, known as alleles, can result in the formation of quite different characteristics. Genetic diagrams such as these Punnett squares help you to explain what is happening and predict what the possible offspring might be like*

Monohybrid crosses

When the genes from two parents are combined, it is called a genetic cross. **Monohybrid crosses** are genetic crosses that involve characteristics which are inherited on single genes. We use the different alleles of the parents to help us predict what the offspring will be like. Mendel showed that the possible offspring between crosses of different types of individuals in monohybrid inheritance fall into clear patterns that you can learn and predict.

The Punnett squares in Figure 2 help us to predict the possible offspring from a monohybrid cross. A Punnett square shows us:

● the alleles for a characteristic carried by the parents (the genotype of the parents)
● the possible gametes which can be formed from these
● how these could combine to form the characteristic in their offspring.

The genotype of the offspring allows you to work out the possible phenotypes too.

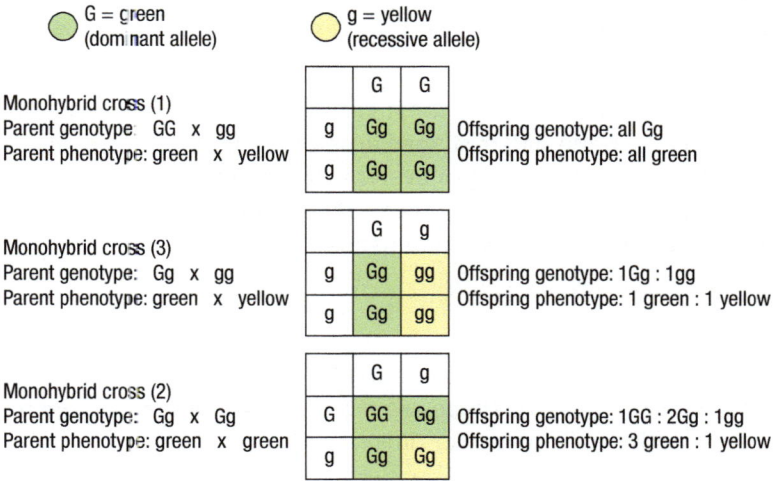

Figure 2 *You can use Punnett squares to show the patterns of monohybrid inheritance. These are some of the crosses originally carried out by Mendel on green and yellow peas*

Summary questions

1 a Define the term dominant allele?
   b Define the term recessive allele?
   c Explain what is meant if an individual is homozygous recessive for a characteristic.

2 Draw a Punnett square similar to the ones in Figure 2 to show the possible offspring from a cross between two people who both have dimples and the genotype Dd.

3 The characteristic of having round peas or wrinkled peas is a monohybrid characteristic in peas. The allele for round peas is dominant to the allele for wrinkled peas.
   a Choose a suitable letter to represent the alleles for round and wrinkled peas.
   b What genotypes and phenotypes might you expect to see in the offspring if two heterozygous round pea plants are crossed? Use a Punnet square to help show your answer.
   c What genotypes and phenotypes might you expect to see in the offspring if a heterozygous round pea plant is crossed with a wrinkled pea plant? Use a Punnett square to help show your answer.

Ext

Key points

● Some characteristics are controlled by a single gene.

● Genes can have different forms called alleles.

● Some alleles are dominant – they always result in the development of a characteristic if present.

● Some alleles are recessive – they only result in the development of a characteristic if no dominant alleles are present.

● If both chromosomes in a pair contain the same allele of a gene, the individual is homozygous. If they contain different alleles of the same gene, the individual is heterozygous.

● You can interpret genetic diagrams of monohybrid inheritance and sex inheritance.

● You can construct Punnett squares to predict inherited characteristics.

Ext

10.6

DNA and family trees

Learning objectives

After this topic, you should know:

- what DNA is

- how the information in the DNA results in different proteins in your body

- how family trees can be used with genetic diagrams to predict the outcomes of genetic crosses.

?? *Did you know …?*

Unless you have an identical twin, your DNA is unique to you. Scientists have used this to develop **DNA fingerprinting**, a way of identifying people from tiny traces of skin cells or body fluids.

Study tip

Three bases on DNA code for one amino acid. Amino acids are joined together to make a protein. It is the particular sequence of amino acids that gives each protein a specific shape and function.

The work of Gregor Mendel was just the start of our understanding of inheritance. Today, you know that your features are inherited on genes carried on the chromosomes that are found in the nuclei of your cells.

DNA – the molecule of inheritance

The chromosomes are made up of long molecules of a chemical known as **DNA** (deoxyribonucleic acid). These very long strands of DNA twist to form a double helix structure. Your genes are small sections of this DNA.

The DNA carries the instructions to make the proteins that form most of your cell structures. These proteins include the enzymes that control your cell chemistry. This is how the relationship between the genes and the whole organism builds up.

The genes make up the chromosomes in the nucleus of the cell. They code for the proteins, which make up the different specialised cells that form tissues. These tissues then form organs and organ systems that make up the whole body. Different alleles of the same gene code for different proteins, which is why they end up coding for different characteristics.

The genetic code

Ext

The long strands of your DNA are made up of combinations of four different compounds called **bases** (Figure 1). These are grouped into threes, and each group of three bases codes for a particular amino acid.

Each gene is made up of hundreds or thousands of these bases. The order of the bases controls the order in which the amino acids are assembled to produce a particular protein for use in your body cells. Each gene codes for a particular combination of amino acids, which make a specific protein. This is sometimes referred to as the 'one gene, one protein' principle.

A change or mutation in a single group of bases can be enough to change or disrupt the whole protein structure and the way it works.

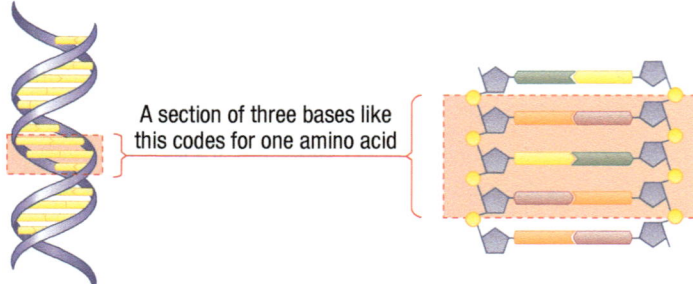

A section of three bases like this codes for one amino acid

Figure 1 *DNA codes for the amino acids that make up the proteins that make up the enzymes and other protein components that make each individual*

Family trees

You can trace genetic characteristics through a family by drawing a family tree (Figures 2 and 3). Family trees show males and females and can be useful for tracing family likenesses. They can also be used for tracking inherited diseases, showing a physical characteristic, or showing the different alleles people or animals have inherited. Family trees can be used to work out where an individual is likely to be homozygous or heterozygous for particular alleles.

On a family tree, males are usually shown as squares and females as circles. Individuals with a particular characteristic are shaded. A family tree can show where a mutation may have taken place, and enables us to work out the possible genotypes of many of the family members. People often build up their family tree. They are also very important for domestic animals, racehorses, and animals which are kept and bred in zoos.

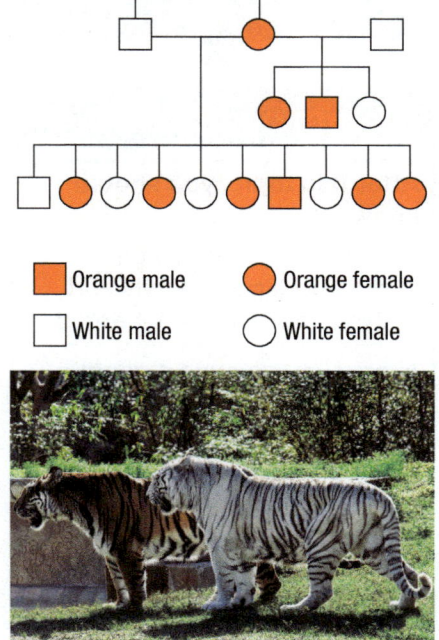

Figure 2 *This animal family tree shows the inheritance of orange and white coat colours in tigers*

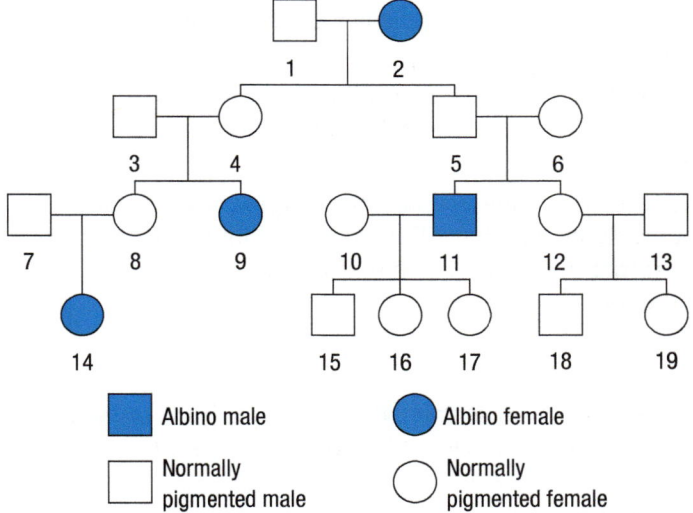

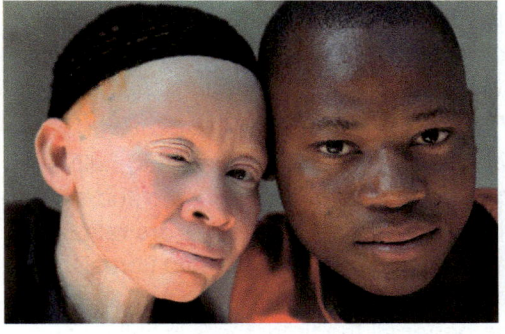

Figure 3 *This human family tree shows the inheritance of albinism in people. Albinism is a recessive characteristic and the normal pigment of the skin, hair, and eyes is missing in affected people (albinos)*

Summary questions

1 **a** What is DNA?
   b Draw and label a diagram to show the structure of DNA.

2 Copy the family tree shown in Figure 3. For each individual shown, write down their possible genotype. If there are two possibilities, put both down.

Ext

3 Draw Punnet squares to show the possible genetic crosses between the following couples in Figure 3:
   a 1 and 2
   a 7 and 8.

Key points

- Chromosomes are made up of large molecules of DNA. DNA is made of long strands twisted to form a double helix, which contains combinations of four different compounds called bases.

Ext

- A gene is a small section of DNA that codes for a particular combination of amino acids which makes a specific protein.

- You can interpret genetic diagrams, including family trees, to predict inherited characteristics.

10.7 Inherited conditions in humans

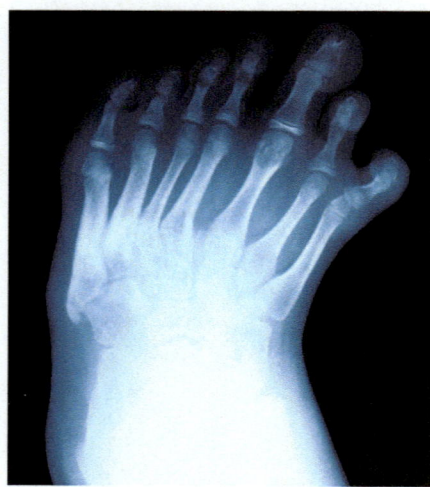

An X-ray of the hand of someone with polydactyly – count the number of fingers!

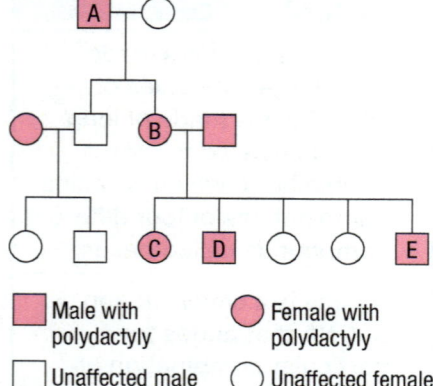

| Male with polydactyly | Female with polydactyly |
| Unaffected male | Unaffected female |

Figure 1 *Polydactyly is passed through a family by a dominant allele*

Not all diseases are infectious. Sometimes diseases are the result of a change in the bases or coding of our genes and can be passed on from parent to child. These diseases are known as **genetic disorders** or **inherited disorders**.

You can use your knowledge of dominant and recessive alleles to work out the risk of inheriting a genetic disorder.

Polydactyly

Sometimes babies are born with extra fingers or toes. This is called **polydactyly**. The most common form of polydactyly is caused by a dominant allele. It can be inherited from one parent who has the condition. People often have their extra digit removed, but some people live quite happily with them.

If one of your parents has polydactyly and is heterozygous, you have a 50% (one in two) chance of inheriting the disorder. That's because half of their gametes will contain the faulty dominant allele. If they are homozygous, you will definitely have the condition (Figure 2).

Some dominant genetic disorders have a much more widespread effect on the way the body works than polydactyly does. For example, Huntington's disease is a dominant genetic disorder that develops in middle age. It affects the nervous system and eventually leads to death.

Cystic fibrosis

Cystic fibrosis (CF) is a genetic disorder that affects many organs of the body, particularly the lungs and the pancreas. Over 8500 people in the UK have cystic fibrosis.

Cystic fibrosis is a disorder of the cell membranes which means affected people are unable to move certain substances from one side of the membrane to the other. As a result, the mucus made by cells in many areas of the body becomes thick and sticky, causing a number of problems. The lungs and parts of the digestive system and reproductive system become clogged up by the thick, sticky mucus, which stops them working properly.

Treatment for cystic fibrosis includes physiotherapy and antibiotics to help keep the lungs clear of mucus and infections. The pancreas cannot make and secrete enzymes properly because the tubes through which the enzymes are released into the small intestine are blocked with mucus. People with cystic fibrosis are given enzymes to replace the ones the pancreas cannot produce and to thin the mucus. Although treatments are getting better all the time, there is still no cure.

Cystic fibrosis is caused by a recessive allele so it must be inherited from both parents if offspring are to suffer from the disease. Children affected by cystic fibrosis are usually born to parents who do not suffer from the disorder. They have a dominant, healthy allele, so their bodies work normally. However, they also carry the recessive cystic fibrosis allele. Because it gives them no symptoms, they have no idea it is there. They are known as **carriers**.

The incidence of cystic fibrosis varies around the world, for example, in the UK, one person in 25 carries the cystic fibrosis allele. Most of them will never be aware of it. The only situation in which it may become obvious is if they have children with a partner who also carries the allele. Then there is a 25% (one in four) chance that any child they have will be affected (Figure 3).

links
You can find out more about how to construct and use genetic diagrams and the patterns of Mendelian inheritance in 10.4 From Mendel to modern genetics and 10.5 Inheritance in action.

The genetic lottery

When looking at the possibility of inheriting genetic disorders, it is important to remember that every time an egg and a sperm fuse it is down to chance which alleles combine. So, if two parents who are heterozygous for the cystic fibrosis allele have four children, there is a 25% (one in four) chance that each child will have the disorder. However, it may be that all four children have cystic fibrosis, or none of them might be affected. They might all be carriers, or none of them might inherit the faulty alleles (Figure 3). It's all down to chance!

N = dominant allele (normal metabolism)
n = recessive allele (cystic fibrosis)

Both parents are carriers (Nn)

	N	n
N	NN	Nn
n	Nn	nn

Genotype of offspring:
25% normal (NN)
50% carriers (Nn)
25% affected by cystic fibrosis (nn)

Phenotype of offspring:

3/4, or 75% chance normal
1/4, or 25% chance cystic fibrosis

Figure 3 *A genetic diagram for cystic fibrosis*

F = dominant allele (polydactyly)

f = recessive allele (normal number of fingers and toes)

Parent with polydactyly (Ff)

		F	f
Normal parent (ff)	f	Ff	ff
	f	Ff	ff

Genotype of offspring:
50% affected (Ff)
50% normal (ff)

Phenotype of offspring:

1/2, or 50% chance polydactyly
1/2, or 50% chance normal

Figure 2 *A genetic diagram for polydactyly*

Curing genetic disorders

So far doctors have no way of curing genetic disorders. In some cases the disorders are very minor, for example, colour blindness and most cases of polydactyly. However, some genetic disorders are very serious and can even shorten lives. Scientists hope that **genetic engineering** could be the answer. It should be possible to add healthy genes which would function normally in the cells. They have tried this in people affected by cystic fibrosis. Unfortunately, so far they have not yet managed to cure anyone with an inherited genetic disorder, but they remain very hopeful of this possibility.

links
Learn more about how scientists can change the genes in the cells of an organism in 11.3 Genetic engineering.

Summary questions

1 a What is polydactyly?
   b Why can one parent with the allele for polydactyly pass the condition on to their children even though the other parent is not affected?
   c Look at the family tree in Figure 1. For each of the five people labelled A to E, give their possible alleles and explain your answers.

2 a Why are carriers of cystic fibrosis not affected by the disorder themselves?
   b Why must both of your parents be carriers of the allele for cystic fibrosis if you are to inherit the disease?

Ext

3 A couple have a baby who has cystic fibrosis. Neither the couple, nor their parents, have any signs of the disorder.
   Draw genetic diagrams showing the possible genotypes of the grandparents and the parents to explain how this could happen.

Key points

- Some disorders are inherited.

- Polydactyly is caused by a dominant allele of a gene and can be inherited from either parent.

- Cystic fibrosis is caused by a recessive allele of a gene so must be inherited from both parents.

- You can use genetic diagrams to predict how genetic disorders might be inherited.

10.8

More inherited conditions in humans

Learning objectives

After this topic, you should know:

- that some inherited conditions give heterozygous individuals protection against other diseases

- that some inherited conditions are the result of inheriting abnormal numbers of chromosomes rather than a fault in an individual gene.

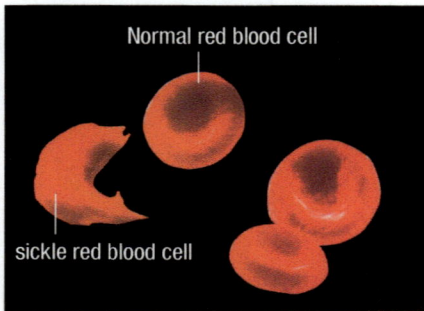

R = normal red blood cells (rbc)
r = sickle red blood cells

Both parents are heterozygous carriers (Rr × Rr)

	R	r
R	RR	Rr
r	Rr	rr

Offspring potential genotypes and phenotypes
1 homozygous dominant (Rr) – normal red blood cells
2 heterozygous (Rr) – some normal and some sickle red blood cells
1 homozygous recessive (rr) – sickle-cell anaemia

Figure 1 *The inheritance and effects of sickle-cell anaemia*

There are many different genetic disorders, some of which are extremely rare, with only a few families around the world affected. Others are relatively common. Some genetic disorders can even carry surprising advantages.

Sickle-cell anaemia

Sickle-cell anaemia is a genetic disorder that affects millions of people around the world. It affects the red blood cells that carry oxygen from your lungs to all the cells of your body. In sickle-cell anaemia, the red blood cells become sickle-shaped. When this happens, they don't carry oxygen effectively so you feel breathless, lack energy, and are tired. Affected children often fail to grow. The sickle-shaped cells also block small blood vessels. This can cause great pain and death of the tissue, leading to severe infections.

Sickle-cell anaemia causes the deaths of up to 2 million people a year. It is particularly common in people originating from Africa, the Mediterranean, India, and Spanish-speaking areas of the Americas. Sickle-cell anaemia is caused by a recessive allele, so you need to inherit one from each parent in order to be fully affected by the disease. However, even heterozygous people are partly affected and have some sickle cells in their blood.

In some ways, it is surprising that the sickle-cell allele survives and is so widespread as it has such a damaging impact on people who are homozygous. However, scientists have discovered that people who are heterozygous for sickle-cell anaemia are less likely to get malaria than people who are homozygous for the dominant normal allele. Malaria is another killer disease which affects the lives of millions of people worldwide. It is caused by a **parasite** passed from the blood of one person to another by mosquito bites.

People who are heterozygous for sickle-cell anaemia have some problems with their blood cells and it seems that this change in shape is enough to protect them against the malaria parasites. This gives them an advantage – and makes sure they survive to pass on their sickle-cell alleles to another generation. This means that sickle-cell anaemia persists at least partly because heterozygotes are protected against malaria. In parts of the world where there is no malaria, sickle-cell anaemia is very rare and is usually only found in families who originate from malarial areas.

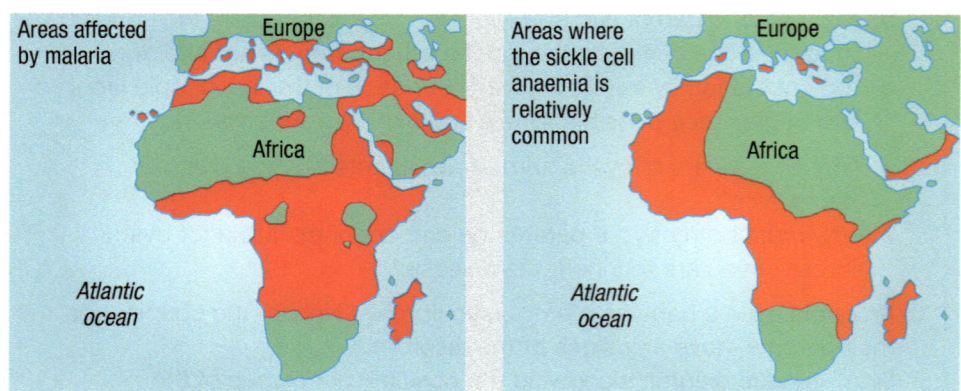

Figure 2 *This diagram shows the distribution of malaria (left) and sickle-cell anaemia (right) in Africa*

Whole chromosome disorders

Some inherited disorders are not the result of a change in a single gene. They are caused by the inheritance of abnormal numbers of chromosomes. As the cells divide by meiosis to form the gametes, sometimes things go wrong. One of the gametes may end up with too few chromosomes or with an extra chromosome. Often these gametes cannot survive, but sometimes they do.

One of the most common inherited conditions resulting from the inheritance of an abnormal number of chromosomes is Down's syndrome. This is caused by an extra copy of chromosome number 21, so the baby has 47 chromosomes instead of 46. This can cause a number of developmental problems in many different areas of the body, including the brain, the heart, and the muscles.

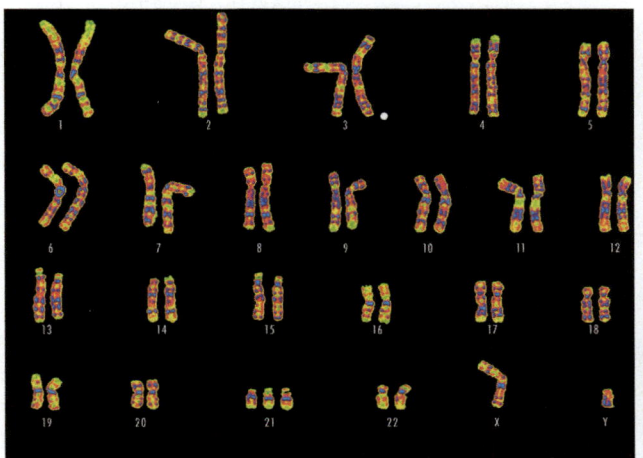

Figure 3 *Inheriting an extra chromosome can affect many systems in the body, which is seen clearly when someone is affected by Down's syndrome. Compare this karyotype with the normal human karyotype on Page 134*

⌾ links

To find out more about meiosis, look back to 2.2 Cell division in sexual reproduction.

Study tip

Learn which type of genes cause genetic disorders:
Dominant – polydactyly and Huntington's disease
Recessive – cystic fibrosis and sickle-cell anaemia.

Summary questions

1 a Sickle-cell anaemia is a recessive inherited disorder. Explain carefully what this means.
   b What are the main symptoms of sickle-cell anaemia? Explain how they are caused.

Ext

2 Using the letter R for the dominant normal allele and r for the recessive sickle-cell allele, carry out the following genetic crosses. For each cross state the likelihood of having a child with sickle-cell anaemia and a child carrying the sickle-cell allele.
   a One parent is homozygous dominant and the other is heterozygous.
   b Both parents are heterozygous.
   c One parent is homozygous dominant and the other has sickle-cell anaemia.

3 a What is Down's syndrome?
   b How does the inheritance of Down's syndrome differ from the inheritance of genetic disorders such as cystic fibrosis and sickle-cell anaemia?

Key points

- Sickle-cell anaemia is an inherited condition that affects the red blood cells and is caused by a recessive allele.

- Some inherited conditions are caused by the inheritance of abnormal numbers of chromosomes. For example, Down's syndrome is caused by the presence of an extra chromosome.

Summary questions

1 If you dig up a strawberry plant, it will often have many other small strawberry plants attached to it.

  a How are these small plants produced?

  b What sort of reproduction is this?

  c Strawberry plants also produce seeds. How are these produced and what sort of reproduction is this?

  d How are the new plants that you would grow from seeds different from the new plants grown directly from the parent plant?

2 **a** What is a gamete?

  b How do the chromosomes in a gamete differ from the chromosomes in a normal body cell?

3 **a** What are chromosomes made of?

  b What is a gene?

  c Explain carefully how the information carried in a gene is expressed in the phenotype of an organism.

4 Explain what Mendel did, what was unusual about his work, and why people initially rejected his ideas. What developments in scientific ideas and technology helped people eventually to understand and accept Mendel's ideas?

5 Whether you have a straight thumb or a curved thumb is decided by a single gene with two alleles. The allele for a straight thumb, S, is dominant to the curved allele, s. Use this information to help you to answer the following questions:

Josh has straight thumbs but Sami has curved thumbs. They are expecting a baby.

  a You know exactly what Sami's thumb alleles are. What are they, and how do you know?

  b i If the baby has curved thumbs, what does this tell you about Josh's thumb alleles?

  Ext   **ii** Draw and complete a Punnett square to show the genetics of your explanation.

  c i If the baby has straight thumbs, what does this tell us about Josh's thumb alleles?

  Ext   **ii** Use Punnett squares to show the genetics of your explanation.

6 Amjid grew some purple-flowering pea plants from seeds he had bought at the garden centre. He planted them in his garden. Here are his results:

Seeds planted	247
Purple-flowered plants	242
White-flowered plants	1
Seeds not growing	4

  a How would you explain these results?

Ext   **b** Amjid was interested in these plants, so he collected the seed from some of the purple-flowered plants and used them in the garden the following year. He made a careful note of what happened.

Here are his results

Seeds planted	406
Purple-flowered plants	295
White-flowered plants	102
Seeds not growing	6

Amjid was slightly surprised. He did not expect to find that a third of his flowers would be white.

  i The purple allele, P, is dominant and the allele for white flowers, p, is recessive. Draw a genetic diagram that explains Amjid's numbers of purple and white flowers.

  ii How accurate were Amjid's results compared with the expected ratio?

  c Suggest another genetic cross that would confirm the genotype of the purple plants. Produce a genetic diagram to show the results you would expect.

7 Many human characteristics are the result of different genes interacting, but there are some characteristics which are the result of monohybrid inheritance.

  a What is monohybrid inheritance?

  b What is meant by the terms dominant allele and recessive allele?

  c Describe two normal human characteristics that are inherited by monohybrid inheritance.

Ext   **d** Cystic fibrosis is a disorder that is inherited on a recessive allele. Explain what this means, and draw a genetic diagram to show how two healthy parents could have a child with cystic fibrosis.

  e Huntington's disease is a serious human genetic disorder carried on a dominant allele. The problems it causes do not show up until the sufferer is middle-aged. Draw a genetic diagram to show how an affected individual could pass this disease to their offspring.

Practice questions

1 Copy and complete sentences **a–d** using the following words.

> *alleles   heterozygous   amino acid   genes   DNA*
> *skin   chromosomes   homozygous   gamete*

a Inherited characteristics are controlled by .... made up of the chemical substance ....

b Human body cells have 46 ...., but the .... cells have only 23.

c Different forms of one gene are called ....

d If an individual inherits two different forms of one gene they are .... for that gene, but if they have two of the same form of the gene they are .... for the gene.   (7)

2 A teacher wanted to grow new geranium plants to use for an investigation in science. She took five identical cuttings and planted them in identical pots of compost.

Figure 1

She gave the five pots to different students to take home for the long summer holiday.

a What type of reproduction is taking cuttings?   (1)

b The parent plant produced red flowers. What colour flowers will the cuttings produce? Why?   (2)

c When the plants were returned after the holiday, they were different heights and had different numbers and sizes of leaves. Name **two** factors that may have caused this variation.   (2)

d The teacher wanted to use these plants as she said it would make the investigation more reliable than buying five new plants. Suggest reasons why.   (2)

3 Copy and complete the passage by inserting the correct scientific term in each space provided:

A gene is a small section of the chemical .... Each gene codes for the synthesis of a specific .... made up of .... joined in the correct sequence. The .... is made up of long strands twisted into a .... shape. On these strands .... bases code for each ....   (7)

4 *In this question you will be assessed on using good English, organising information clearly, and using specialist terms where appropriate.*

A class in school included identical twin girls.

Twin A was 150 cm tall and weighed 51 kg. She had long, blonde, straight hair and blue eyes.

Twin B was 151 cm tall and weighed 63 kg. She had short, black, curly hair and blue eyes.

Use your knowledge and understanding of inheritance and variation due to genetic and environmental factors to evaluate the characteristics seen in these two girls.   (6)

5 Figure 2 shows the present distribution of malaria and the sickle-cell allele in Africa.

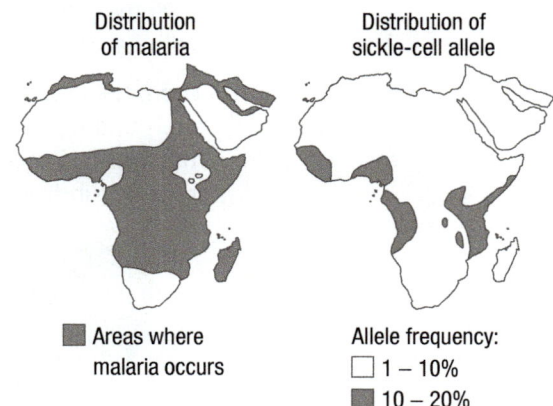

Distribution of malaria   Distribution of sickle-cell allele

■ Areas where malaria occurs

Allele frequency:
□ 1 – 10%
■ 10 – 20%

Figure 2

Ext

a Draw a genetic diagram to show how sickle-cell anaemia can be inherited from parents who do not have the condition. Use the following key to symbols for alleles:

HbA   Normal adult haemoglobin

Hbs   Sickle-cell haemoglobin   (4)

b i Explain the link between sickle-cell anaemia, resistance to malaria, and the frequency of the Hbs allele.   (3)

ii Select and evaluate the evidence from the maps that accounts for the distribution of the sickle-cell allele and the resistance to malaria in parts of Africa.   (2)

11.1

Cloning

Learning objectives

After this topic, you should know:

- different ways of creating clones

- why clones are useful.

A **clone** is an individual that has been produced asexually and is genetically identical to the parent. Many plants reproduce naturally by cloning. This has been used by farmers and gardeners for many years.

Cloning plants

Taking cuttings is a form of artificial asexual reproduction or cloning that has been carried out for hundreds of years.

In recent years, scientists have come up with a more modern way of cloning plants called **tissue culture**. It is more expensive, but it allows you to make thousands of new plants from one tiny piece of plant tissue.

The first step is to add a mixture of plant hormones to a small group of cells from the plant you want to clone, to produce a big mass of identical plant cells called a **callus**.

Then, using a different mixture of hormones and conditions, you can stimulate each of these cells to form a tiny new plant. This type of cloning guarantees that you can produce thousands of offspring with the characteristics you want from one individual plant.

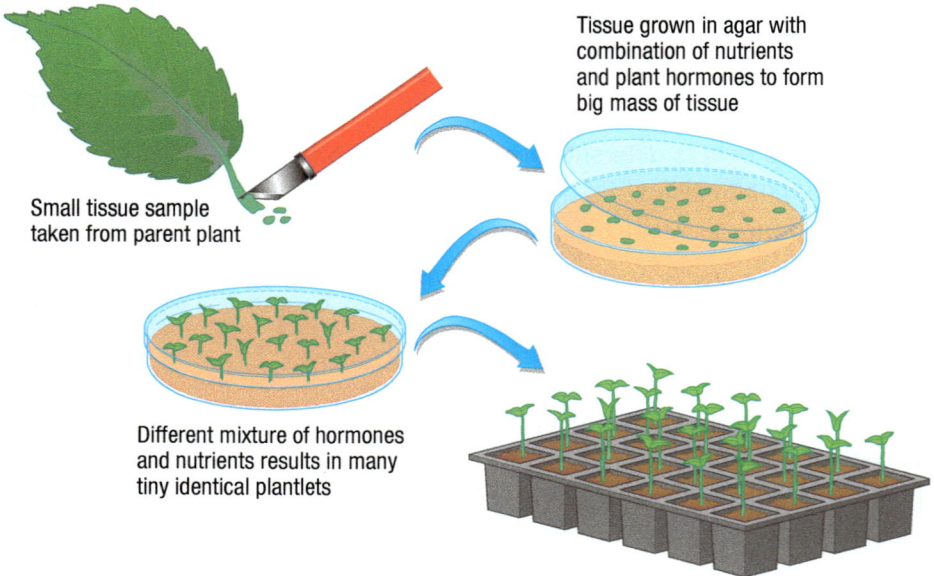

Small tissue sample taken from parent plant

Tissue grown in agar with combination of nutrients and plant hormones to form big mass of tissue

Different mixture of hormones and nutrients results in many tiny identical plantlets

Plantlet clones grown on

Figure 2 *Tissue culture makes it possible to produce thousands of identical plants quickly and easily from one small tissue sample*

Cloning animals

In recent years, cloning animals has become quite common in farming, particularly transplanting cloned cattle embryos. Cows normally produce only one or two calves at a time. If you use embryo cloning, your best cows can produce many more top-quality calves each year.

How does embryo cloning work? First, you give a top-quality cow fertility hormones so that it produces a lot of eggs. You then fertilise these eggs using sperm from a really good bull. Often, this is done inside the cow and the embryos that are produced are then gently washed out of her womb. Sometimes the eggs are collected and you add sperm in a laboratory to produce the embryos. Each embryo is not genetically the same as its mother as it is formed when the egg and sperm fuse.

At this very early stage of development, every cell of the embryo can still form all of the cells needed for a new cow. This is because the cells have not become specialised. These cells can be split apart and then transplanted into host mothers (Figure 3). The split cells are effectively identical twins.

Study tip

Remember that clones have identical genetic information to each other.

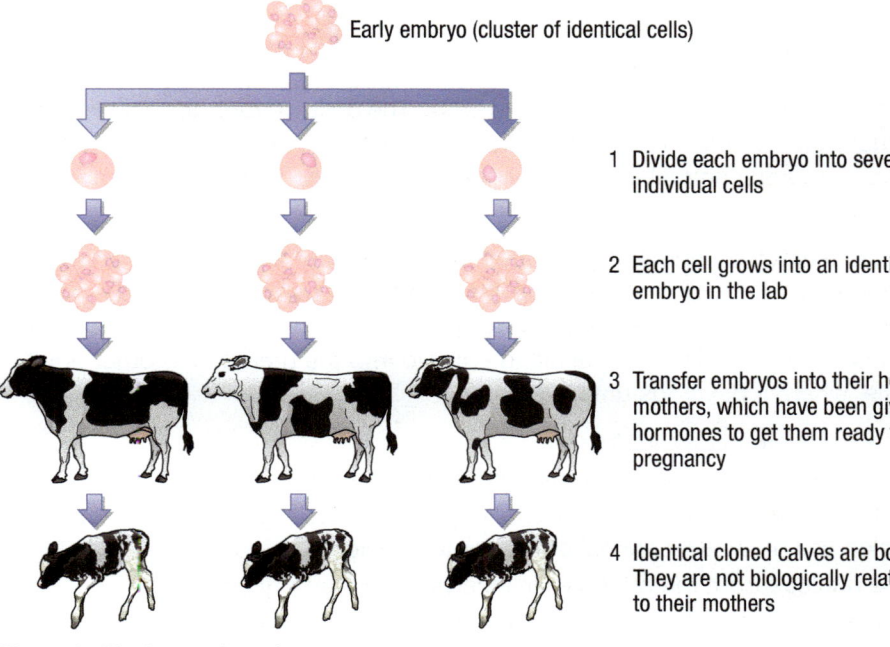

Early embryo (cluster of identical cells)

1 Divide each embryo into several individual cells

2 Each cell grows into an identical embryo in the lab

3 Transfer embryos into their host mothers, which have been given hormones to get them ready for pregnancy

4 Identical cloned calves are born. They are not biologically related to their mothers

Figure 3 *Cloning cattle embryos*

links

For more information on cloning embryos, see 11.2 Adult cell cloning.

Cloning cattle embryos and transferring them to host cattle is skilled and expensive work. However, it is worth it because a top cow might produce 8–10 calves through normal reproduction during her working life. By using embryo cloning, the same cow can produce 30 or more calves in a single year.

Cloning embryos also means that high-quality embryos can be transported all around the world. They can be carried to places where cattle with a high milk yield or lots of meat are badly needed for breeding with poor local stock. Embryo cloning is also used to make lots of identical copies of embryos that have been **genetically modified** to produce medically useful compounds.

Summary questions

1 Define the following terms:
  a embryo cloning   **c** asexual reproduction
  b tissue cloning

2 **a** Cloning cattle embryos is very useful. Why?
  b Draw a flow chart to show the stages in the embryo cloning of cattle.
  c Suggest some of the economic and ethical issues raised by embryo cloning in cattle.

3 Make a table to compare the similarities and differences between tissue cloning in plants and embryo cloning in cattle.

Key points

- A modern technique for cloning plants is tissue culture, using small groups of cells taken from part of the original plant.

- Embryo cloning involves splitting apart cells from a developing animal embryo before they become specialised and then transplanting the identical embryos into host mothers.

11.2 | Adult cell cloning

Many small invertebrate animals such as *Hydra* reproduce asexually in a way which is quite similar to asexual reproduction in plants. Even some reptiles can reproduce asexually with eggs that do not need to be fertilised. However, mammals never reproduce asexually. True cloning of large mammals, without sexual reproduction, has been a major scientific breakthrough. It is the most complicated form of asexual reproduction you can find.

Adult cell cloning

To clone a cell from an adult animal is easy. The cells of your body reproduce asexually all the time to produce millions of identical cells. However, to take a cell from an adult animal and make an embryo or even a complete identical animal is a very different thing.

When a new whole animal is produced from the cell of another adult animal, it is known as **adult cell cloning**. This is still relatively rare. You place the nucleus of one normal body cell into the empty egg cell of another animal of the same species. Then you place the resulting embryo into the uterus of another adult female, where it develops until it is born.

Here are the steps involved:

● The nucleus is removed from an unfertilised egg cell.
● At the same time, the nucleus is taken from an adult body cell, for example, a skin cell of another animal of the same species.
● The nucleus from the adult cell is inserted in the empty egg cell.
● The new egg cell is given a tiny electric shock, which stimulates it to start dividing to form embryo cells. These contain the same genetic information as the original adult cell and the original adult animal.
● When the embryo has developed into a ball of cells, it is inserted into the womb of an adult female to continue its development.

Adult cell cloning has been used to produce a number of whole animal clones. The first large mammal ever to be cloned from the cell of another adult animal was Dolly the sheep, born in 1997.

Figure 1 *Dolly the sheep went on to have lambs of her own in the normal way*

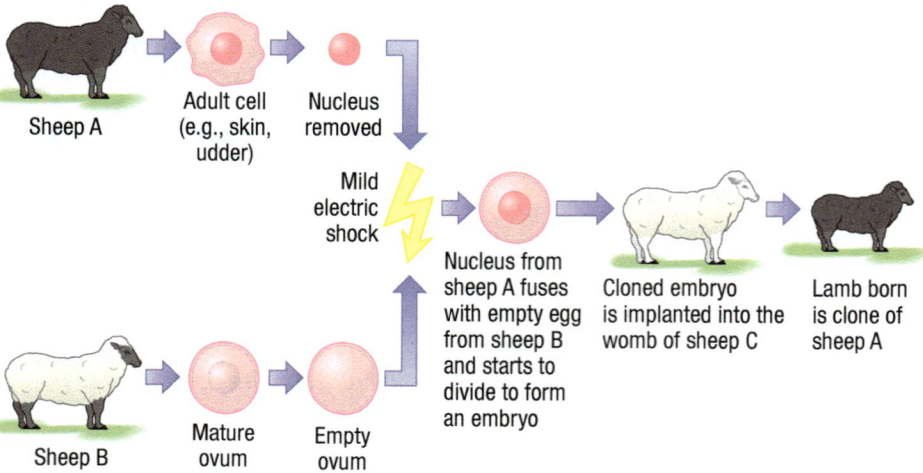

Figure 2 *Adult cell cloning remains a very difficult technique, but scientists hope it may bring benefits in the future*

The birth of Dolly was the only success from hundreds of attempts at adult cell cloning. The cloning technique is tricky and unreliable, but scientists hope that it will become easier in future.

The benefits and disadvantages of adult cell cloning

One big hope for adult cell cloning is that it can be used to clone animals that have been genetically engineered to produce useful proteins in their milk. This would give us a good way of producing large numbers of cloned, medically useful animals.

This technique could also be used to help save animals from extinction, or even to bring back species of animals that died out years ago. The technique could be used to clone pets or prized animals, so that they continue even after the original has died. However, some people are not happy about this idea.

There are some disadvantages to this exciting science as well. Many people fear that the technique could lead to the cloning of human babies. This could be used to help infertile couples, but it could also be abused. Cloning of humans is not possible at the moment, but who knows what might be possible in the future?

Another problem is that modern cloning techniques produce lots of plants or animals with identical genes. In other words, cloning reduces variation in a population. This means the population is less able to survive any changes in the environment that might happen in the future. That's because if one of the individuals in the population does not contain a useful characteristic, none of them will.

In a more natural population, at least one or two individuals can usually survive change. They go on to reproduce and restock. Inability to survive environmental changes could be a problem in the future for cloned crop plants or for cloned farm animals.

??? Did you know … ?

The only human clones alive at the moment are natural ones known as identical twins! However, the ability to clone mammals such as Dolly the sheep has led to fears that some people may want to have a clone of themselves produced – whatever the cost.

Study tip

Plants can be cloned by tissue culture or by taking cuttings. Animals can be cloned by dividing embryos or by adult cell cloning.

⬭ links

For more information on adult cell cloning, see 11.4 Making choices about genetic technology.

Key points

- In adult cell cloning, the nucleus of a cell from an adult animal is transferred to an empty egg cell from another animal. A small electric shock causes the egg cell to begin to divide and starts embryo development. The embryo is then placed in the womb of a third animal to develop.

- The animal that is born is genetically identical to the animal that donated the nucleus from the original adult body cell.

Summary questions

1 Make a flow chart to show how adult cell cloning works.

2 Explain clearly the differences between natural mammalian clones (identical twins), embryo clones, and adult cell clones.

3 What are the main advantages and disadvantages of the development of adult cell cloning techniques? How valid do you consider the main concerns expressed?

11.3 Genetic engineering

Learning objectives

After this topic, you should know:

- how genes are transferred from one organism to another in genetic engineering

- the potential benefits and problems associated with genetically modified crops.

What is genetic engineering?

Genetic engineering involves changing the genetic material of an organism using the following process:

- You take a gene from one organism and transfer it to the genetic material of a completely different organism. Enzymes are used to isolate and cut out the required gene.

- The gene is then inserted into a vector using more enzymes. The vector – usually a bacterial plasmid or a virus – carries the new gene to another organism.

- The vector is then used to insert the gene into the required cells, which may be bacterial, animal, fungal, or plant cells.

So, for example, genes from the chromosomes of a human cell can be cut out using enzymes and transferred to the cell of a bacterium. The gene carries on making a human protein, even though it is now in a bacterium.

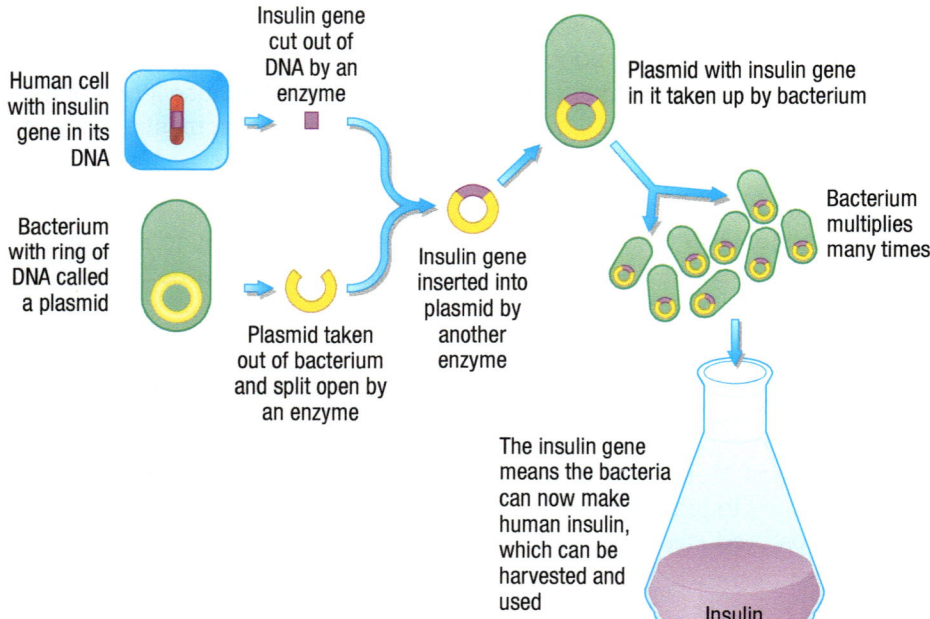

Figure 1 *The principles of genetic engineering. A bacterial cell receives a gene from a human so it makes a human protein – in this case, the human hormone insulin*

⬭ **links**

To find out more about the use of insulin to treat diabetes, see 7.4 Treating diabetes.

Study tip

Cloning and genetic engineering are different! Learn the techniques for both processes.

If genetically engineered bacteria are cultured on a large scale, they can make huge quantities of protein from other organisms. Humans now use them to make a number of drugs and hormones used as medicines, for example, genetically engineered bacteria are used to make human insulin (Figure 1).

Transferring genes to animal and plant cells

There is a limit to the types of proteins that bacteria are capable of making. As a result, genetic engineering has moved on. Scientists have found that genes from one organism can be transferred to the cells of another type of animal or plant at an early stage of their development. As the animal or plant grows, it develops with the new desired characteristics from the other organism. For example, glowing genes from jellyfish have been used to produce crop plants that glow in the dark when they are lacking in water. The farmer can then tell when the crops need irrigation.

Animals have been genetically engineered in a number of ways, but it is with plants that most progress has been made.

Genetically modified crops

Crops that have had their genes modified by genetic engineering techniques are known as **genetically modified crops (GM crops)**. Genetically modified crops often show increased yields. For example, genetically modified crops include plants that are resistant to attack by insects because they have been modified to make their own **pesticide**. This means that more of the crops survive to provide food for people. Growing GM plants that are more resistant than usual to **herbicides** means farmers can spray and kill weeds more effectively without damaging their crops. Again, this increases the crop yield.

Increasing crop yields is extremely important in providing food security for the world's human population, which is growing all the time. For example:

- Genetic engineering has resulted in many cereal plants with much shorter stems than the original plants. This means they are much less likely to be damaged by wind or storms, which increases the crop yield.
- Recent work on rice plants has produced crops that can withstand being completely covered in water for up to three weeks during flooding and still produce a high-yielding crop of rice (Figure 2). As more than 3.3 billion people worldwide rely on rice for a large proportion of their daily calorie intake, and climate change seems to be bringing much more severe flooding in many rice-growing countries, genetic modifications such as these could save millions of people from starvation.

Sometimes genetically modified crops contain genes from a completely different species, such as the jellyfish genes added to crop plants described earlier. Sometimes genetic modification simply speeds up normal selective breeding, by taking a gene from another closely related plant and inserting it into the genome. This has been done in the example of the flood-resistant rice.

Figure 2 *Rice is a vital staple food for millions of people. This genetically modified strain is still growing when traditional strains have been destroyed by prolonged flooding*

Key points

- Genes can be transferred to the cells of animals and plants at an early stage of their development so they develop desired characteristics. This is genetic engineering.

- In genetic engineering, genes from the chromosomes of humans and other organisms can be cut out using enzymes and transferred to the cells of bacteria and other organisms using a vector, which is usually a bacterial plasmid or a virus. **Ext**

- Crops that have had their genes modified are known as genetically modified (GM) crops. GM crops often have improved resistance to insect attack or herbicides and they generally produce a higher yield.

Summary questions

1 Early genetic engineering was carried out almost entirely in bacteria. Now, many different species of animals and plants have been genetically engineered. Explain why it is important to be able to modify the genomes of these different organisms.

Ext

2 Make a flow chart that explains the stages of transferring a gene for a shorter stem from one plant to another, using a bacterial plasmid as a vector.

3 Give three examples of ways in which food crops have been genetically modified. For each, explain how the change has increased the yield of the crop, and why this is important.

11.4 Making choices about genetic technology

Learning objectives

After this topic, you should know:

- some of the concerns and uncertainties about new genetic technologies such as cloning and genetic engineering.

Did you know …?

Short-stemmed flood- and drought-resistant plants are among the GM crops that are already helping to solve the problems of world hunger.

🔗 links

To find out more about GM crops, look back to 11.3 Genetic engineering.

Did you know …?

A great deal of work is currently being done on the lungs of people with cystic fibrosis and children with SCID (severe combined immunodeficiency). There have been some successes in using gene therapy to cure SCID, although there were also major problems.

Figure 1 *Yellow beta carotene is needed to make vitamin A in the body. The amount of beta carotene in golden rice and golden rice 2 is reflected in the depth of colour of the rice*

The benefits of genetic engineering and cloning are becoming more apparent all the time. However, there are some concerns about the use of these new technologies. More research is needed before scientists can fully understand any long-term impact they may have on individuals or on the environment.

Benefits of genetic engineering

People are already seeing many benefits from genetic engineering. Genetically engineered bacteria can make exactly the proteins humans need, in exactly the amounts needed, and in a very pure form. For example, pure human insulin is mass-produced using genetically engineered bacteria.

Some of the advantages of genetic engineering are:

- improved growth rates of plants and animals
- increased food value of crops, as genetically modified (GM) crops usually have much bigger yields than ordinary crops
- crops can be designed to grow well in dry, hot, or cold parts of the world
- crops can be engineered to produce plants that make their own pesticide or that are resistant to herbicides used to control weeds.

Human engineering

One huge potential benefit of genetic engineering that has still to be fully realised is curing human genetic conditions. Scientists are getting close to curing several conditions by putting healthy genes into affected cells using genetic engineering, so the cells work properly. Perhaps the cells of an early embryo can be engineered so that the individual develops into a healthy person. If these treatments become possible, many people would have new hope of a normal life for themselves or for their children. However, there is a lot of work to be done before gene therapy is widely used in people.

Concerns about genetic engineering

Genetic engineering is still a very new science, and no one can be sure what the long-term effects might be. For example, insects may become pesticide-resistant if they eat a constant diet of pesticide-producing plants.

Some people are concerned about the effect of eating GM food on human health. However, people eat a wide range of organisms with many different types of DNA every day as part of a normal diet, so this concern may well be unfounded. Others feel that farmers have enough varieties of crop plants without using GM crops.

Another concern is that genes from genetically modified plants and animals might spread into the wildlife of the countryside. Some people are very anxious about the effect these GM organisms might have on populations of wild flowers and insects. GM crops are often made infertile, which means that farmers in poor countries have to buy new seed each year. Many people are unhappy with this practice. If these infertility genes spread into wild populations, it could cause major problems in the environment – although scientists are working hard to prevent this.

Ever since genetically modified foods were first introduced, there has been controversy and discussion about them. For example, varieties of GM rice known as 'golden rice' and 'golden rice 2' have been developed. These varieties of rice produce large amounts of vitamin A. Up to 500 000 children go blind each year as a result of a lack of vitamin A in their diets. In theory, golden rice offers a solution to this problem.

In fact, some people objected to the way in which the trials of the rice were run and the cost of the product, even though the rice is fertile and farmers can grow their own after they buy the original seed. No golden rice is yet being grown in countries affected by vitamin A blindness, and children continue to go blind due to vitamin A deficiency. However, the majority of plant scientists around the world believe that GM crops are the way forward in solving the problem of feeding the world's expanding population.

Another big concern is that people might want to manipulate the genes of their future children. This may be to make sure that they are born healthy, but there are also concerns that people might want to use it to have 'designer' children with particular characteristics such as high intelligence. Genetic engineering raises issues for us all to think about.

Cloning pets

Cloning plants and even some animals is widely used, but cloning mammals has also led to some ethical issues. Most of the research into cloning has been focused on farm and research animals, but some companies are hoping to be able to clone people's dying or dead pets for them. It has already been shown that a successful clone can be produced from a dead animal, as beef from a slaughterhouse has been used to create a live cloned calf.

Cloning your pet won't be easy or cheap. The issue is – should people be cloning their dead cats and dogs when there are thousands of unwanted animals already in existence? Even if a favourite pet cat is cloned, it may look nothing like the original because the coat colour of many cats is the result of genes switching on and off at random in the skin cells (Figure 2). A cloned pet will develop and grow in a different environment to the original animal as well. This means other characteristics that are affected by the environment will probably be different, too.

Figure 2 *The cat on the left is Rainbow. The cat on the right is Cc, Rainbow's clone. Rainbow and Cc share the same DNA, but they don't look the same*

To some people these are exciting events. To others they are a waste of time, money, and the lives of all the embryos that don't make it.

Figure 3 *Lancelot Encore, an adult cell clone, with his owners and a portrait of the original dog*

Study tip

Think about the pros and cons of these new techniques and be ready to discuss them in your exams.

Summary questions

1 Describe the main advantages and disadvantages of genetically modifying plants.

2 People get very concerned about cloning. Do you think these fears are justified? Explain your answer.

3 Summarise the use of genetic engineering and cloning in the treatment of human diseases so far.

Key points

- Concerns about GM crops include their effects on populations of wild flowers and insects, and uncertainty about the effects of eating GM crops on human health.

Summary questions

1 Tissue culture techniques mean that 50000 new raspberry plants can be grown from one old plant, instead of two or three by taking cuttings. Cloning embryos from the best-bred cows means that they can be genetically responsible for 30 or more calves each year, instead of two or three.

   a How can one cow produce 30 or more calves in a year?

   b What are the similarities between cloning animals in this way and tissue culture in plants?

   c What are the differences between the techniques for cloning animals and plants?

   d Why do you think there is so much interest in finding different ways to make the breeding of farm animals and plants increasingly efficient?

2 a Describe the process of adult cell cloning.

   b There has been a great deal of media interest and concern about cloning animals, but very little about cloning plants. Why do you think there is such a difference in the way people react to these two different technologies?

3 Human growth is usually controlled by growth hormones produced by the pituitary gland in the brain. If you don't make enough hormones, you don't grow properly and remain very small. This condition affects 1 in every 5000 children. Until recently, the only way to get human growth hormone was from the pituitary glands of dead bodies. Genetically engineered bacteria can now make plenty of pure human growth hormone.

   a Draw and label a diagram to explain how a healthy human gene for making growth hormone can be taken from a human chromosome and put into a working bacterial cell.

   b What are the advantages of producing substances such as growth hormone using genetic engineering?

4 In 2003, two mules called Idaho Gem and Idaho Star were born in America. They were clones of a famous racing mule. They were separated and sent to different stables to be reared and trained for racing. They both seem very healthy. So far Idaho Gem has been more successful than his cloned brother, winning several races against ordinary racing mules. There is a third clone, Utah Pioneer, which has not been raced.

   a The mules are genetically identical. How do you explain the fact that Idaho Gem has beaten Idaho Star in several races?

   b Why do you think one of the clones is not being raced?

   c The clones' progress is being carefully monitored by scientists. What type of data do you think will be available from these animals?

5 a What is meant by the term GM crops?

   b Explain the main concerns of people about the use of GM crops around the world.

   c Most plant scientists believe that GM technology will be the key to producing enough food to feed the world population. How can it be used to do that?

   d One concern people have about GM crops is that they might cross-pollinate with wild plants. Scientists need to find out how far pollen from a GM crop can travel to be able to answer these concerns.

   Describe how a trial to investigate this might be set up.

Practice questions

1 a What is a clone? (1)

b Clones of plants can be created by taking small blocks of stem, leaf, or root cells and growing them in a special medium. Name this type of cloning. (1)

c Animal clones can be created from embryos or by adult cell cloning. Briefly describe the steps in these two methods. (7)

d Suggest which method would be most suitable for creating a herd of cows that produce high yields of milk. Give reasons for your choice. (4)

2 The use of cloned animals in food production is controversial.

It is now possible to clone champion cows. Champion cows produce large quantities of milk.

a Describe how adult cell cloning could be used to produce a clone of a champion cow. (4)

b Read the passage below about cloning cattle:

The government has been accused of 'inexcusable behaviour' because a calf of a cloned American champion cow has been born on a British farm. Campaigners say it will undermine trust in British food because the cloned cow's milk could enter the human food chain.

However, supporters of cloning say that milk from clones and their offspring is as safe as the milk consumers drink every day.

Those in favour of cloning say that an animal clone is a genetic copy. It is not the same as a genetically engineered animal. Opponents of cloning say that consumers will be uneasy about drinking milk from cloned animals.

Use the information in the passage and your own knowledge and understanding to evaluate whether or not the government should allow the production of milk from cloned 'champion' cows.

Remember to give a conclusion to your evaluation. (5)

3 Read the passage below. Use the information and your own knowledge to answer the questions that follow.

At one time, the boll weevil destroyed cotton crops. Farmers sprayed the crops with a pesticide.

The weevil died out but another insect, the bollworm moth, became resistant to this pesticide.

In the 1990s, large crops of the cotton plant were destroyed by the bollworm moth. The pesticides then used to kill the moth were expensive and very poisonous, resulting in human deaths.

Scientists investigated alternative ways to control the bollworm moth. They found out that a type of bacterium produced a poison that killed bollworm larvae (grubs).

A GM cotton crop plant was developed that produced the poison to kill bollworms. This proved to be very effective and farmers were able to stop using pesticide sprays.

Now farmers have another problem. Large numbers of other insects have multiplied because they were not killed when the farmers stopped using pesticides. Some of these insects have started to destroy the GM cotton and farmers are beginning to use pesticides again!

a i Give **one** advantage of spraying crops with pesticides. (1)

ii Give **two** disadvantages of spraying crops with pesticides. (2)

iii Give **one** economic advantage of using GM cotton. (1)

iv Some people object to using GM crops. Suggest **one** reason why. (1)

b *In this question you will be assessed on using good English, organising information clearly, and using specialist terms where appropriate.*

The GM cotton was genetically engineered to produce the same poison as the bacterium.

Describe fully how this is done. (6)

12.1 Theories of evolution

Learning objectives

After this topic, you should know:
- the theory of evolution
- some of the evidence for evolution discovered by Darwin.

You are surrounded by an amazing variety of life on planet Earth. Questions such as 'Where has it all come from?' and 'When did life on Earth begin?' have puzzled people for many generations.

Darwin's theory of **evolution** by **natural selection** tells us that all the species of living things alive today have evolved from the first simple life forms. Scientists think these early forms of life developed on Earth more than 3 billion years ago. Most of us take these ideas for granted, but they are really quite new.

Until the 18th century, most people in Europe believed that the world had been created by God. They thought it was made, as described in the Christian Bible, a few thousand years ago. However, by the beginning of the 19th century, scientists were beginning to come up with new ideas.

Lamarck's theory of evolution

Jean-Baptiste Lamarck was a French biologist. He thought that all organisms were linked by what he called a 'fountain of life'. He made the great step forward of suggesting that individual animals adapted and evolved to suit their environment. His idea was that every type of animal evolved from primitive worms. The change from worms to other organisms was caused by the **inheritance of acquired characteristics**.

Lamarck's theory proposed that the way organisms behaved affected the features of their body – a case of 'use it or lose it'. If animals used something a lot over a lifetime, Lamarck thought this feature would grow and develop. Any useful changes that took place in an organism during its lifetime would be passed from a parent to its offspring. The neck of the giraffe is a good example (Figure 1). If a feature wasn't used, Lamarck thought it would shrink and be lost.

Lamarck's theory influenced the way **Charles Darwin** thought. However, there were several problems with Lamarck's ideas. There was no evidence for his 'fountain of life' and people didn't like the idea of being descended from worms. People could also see quite clearly that changes in their bodies – such as big muscles, for example – were not passed on to their children.

People now know that, in the great majority of cases, Lamarck's idea of inheritance cannot happen. However, his ideas paved the way for the scientists such as Darwin who followed him.

Figure 1 *In Lamarck's model of evolution, giraffes have long necks because each generation stretched up to reach the highest leaves. So each new generation had a slightly longer neck*

Charles Darwin and the origin of species

Our modern ideas about evolution began with the work of one of the most famous scientists of all time – Charles Darwin. Darwin set out in 1831 as the captain's companion and ship's naturalist on HMS *Beagle*. He was only 22 years old at the start of the voyage to South America and the South Sea Islands.

Darwin planned to study geology on the trip. Yet as the voyage went on, he became as excited by his collection of animals and plants as by his rock samples.

In South America, Darwin discovered a new form of the common rhea, an ostrich-like bird – although he had almost finished eating it before he noticed the differences! When he observed two different types of the same bird living in slightly different areas, this set Darwin thinking.

On the Galapagos Islands, Darwin was amazed by the variety of species. He noticed that they varied from island to island. Darwin found strong similarities between types of finches, iguanas, and tortoises on the different islands. Yet each was different and adapted to make the most of local conditions.

Darwin collected huge numbers of specimens of animals and plants during the voyage. He also made detailed drawings and kept written observations. The long journey home gave him plenty of time to think about what he had seen. Charles Darwin returned home after five years with some new ideas forming in his mind.

After returning to England, Darwin spent the next 20 years working on his ideas. Darwin's theory is that all living organisms have evolved from simpler life forms. This evolution has come about by a process of natural selection.

Reproduction always gives more offspring than the environment can support. Only those that have inherited features most suited to their environment – the 'fittest' – will survive. When they breed, they pass on the alleles for those useful inherited characteristics to their offspring. This is natural selection.

When Darwin suggested how evolution took place, no one knew about genes. He simply observed that useful inherited characteristics were passed on. Today, scientists know that it is useful genes/alleles that are passed from parents to their offspring in natural selection.

Figure 2 *Darwin was impressed by the marine iguanas he found on the Galapagos Islands and he studied them very carefully, comparing them in detail to land-dwelling iguanas*

Study tip

Remember the key steps in natural selection:

mutation of gene → advantage for survival → breed → pass on useful allele

Figure 3 *Darwin worked here in his study for around 20 years, carrying out experiments and organising his ideas on evolution by natural selection*

Summary questions

1 Define the following terms:
   a evolution
   b natural selection.

2 How did Jean-Baptiste Lamarck affect the development of ideas about evolution?

3 Explain the importance of the following in the development of Darwin's ideas:
   a South American rheas
   b Galapagos tortoises, iguanas, and finches
   c the long voyage of HMS *Beagle*
   d the 20 years from Darwin's return to the publication of his book, *The Origin of Species*.

Key points

- The theory of evolution states that all the species that are alive today – and many more which are now extinct – evolved from simple life forms that first developed more than 3 billion years ago.

- Darwin's theory is that evolution takes place through natural selection.

- Other theories, including that of Lamarck, are based mainly on the idea that changes that occur in an organism over its lifetime can be inherited. It is now understood that in the great majority of cases this is not true.

12.2 Natural selection

Learning objectives

After this topic, you should know:
- how natural selection works
- the timescales involved in evolution.

Figure 1 *The tiny number of thistles that will survive and grow into adults from this mass of floating seeds will have combinations of alleles that give them an advantage over all the others*

∞ links

For more information on the competition between plants and animals in the natural world, look back to 12.7 Competition in animals and 12.8 Competition in plants.

Figure 2 *The natural world is often brutal. Foxes aren't the only animals to hunt rabbits! Only the best-adapted predators capture prey and survive to breed – and only the best-adapted prey animals escape to breed as well*

Scientists explain the variety of life today as the result of a process called natural selection. The idea was first suggested about 150 years ago by Charles Darwin.

Animals and plants are always in competition with each other. Sometimes an animal or plant gains an advantage in the competition. This might be against other species or against other members of its own species. That individual is more likely to survive and breed. This is known as natural selection.

Survival of the fittest

Charles Darwin was the first person to describe natural selection as the 'survival of the fittest'. Reproduction is a very wasteful process. Animals and plants always produce more offspring than the environment can support.

Genetic variation

The individual organisms in any species may show a wide range of variation. This is because of differences in the genes they inherit. Differences in the genes can arise as a result of mutation (see below). The offspring with the alleles that produce the characteristics best suited to the environment are more likely to survive to breed successfully. The alleles that have enabled these individuals to survive are then passed on to the next generation. Less well-adapted alleles will be lost. This is natural selection at work.

Think about rabbits. The rabbits with the best all-round eyesight, the sharpest hearing, and the longest legs will be the ones that are most likely to escape being eaten by a fox. They will be the ones most likely to live long enough to breed. What's more, they will pass those useful genes on to their babies. The slower, less alert rabbits will get eaten and their genes are less likely to be passed on.

The part played by mutation

New forms of genes (alleles) result from changes in existing genes. These changes are known as mutations. They are tiny changes in the long strands of DNA. Mutations occur quite naturally all the time through mistakes made in copying DNA when the cells divide. Mutations introduce more variation into the genes of a species as new and different alleles are produced. In terms of survival, this is very important.

Many mutations have no effect on the characteristics of an organism, and some mutations are harmful. However, just occasionally a mutation has a good effect. It produces an adaptation that makes an organism better suited to its environment. This makes it more likely to survive and breed. The mutant allele will gradually become more common in the population and will cause the species to evolve.

 Did you know … ?

Fruit flies can produce 200 offspring every two weeks. The yellow star thistle, an American weed, produces around 150 000 seeds per plant, per year. If all those offspring survived, we'd be overrun with fruit flies and yellow star thistles!

Natural selection in action

When new forms of a gene arise from mutation, there may be a relatively rapid change in a species. This is particularly true if the environment changes. If the mutation gives the organism an advantage in the changed environment, it will soon become common.

Malpeque Bay in Canada has some very large oyster beds. In 1915, the oyster fishermen noticed a few small, flabby oysters with pus-filled blisters among their healthy catch. By 1922, the oyster beds were almost empty. The oysters had been wiped out by a destructive new disease (soon known as Malpeque disease).

Fortunately, a few of the oysters had a chance mutation that made them resistant to the disease. These were the only ones to survive and breed. The oyster beds filled up again, and by 1940 they were producing more oysters than ever.

A new population of oysters had evolved. As a result of natural selection, almost every oyster in Malpeque Bay now carries an allele that makes it resistant to Malpeque disease. So the disease is no longer a problem.

Figure 3 *People will pay a lot of money for healthy oysters like these, so the evolution of a disease-resistant strain in Malpeque Bay allowed a new oyster business to emerge from the ruins of the old one*

Timescales of evolution

Natural selection can bring about change very quickly. In bacteria, it can take a matter of days for the genetic make-up of a population to change. In the Malpeque Bay oysters, the population changed over about 20 years. However, to produce an entire new species rather than just a different population usually takes much longer. It has taken millions of years for the organisms present on Earth in the 21st century to evolve. There have been many different species that no longer exist, which lived on Earth many millions of years ago. The descendants of some of those species are the animals and plants you see around you.

Summary questions

1 Many features that help animals and plants survive are the result of natural selection. Give **three** examples, for example, all-round eyesight in antelope, and use them to explain what is meant by natural selection.

2 **a** What is mutation?
   b Why is mutation important in natural selection?

3 Explain how the following characteristics of animals and plants have come about in terms of natural selection.
   a Male red deer have large sets of antlers.
   b Cacti have spines instead of leaves.
   c Camels can tolerate their body temperature rising far higher than most other mammals.

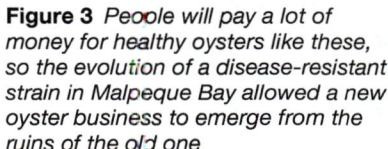

links

For more information on genes, look back to 10.1 Inheritance.

links

You can find out about the evolution of antibiotic-resistant bacteria by natural selection in 8.5 Changing pathogens.

Study tip

Remember – mutations happen all the time by chance. Sometimes factors in the environment such as excess UV radiation can increase the rate of mutation.

Key points

- Natural selection works by selecting the organisms best adapted to a particular environment.

- Different organisms in a species may show a wide range of variation because of differences in their genes.

- The individuals with the characteristics most suited to their environment are most likely to survive and breed successfully.

- The alleles that have produced these successful characteristics are then passed on to the next generation.

- The timescales of evolution vary depending on the complexity and life cycle of organisms. For example, simple organisms such as bacteria evolve much faster than complex multicellular organisms such as mammals.

12.3

Isolation and the evolution of new species

 links

Revisit your learning on genetic variation and natural selection by looking back to 12.2 Natural selection.

??? Did you know ...?

Sometimes organisms are separated by **environmental isolation**. This is when the climate changes in one area where an organism lives, but not in other areas. For example, if the climate becomes warmer in one area, plants will flower at a different time of year. The breeding times of the plants and the animals linked with them will change and so new species will emerge.

Figure 1 *Both the marsupial koala and the eucalyptus tree it feeds on have evolved in geographical isolation in Australia*

Some of the best evidence scientists have about the history of life on Earth comes from **fossils**. Fossils are the remains of organisms from many hundreds of thousands or millions of years ago, that are found preserved in rocks, ice, and other places. The fossil record shows that many different **species** have appeared and then died out over millions of years. This is evolution in action. Natural selection takes place and new organisms adapted to the different conditions evolve. However, evolution is happening all the time. There is a natural cycle of new species appearing whilst others become extinct.

Isolation and evolution

You have already learnt about the role of genetic variation and natural selection in evolution. Any population of living organisms contains genetic variety. If one population becomes isolated from another, the conditions they are living in are likely to be different. This means that, within each group, different characteristics will be selected as preferable over time. The two populations might change so much over time that they cannot interbreed successfully and then a new species evolves. A species is a group of organisms with many features in common that can breed successfully producing fertile offspring.

How do populations become isolated?

The most common way in which populations become separated is by **geographical isolation**. This is when two populations become physically isolated by a geographical feature, for example, a new mountain range, a new river, or an area of land becoming an island. Earthquakes can separate areas of land, and volcanoes can produce completely new islands.

There are some well-known examples of populations becoming isolated. Australia separated from the other continents over 5 million years ago. That's when the Australian populations of marsupial mammals that carry their babies in pouches became geographically isolated. As a result of natural selection, many different species of marsupials evolved. Organisms as varied as kangaroos and koala bears appeared. Across the rest of the world, competition resulted in the evolution of other mammals with more efficient reproductive systems. In Australia, marsupials remain dominant.

Organisms in isolation

Organisms on islands are geographically isolated from the rest of the world. The closely related but very different species on the Galapagos Islands helped Darwin form his ideas about evolution.

When a species evolves in isolation and is found in only one place in the world, it is said to be **endemic** to that area. One area where scientists are finding many new endemic species is Borneo. It is one of the largest islands in the world. Borneo still contains huge areas of tropical rainforest.

Between 1994 and 2006, scientists discovered over 400 new species in the Borneo rainforest. There are more than 25 species of mammal found only on the island. All of these organisms have evolved through geographical isolation.

Speciation

Any population will contain natural genetic variety. This means it contains a wide range of alleles controlling its characteristics that result from sexual reproduction and mutation. This is genetic variation.

In each population, the alleles that are selected will control characteristics that help the organism to survive and breed successfully. This is natural selection.

Sometimes, part of a population becomes isolated with new environmental conditions. Alleles for characteristics that enable organisms to survive and breed successfully in the new conditions will be selected. These are likely to be different from the alleles that gave success in the original environment. As a result of the selection of these different alleles, the characteristic features of the isolated organisms will change. Eventually, they can no longer interbreed with the original organisms and a new species forms. This is known as **speciation**.

This is what has happened on the island of Borneo, in Australia, and on the Galapagos Islands. If conditions in these isolated places are changed or the habitat is lost, the species that have evolved to survive within it could easily become extinct.

However, a similar version of speciation has also happened all over the world. Geographical isolation may involve very large areas, such as Borneo, or very small regions. Mount Bosavi is the crater of an extinct volcano in Papua New Guinea (Figure 3). It is only 4 km wide and the walls of the crater are 1 km high. The animals and plants trapped within the crater have evolved in different ways to those outside it.

Very few people have been inside the crater. During a three-week expedition in 2009, scientists discovered around 40 new species. These included mammals, fish, birds, reptiles, amphibians, insects, and plants. All of these species are the result of natural selection caused by the specialised environment of the isolated crater. They include an enormous 82 cm-long rat that weighs 1.5 kg!

Figure 2 *Lemurs are endemic to the island of Madagascar. Some 99 species and subspecies of lemur have evolved there, such as this ring-tailed lemur*

Figure 3 *Mount Bosavi in Papua New Guinea is a small, geographically isolated environment where many new species have evolved*

Study tip

Speciation is a difficult concept. If you cannot remember all the steps, at least remember that two groups of the same species become separated and different natural selection takes place in each group, so eventually they become so different that they are no longer able to interbreed.

Key points

- New species arise as a result of:
  - Isolation – two populations of a species become separated (e.g., geographically).
  - Genetic variation – each population has a wide range of alleles that control their characteristics.
  - Natural selection – in each population alleles are selected that control the characteristics which help the organism to survive and breed.
  - Speciation – the populations become so different that successful interbreeding is no longer possible.

Summary questions

1 a How might populations become isolated?
  b Why does this isolation lead to the evolution of new species?

2 Islands often have their own endemic organisms.
  a What is an endemic organism?
  b Why are endemic organisms common on islands?
  c How does this act as evidence for our current model of speciation and evolution?

3 Explain how genetic variation and natural selection result in the formation of new species in isolated populations.

12.4

Adapt and survive

🔗 **links**

For more information on plant adaptation, see 12.6 Adaptations in plants.

Figure 1 *Mangroves are trees that live in soil with very little oxygen, often with their roots covered by salty water. They have special adaptations to get rid of the salt through their leaves, and roots that grow in the air to get oxygen*

🔗 **links**

For more information on animal adaptations, see 12.5 Adaptations in animals.

The variety of conditions on the surface of the Earth is huge. They range from hot, dry deserts to permanent ice and snow. There are deep, saltwater oceans and tiny, freshwater pools. Whatever the conditions, almost everywhere on Earth you will find living organisms that are able to survive and reproduce.

Survive and reproduce

Living organisms need a supply of materials from their surroundings and from other living organisms so that they can survive and reproduce successfully. What they need depends on the type of organism:

- Plants need light, carbon dioxide, water, oxygen, and nutrients to produce glucose and other larger molecules and to provide them with the energy they need to survive.
- Animals need food from other living organisms, water, and oxygen.
- Microorganisms need a range of things. Some are similar to plants, whilst some are similar to animals. Some don't need oxygen or light to survive.

Living organisms have special features known as **adaptations**. These features make it possible for them to survive in their particular habitat, even when the conditions are very extreme.

Plant adaptations

Plants need to photosynthesise to produce the glucose needed for energy and growth. They also need to have enough water to maintain their cells and tissues. They have adaptations that enable them to live in many different places. For example, most plants get water and mineral nutrients from the soil through their roots.

Epiphytes are found in rainforests. They have adaptations that allow them to live high above the ground attached to other plants. They collect water and nutrients from the air and in their specially adapted leaves.

Some plant adaptations are all about reproduction. For example, the South African sausage tree (*Kigelia pinnata*) is one of a relatively small number of plants that rely on bats to pollinate their flowers. The flowers open at night, have a strong perfume, and produce lots of nectar. They hang down below the branches and leaves, which makes it as easy as possible for the bats to approach and feed from them – and at the same time transfer pollen from one flower to another on their fur.

Animal adaptations

Animals cannot make their own food – they have to eat plants or other animals. This type of feeding is known as heterotrophic. Many of the adaptations of animals help them to get the food they need. This means that you can tell what a mammal eats by looking at its teeth.

- **Herbivores** have teeth adapted for grinding up plant cells.
- **Carnivores** have teeth adapted for tearing flesh or crushing bones.

Animals also often have adaptations to help them find and attract a mate.

Adapting to the environment

Some of the adaptations seen in animals and plants help them to survive in a particular environment. For example:

- Some sea birds get rid of all the extra salt they take in from the seawater by 'crying' very salty tears from a special salt gland.
- Animals that need to survive extreme winter temperatures often produce a chemical in their cells which acts as antifreeze. It stops the water in the cells from freezing and destroying the cell.
- Plants such as water lilies have lots of big air spaces in their leaves. This adaptation enables them to float on top of their watery environment and make food by photosynthesis.

Living in extreme environments

Organisms that survive and reproduce in the most difficult conditions are known as **extremophiles**. Many extremophiles are microorganisms. Microorganisms are found in more places in the world than any other living thing. These places range from ice packs to hot springs and geysers. Microorganisms have a range of adaptations which make this possible.

Some extremophiles live at very high temperatures. Bacteria known as thermophiles can survive at temperatures of over 45 °C and often up to 80 °C or higher. (In most organisms, enzymes stop working at around 40 °C.) These extremophiles have specially adapted enzymes that do not denature and can work at these high temperatures. In fact, many of these organisms cannot survive and reproduce at lower temperatures. Larger organisms that live in very hot environments have similarly adapted enzymes.

Other bacteria have adaptations enabling them to grow and reproduce at very low temperatures, down to −15 °C. They are found in ice packs and glaciers around the world.

Most living organisms struggle to survive in a very salty environment because of the problems it causes with water balance. However, there are species of extremophile bacteria that can only live in extremely salty environments, such as the Dead Sea and salt flats. These bacteria have adaptations to their cytoplasm so that water does not move out of their cells into their salty environment. However, in ordinary sea-water they would swell up and burst!

Figure 3 *Black smoker bacteria live in deep ocean vents, 2500 m down, at temperatures of well over 100 °C, with enormous pressure, no light, and an acid pH of about 2.8. They have adaptations to cope with some of the most extreme conditions on Earth*

Figure 2 *Animals from the deep oceans are adapted to cope with enormous pressure, no light, and very cold, salty water. If these extremophiles are brought to the surface too quickly, they explode because of the rapid change in pressure*

Key points

- Organisms need a supply of materials from their surroundings and from other living organisms to survive and reproduce.

- Organisms, including microorganisms, have features (adaptations) that enable them to survive in the conditions in which they normally live.

- Extremophiles have adaptations that enable them to live in environments with extreme conditions of salt, temperature, or pressure.

Summary questions

1 Describe what plants and animals need from their surroundings to survive and reproduce, and explain the differences between them.

2 **a** What is an extremophile?
  b Give two examples of adaptations found in different extremophiles.

3 Explain what is meant by an adaptation, and give three examples of adaptations in either animals or plants to a particular environment or way of life.

12.5 Adaptations in animals

∞ **links**

You can see a diagram explaining surface area to volume ratio in 3.1 Exchanging materials.

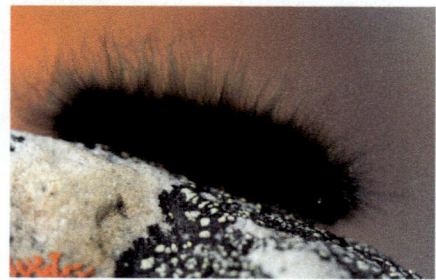

Figure 1 *The Arctic woolly bear moth caterpillar is adapted to survive up to 14 years of freezing and thawing before it becomes an adult moth. It contains a special chemical that acts like antifreeze and protects the cells from damage in the extreme conditions of the Arctic*

Study tip

Remember that animals living in very cold conditions often have a low surface area to volume ratio. This means that there is less area for energy to be transferred to the environment. The opposite can be true in hot climates.

??? Did you know …?

Polar bears don't change colour. They have no natural predators on the land. They hunt seals all year round in the sea, where their white colour makes them less visible amongst the ice.

Animals have adaptations that help them to get the food and mates they need to survive and reproduce. They also have adaptations for survival in the conditions where they normally live. These include:

- Structural adaptations such as the shape or colour of the organism or part of the organism. Examples include the large surface area of an elephant's ears, which are an adaptation to help transfer energy to the surroundings and cool the animal.
- Behavioural adaptations such as migration, basking, and huddling together. For example, mammals such as wildebeest and zebra, and birds such as Arctic terns and swallows, migrate hundreds or even thousands of miles every year in search of food or ideal breeding conditions.
- Functional adaptations related to processes such as reproduction and metabolism. For example, the reproduction of many organisms is coordinated with when there will be plenty of food available for the offspring. And some seals that dive to great depths collapse their lungs to cope with the high pressures deep underwater. They have special chemicals in their muscles which store the oxygen they need until they return to the surface, reinflate their lungs, and breathe again.

Animals in cold climates

To survive in a cold environment, you must be able to keep yourself warm. Animals that live in very cold places, such as the Arctic, are adapted to reduce the energy transferred from their bodies to the environment.

You transfer energy to the environment through your body surface (mainly your skin), cooling you down. The amount of energy you transfer is closely linked to your surface area to volume ratio (SA:V). The larger the surface area to volume ratio, the larger the rate of energy transfer.

The ratio of surface area to volume falls as objects get bigger. This is why mammals in a cold climate grow to a large size – it keeps their surface area to volume ratio as small as possible, which helps them to keep warm (reduces energy transfers to the surroundings). The surface area to volume ratio is very important when you look at the adaptations of animals that live in cold climates. It explains why so many Arctic mammals, such as seals, walruses, whales, and polar bears, are relatively large.

Animals in very cold climates often have other adaptations, too. The surface area of the thinly skinned areas of their bodies, like their ears, is usually very small. This reduces the rate at which energy is transferred to the environment.

Many Arctic mammals have plenty of insulation, both inside and out. Inside they have blubber (a thick layer of fat that builds up under the skin). On the outside they have a thick fur coat that provides very effective insulation. These adaptations really reduce the rate of energy transfer to the environment through their skin.

The fat layer also provides a food store. Animals often build up their fat in the summer. Then they can live off their body fat through the winter when there is almost no food.

Penguins are well-known for huddling together to help them survive in very cold environments. This reduces the amount of exposed surface area of every bird. It reduces energy transfers to the environment and helps keep the birds warm.

Camouflage

Camouflage is a form of structural adaptation that is important both to **predators** (so their prey doesn't see them coming) and to prey (so they can't be seen) (Figure 2).

The colours that would camouflage an Arctic animal against plants in summer would stand out against the snow in winter. Many Arctic animals, including the Arctic fox, the Arctic hare, and the stoat, therefore have grey or brown summer coats that change to pure white in the winter.

The colour of the coat of a lioness is another example of effective camouflage. The sandy brown colour matches perfectly with the dried grasses of the African savannah. Her colour hides the lioness from the grazing animals that are her prey.

Surviving in dry climates

Dry climates are often also hot climates. For example, hot deserts are very difficult places for animals to live. There is scorching heat during the day, often followed by severe cold at night. Water is also in short supply. Cold deserts can also be dry because the water is frozen and unavailable to plants and animals.

The biggest challenges if you live in a desert are:
- coping with the lack of water
- stopping your body temperature from getting too high.

Many desert animals are adapted to need little or nothing to drink. They get the water they need from the food they eat. Mammals need to keep their body temperature the same all the time. So as the environment gets hotter, they have to find ways of keeping cool. Sweating means they lose water, which is not easy to replace in the desert. Animals that live in hot conditions adapt their behaviour to keep cool. They are often most active in the early morning and late evening, when it is not so hot. During the cold nights and the hottest times of the day, they rest in burrows or shady areas where the temperature doesn't change much.

Many desert animals are quite small, so their surface area is large compared to their volume. This helps them to transfer energy to the environment. They also often have large, thin ears to increase their surface area. These adaptations help them to transfer energy to their surroundings, cooling their bodies.

Another adaptation of many desert animals is to have thin fur. Any fur they do have is fine and silky. They also have relatively little body fat stored under the skin. These features all make it easier for them to transfer energy to the environment through the surface of the skin and so cool down.

Figure 2 *This moth in an English woodland needs to avoid the birds that might eat it, so camouflage is an important adaptation for survival*

Figure 3 *Bactrian camels have to survive extremes of both heat (38 °C) and cold (–29 °C) in the rocky deserts of East and Central Asia. Adaptations that help them to survive include a thick winter coat, very little sweating, tissues that can cope with big fluctuations in their core temperature, a large store of fat in the humps, the ability to take water from their food, and drinking large amounts of water (around 135 litres) at a time when it becomes available*

Summary questions

1 **a** List the main problems that face animals living in cold conditions such as in the Arctic.
   b List the main problems that face animals living in the desert.
2 Animals that live in the Arctic are adapted to keep warm through the winter. Describe three of these adaptations and explain how they work.
3 **a** Describe the visible adaptations of an elephant that enable it to keep cool in hot conditions.
   b Suggest other ways in which animals might be adapted to survive in hot, dry conditions.
   c Describe and explain at least **two** adaptations that enable marine mammals such as whales and seals to survive in the seas and oceans of the world.

Key points

- All living things have adaptations that help them to survive in the conditions where they live.

- Animal adaptations include:
  - structural adaptations such as the shape and colour of the organism
  - behavioural adaptations such as migration
  - functional adaptations of processes such as reproduction and metabolism.

12.6 Adaptations in plants

∞ **links**

To find out more about transport in plants and transpiration, look back to 9.4 Transport systems in plants.

Plants need light, water, space, and nutrients to survive. There are some places where plants cannot grow. In deep oceans, no light penetrates, so plants cannot photosynthesise. In the icy wastes of the Antarctic, it is simply too cold for plants to grow.

Almost everywhere else, including the hot, dry areas of the world, you find plants growing. Without them there would be no food for animals. However, plants need water for photosynthesis and to keep their tissues supported. If a plant does not get the water it needs, it wilts and eventually dies.

Plants take in water from the soil through their roots. It moves up through the plant and into the leaves. There are small openings called stomata in the leaves of a plant. These open to allow gases in and out for photosynthesis and respiration. At the same time, water vapour is lost through the stomata by diffusion after it evaporates from the surface of the cells into the air spaces in the leaves.

The rate at which a plant loses water is linked to the conditions in which it grows. When a plant grows in hot and dry conditions, photosynthesis and respiration take place quickly. As a result, plants lose water vapour very quickly. Plants that live in very hot, dry conditions therefore need special adaptations to survive. Most plants either reduce their surface area so they lose less water or store water in their tissues. Some plants do both!

Changing surface area

When it comes to stopping water loss through the leaves, the surface area to volume ratio is very important to plants. A few desert plants have broad leaves with a large surface area. These leaves collect the dew that forms in the cold evenings. They then funnel the water towards their shallow roots.

Some plants in dry environments have curled leaves to reduce the surface area of the leaf. This also traps a layer of moist air around the leaf to reduce the amount of water that the plant loses by evaporation.

Most plants that live in dry conditions have leaves with a very small surface area. This adaptation cuts down the area from which water can be lost. Some desert plants have small, fleshy leaves with a thick cuticle to keep water loss down. The cuticle is a waxy covering on the leaf that stops water evaporating.

Some examples of plant adaptations are:

- Marram grass, which grows on sand dunes. It has tightly curled leaves to reduce the surface area for water loss so it can survive the dry conditions.
- Butcher's broom, which lives in shady, dry conditions under woodland trees and in hedgerows. To reduce water loss its 'leaves' are really flattened, leaf-like bits of stem. Stems have far fewer stomata than true leaves, so the butcher's broom loses very little water and can survive and reproduce in conditions where there is little competition from other species.

Figure 1 *The leaves of butcher's broom are really stems, not leaves. As a result of this adaptation to reduce water loss, the flowers and berries appear to grow out of the 'leaves'*

∞ **links**

For information on surface area to volume ratio, look back to 12.5 Adaptations in animals.

The best-known desert plants are the cacti. Their leaves have been reduced to spines with a very small surface area. This means that cacti only lose a tiny amount of water. Not only that, their sharp spines also discourage animals from eating them.

Collecting water

Many plants that live in very dry conditions have very large, specially adapted root systems. They may have extensive root systems that spread over a very wide area, roots that go down a very long way, or both. These adaptations allow the plant to take up as much water as possible from the soil. For example, the mesquite tree has roots that grow as far as 50 m down into the soil.

Storing water

Some plants cope with dry conditions by storing water in their tissues. When there is plenty of water after a period of rain, the plant stores it. Some plants use their fleshy leaves to store water, whilst other plants use their stems or roots.

For example, cacti don't just rely on their spiny leaves to help them survive in dry conditions. The fat green body of a cactus is its stem, which is full of water-storing tissue. These adaptations make cacti the most successful plants in a hot, dry climate.

Figure 2 *Plants such as this saguaro cactus (left) and apple tree (right) live in very different environments and have very different adaptations for maintaining their water balance*

??? Did you know … ?

After a storm, a large saguaro cactus in the desert can take in 1 tonne of water in a single day. Its adaptations for water conservation mean that it normally loses less than one glass of water a day even in the desert heat. A UK apple tree can lose a whole bath of water in the same amount of time!

Summary questions

1 a Why do plants need water?
 b How do plants get the water they need?

2 a How do plants lose water from their leaves?
 b Why does this make living in a dry place such a problem?

3 a Plants living in dry conditions have adaptations to reduce water loss from their leaves. Give three of these and explain how they work.
 b Preventing water loss from the leaves is not the only way in which plants can deal with dry conditions. Describe and explain three other adaptations seen in plants to help them survive in dry conditions.

Key points

- Plants lose water vapour from the surface of their leaves.

- Plant adaptations for surviving in dry conditions include reducing the surface area of the leaves, having water-storage tissues, and growing extensive root systems.

12.7 Competition in animals

Did you know ...?

Competition between members of different species for the same resources is known as **inter-specific competition**. Competition between members of the same species is known as **intra-specific competition**.

Figure 1 *Some herbivores only feed on one particular plant, but this approach is risky. Pandas only eat bamboo, so they are vulnerable to competition from other animals that eat bamboo and to diseases that damage bamboo*

Figure 2 *The dramatic colours of this poison dart frog are a clear warning to predators to keep well away*

Animals and plants grow alongside lots of other living things. Some will be from the same species, whilst others will be completely different. In any area there is only a limited amount of food, water, and space, and a limited number of mates. As a result, living organisms have to compete for the things they need.

The best-adapted organisms are those most likely to win the **competition** for resources. They will be most likely to survive and produce healthy offspring.

What do animals compete for?

Animals compete for many things, including:

- food
- **territory**
- mates.

Competition for food

Competition for food is very common. Herbivores sometimes feed on many types of plant, and sometimes on only one or two different sorts. Many different species of herbivores will all eat the same plants. Just think how many types of animal eat grass!

The animals that eat a wide range of plants are most likely to be successful. If you are a picky eater, you risk dying out if anything happens to your only food source. An animal with wider tastes will just eat something else for a while!

Competition is also common among carnivores. They compete for prey. Small mammals such as mice are eaten by animals such as foxes, owls, hawks, and domestic cats. The different types of animals all hunt the same mice. The animals that are best-adapted to finding and catching mice will be most successful.

Carnivores have to compete with their own species and with different species for their prey. Some successful predators are adapted to have long legs for running fast and sharp eyes to spot prey. These features will be passed on to their offspring.

Animals often avoid direct competition with members of other species when they can. It is the competition between members of the same species which is most intense.

Prey animals compete with each other, too – to be the one that *isn't* caught! Their adaptations help to prevent them becoming a meal for a predator. Some animals contain poisons that make anything that eats them sick or even kills them. Very often these animals also have bright warning colours so that predators quickly learn which animals to avoid. Poison dart frogs are a good example (Figure 2).

Competition for territory

For many animals, setting up and defending a territory is vital. A territory may simply be a place to build a nest, or it could be all the space needed for an animal to find food and reproduce (Figure 3). Most animals cannot reproduce successfully if they have no territory, so they will compete for the best spaces.

This helps to make sure that they will be able to find enough food for themselves and for their young. For example, the number of territories of many small birds such as tits found in an area varies with the amount of food available. Many animals mark the boundaries of their territories to keep other competitors out. This is often done using urine or faeces to make a strongly scented boundary.

Competition for a mate

Competition for mates can be fierce. In many species, the male animals put a lot of effort into impressing the females. The males compete in different ways to win the privilege of mating with a female.

In some species – such as deer, lions, and elephant seals – the males fight between themselves. Then the winner gets to mate with several females.

Many male animals display to the females to get their attention. Some birds have spectacular adaptations to help them stand out. Male peacocks have the most amazing tail feathers. They use them for displaying to other males (to warn them off) and to females (to attract them). Birds of paradise also produce spectacular displays to attract a mate (Figure 4).

What makes a successful competitor?

A successful competitor is an animal that is adapted to be better at finding food or a mate than the other members of its own species. It also needs to be better at finding food than the members of other local species. In addition, it must be able to breed successfully.

Many animals are successful because they avoid competition with other species as much as possible. They feed in a way that no other local animals do, or they eat a type of food that other animals avoid. For example, one plant can feed many animals without direct competition. Caterpillars eat the leaves, greenfly drink the sap, butterflies suck nectar from the flowers, and beetles feed on pollen.

Figure 3 *The territory of a gannet pair may be small, but without a space they cannot build a nest and reproduce*

Figure 4 *The male bird of paradise uses a very spectacular display to attract a more camouflaged female*

Study tip

Learn to look at an animal and spot the adaptations that make it a successful competitor.

Summary questions

1 **a** Animals that rely on a single type of food can easily become extinct. Explain why.
   b Give one example of animals competing with members of the same species for food.
   c Give one example of animals competing with members of other species for food.
   d Why is competition between members of the same species often more fierce than competition between different species?

2 **a** Give two ways in which animals compete for mates.
   b Suggest the advantages and disadvantages of the methods chosen in part **a**.

3 Explain some of the adaptations you would expect to find in the following organisms, and the advantages they would give:
   a an animal that hunts small mammals such as mice and voles
   b an animal that eats grass
   c an animal that is hunted by many different predators
   d an animal that feeds on the tender leaves at the top of trees.

Key points

- Animals compete with each other for food, territories, and mates.

- Animals have adaptations that make them successful competitors.

12.8 Competition in plants

Practical

Investigating competition in plants

Carry out an investigation to look at the effect of competition on plants. Set up two trays of seeds – one crowded and one spread out. Then monitor the plants' height and wet mass (mass after watering). Keep all of the conditions – light level, amount of water and nutrients available, and temperature – exactly the same for both sets of plants. The differences in their growth will be the result of overcrowding and competition for resources in one of the groups.

The data shows the growth of tree seedlings. You can get results in days rather than months by using cress seeds.

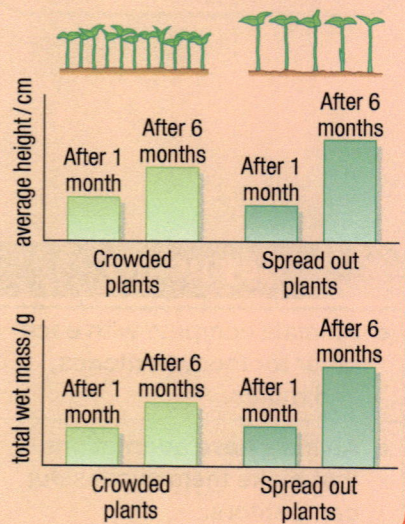

Plants compete fiercely with each other. They compete for:

- light for photosynthesis, to make food using energy from sunlight
- water for photosynthesis and to keep their tissues rigid and supported
- nutrients (ions) from the soil so they can make all the chemicals they need in their cells
- space to grow, allowing their roots to take in water and nutrients and their leaves to capture light.

Why do plants compete?

As with animals, plants are in competition both with other species of plants and with their own species.

Big, tall plants such as trees take up a lot of water and nutrients from the soil. They also prevent light from reaching the plants beneath them. So the plants around them need adaptations to help them to survive.

When a plant sheds its seeds they might land nearby. Then the parent plant will be in direct competition with its own seedlings. As the parent plant is large and settled, it will take most of the water, nutrients, and light. So the plant will deprive its own offspring of everything they need to grow successfully. The roots of some desert plants even produce a chemical that stops seeds from germinating, killing the competition even before it begins to grow!

Sometimes the seeds from a plant will all land close together, a long way from their parent. They will then compete with each other as they grow.

Coping with competition

Plants that grow close to other species often have adaptations which help them to avoid competition.

Small plants found in woodlands often grow and flower very early in the year. This is when plenty of light gets through the bare branches of the trees. The dormant trees take very little water out of the soil. The leaves shed the previous autumn have rotted down to provide nutrients in the soil. Plants such as snowdrops, anemones, and bluebells are all adapted to take advantage of these things. They flower, set seeds, and die back again before the trees are in full leaf.

Another way in which plants compete successfully is by having different types of roots. Some plants have shallow roots taking water and nutrients from near the surface of the soil, whilst other plants have long, deep roots that go far underground. In this way, both types of plants compete successfully for what they need without affecting each other.

Leguminous plants such as peas, beans, and clover all have special bacteria living in nodules on their roots (Figure 1). These bacteria fix nitrogen from the air – in other words, they carry out chemical reactions that produce nitrates. Some of these nitrates are used by the plants for making amino acids, which in turn are used to build up proteins for growth. This gives the plants a real advantage over competing species that have to take their nitrates from the soil.

If one plant is growing in the shade of another, it may grow taller to reach the light. It may also grow leaves with a bigger surface area to take advantage of all the light it does get. Plants may have adaptations such as tendrils or suckers that allow them to climb up artificial structures or large trees to reach the light.

Some plants are adapted to prevent animals from eating them. They may have thorns, like the African acacia or the blackberry, or they may make poisons that mean they taste very bitter or make the animals that eat them ill. Either way, these plants compete successfully because they are less likely to be eaten than other plants without these adaptations.

Spreading the seeds

To reproduce successfully, a plant has to avoid competition with its own seedlings for light, space, water, and nutrients. Many plants use the wind to help them spread their seeds as far as possible. They produce fruits or seeds with special adaptations for flight to carry their seeds away. Examples of this are the parachutes of the dandelion 'clock' (Figure 2) and the winged seeds of the sycamore tree.

Some plants use mini-explosions to spread their seeds. The pods dry out, twist, and pop, flinging the seeds out and away. Gorse bushes and peas are examples of plants that use this method.

Juicy berries such as grapes and blackcurrants and nuts such as hazelnuts and walnuts are adaptations to tempt animals to eat them. The fruit is digested and the tough seeds are deposited well away from the parent plant in their own little pile of fertiliser!

Fruits that are sticky or covered in hooks, such as burrs, get caught up in the fur or feathers of a passing animal. They are carried around until they fall off hours or even days later.

Sometimes, the seeds of several different plants land on the soil and start to grow together. The plants that grow fastest will compete successfully against the slower-growing plants. For example:

- The plants that get their roots into the soil first will get most of the available water and nutrients.
- The plants that open their leaves fastest will be able to photosynthesise and grow faster still, depriving the competition of light.

Figure 1 *Clover has special bacteria in the roots that supply it with nitrates. This helps it to outcompete the grass, which has to take its minerals from the soil*

Figure 2 *The light seeds and fluffy parachutes of dandelion mean they are spread widely and compete very successfully*

Figure 3 *Coconuts will float for weeks or even months on ocean currents, which can carry them hundreds of miles from competition with their parents – or any other coconuts!*

Summary questions

1 a Suggest three ways in which plants can overcome the problems of growing in the shade of another plant.
  b How do snowdrops and bluebells grow and flower successfully in spite of living under large trees in woodlands?

2 a Why do so many plants have adaptations to make sure that their seeds are spread successfully?
  b Give three examples of successful adaptations for spreading seeds.

3 The dandelion is a successful weed. Carry out some research and evaluate the adaptations that make it a better competitor than many other plants on a school field.

Key points

- Plants often compete with each other for light, space, water, and nutrients (ions) from the soil.
- Plants have many adaptations that make them good competitors.

Summary questions

1 Match the following words to their definitions:

a	competition	A	an animal that eats plants
b	carnivore	B	an area where an animal lives and feeds
c	herbivore	C	an animal that eats meat
d	territory	D	the way in which animals compete with each other for food, water, space, and mates

2 Animals such as amphibians and reptiles do not control their own body temperature internally. They need energy transferred from their surroundings and cannot move until they are warm.

  a Why do you think that there are no frogs or snakes in the Arctic?

  b What problems do you think reptiles face in desert conditions, and what adaptations could they have to cope with them?

  c Most desert animals are quite small. Explain how this adaptation helps them survive in the harsh conditions.

3 **a** What are the main problems for plants living in a hot, dry climate?

  b Why does reducing the surface area of their leaves help plants to reduce water loss?

  c Describe **two** ways in which the surface area of the leaves of some desert plants is reduced.

  d Describe other plant adaptations for hot, dry conditions.

  e Why are cacti such perfect desert plants?

4 Bamboo plants all tend to flower and die at the same time. Why is this such bad news for pandas, but doesn't affect most other animals?

5 **a** Why is competition between animals of the same species so much more intense than the competition between different species?

  b How does marking out and defending a territory help an animal to compete successfully?

  c What are the advantages and disadvantages for males of having an elaborate courtship ritual and colouration compared with fighting over females?

6 Use the bar charts from the practical activity in *12.8 Competition in plants* to answer these questions.

  a Describe what happens to the height of both sets of seedlings over the first six months, and explain why the changes take place.

  b The total wet mass of the seedlings after one month was the same whether or not they were crowded. After six months there was a big difference.
    i Why do you think that both sets of seedlings had the same mass after one month?
    ii Explain why the seedlings that were more spread out each had more wet mass after six months.

  c When scientists carry out experiments such as the one described, they try to use large sample sizes. Why?

  d **i** Name a control variable mentioned in the practical.
    ii Why were other variables kept constant?

7 Figure 1 shows the life cycle of the *Plasmodium* parasite.

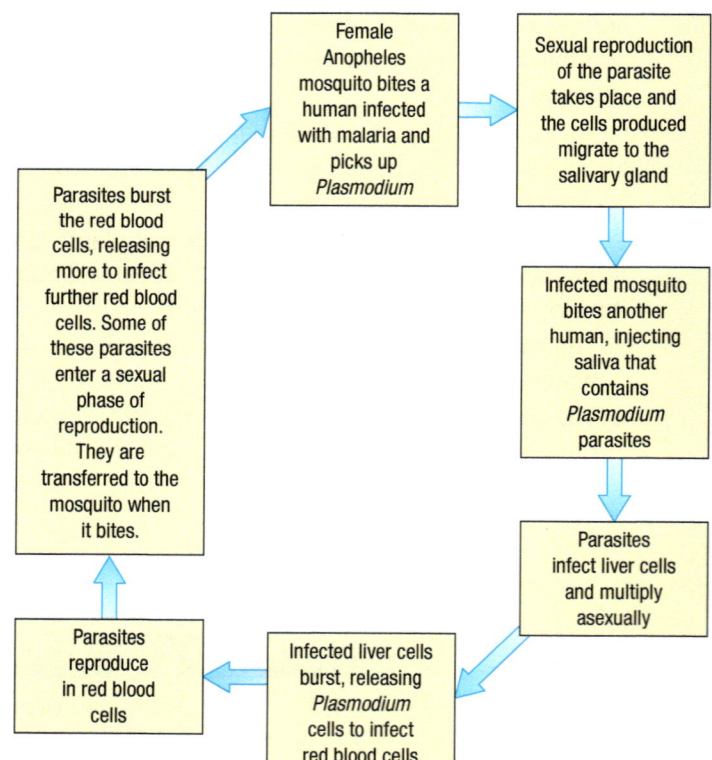

Figure 1

Using Figure 1 to help you, give a clear description of the life cycle of the *Plasmodium* parasite and how it is adapted to survive and be passed on.

Practice questions

1 *In this question you will be assessed on using good English, organising information clearly, and using specialist terms where appropriate.*

Elephants, shown in Figure 1, can survive in hot, dry areas.

Figure 1

Explain how the large, thin ears, lack of a fat layer beneath the skin, and fine bristles instead of fur help the elephant to live in hot, dry areas. Suggest why they move about and feed in the early morning and evening. (6)

2 Figure 2 is a photograph of a flea taken under a microscope. A flea is a parasite that lives in the fur of animals such as cats and dogs.

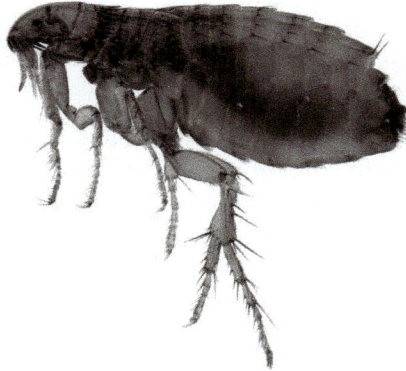

Figure 2

Suggest how each of the following adaptations helps the flea to survive in its habitat:

a no wings

b piercing mouth parts

c body flattened from side to side

d hard external exoskeleton

e long back legs with spring-like mechanism

f covered in bristles and combs. (6)

3 Gardeners may spray their vegetable plots with herbicide in the hope of growing bigger vegetables.

a What is a herbicide? (1)

b Explain, using your knowledge of competition in plants, how using a herbicide might increase the size of the vegetables. (5)

4 The gemsbok is a large herbivore living in dry, desert regions of South Africa. It feeds on grasses that are adapted to the dry conditions by obtaining moisture from the air as it cools at night. The following table shows the water content of these grasses and the feeding activity of the gemsbok over a 24-hour period.

Time of day	% water content of grasses	% of gemsboks feeding
03.00	18	40
06.00	23	60
09.00	25	20
12.00	08	17
15.00	06	16
18.00	05	19
21.00	07	30
24.00	14	50

a i Name the independent variable investigated. (1)
  ii Name a variable that should have been controlled. (1)

b How does the water content of the grasses change throughout the 24-hour period? (1)

c Between which recorded times are more than 30% of the gemsboks feeding? (1)

d Suggest **three** reasons why the gemsboks benefit from feeding at this time. (3)

13.1	Pyramids of biomass

Learning objectives

After this topic, you should know:
- where biomass comes from
- how to construct a pyramid of biomass.

Figure 1 *Plants such as this sugar cane can produce a huge mass of biological material in just one growing season*

Figure 2 *The difference in the water content between fresh plants and dried plants makes a huge difference to the amount of biological material you appear to have*

Study tip

All the energy for life comes from the Sun's radiation.

Radiation from the Sun is the source of energy for most communities of living organisms on Earth. Light pours out continually onto the surface of the Earth. Green plants and algae absorb about 1% of this incident energy from light for photosynthesis. During photosynthesis, some of the energy is transferred to the chemical energy store of glucose molecules that are made. This energy is then stored in the substances that make up the cells of the plants and algae. This new material adds to the **biomass**.

Measuring biomass

Biomass is the mass of material in living organisms. Ultimately, almost all of the biomass on Earth is built up using energy from the Sun.

Biomass is often measured as the dry mass of biological material in grams. The main problem with measuring dry biomass is that you have to kill the living organisms to dry them out.

Wet biomass in grams can be measured instead. This does not involve killing the organisms, but this measurement is less reliable because the amount of water in living organisms can vary throughout the day and depending on conditions, so any results are less repeatable and reproducible than those for dry biomass.

The biomass made by plants is passed on through food chains or food webs. It goes into the animals that eat the plants. It then passes into the animals that eat other animals. No matter how long the food chain or how complex the food web, the original source of all the biomass involved is the Sun.

In a food chain, there are several **trophic levels**. There are usually more **producers** (plants) than **primary consumers** (herbivores). There are also usually more primary consumers than **secondary consumers** (carnivores). However, the number of organisms often does not accurately reflect what is happening to the biomass – the size of the organisms matters as well as the actual numbers. So, measuring the biomass produced is a useful way of looking at the feeding relationships between the different organisms.

Pyramids of biomass

The amount of biomass at each stage of a food chain is less than it was at the previous stage. You can draw the total amount of biomass in the living organisms in the trophic levels at each stage of the food chain. When this biomass is drawn to scale, you can show it as a **pyramid of biomass**.

Only about 10% of the biomass from each trophic level is transferred to the level above it because:
- Not all organisms or parts of organisms at one stage are eaten by the stage above. For example, parts such as plant roots or animal bones may be left behind.
- Some of the materials and energy taken in are passed out and lost in the waste materials of the organism.

- Cellular respiration supplies all the energy needs for living processes in an organism, including movement. Much of the energy is eventually transferred to the surroundings, warming them up. For example, when a herbivore eats a plant, lots of the plant biomass is used in respiration by the animal cells to release energy. Only a relatively small proportion of the plant material is used to build new herbivore biomass by making new cells, building muscle tissue, etc. This means that very little of the plant biomass eaten by the herbivore in its lifetime is available to be passed on to any carnivore that eats it.

So, at each stage of a food chain, the amount of energy in the biomass that is passed on gets less. A large amount of plant biomass supports a smaller amount of herbivore biomass. This in turn supports an even smaller amount of carnivore biomass.

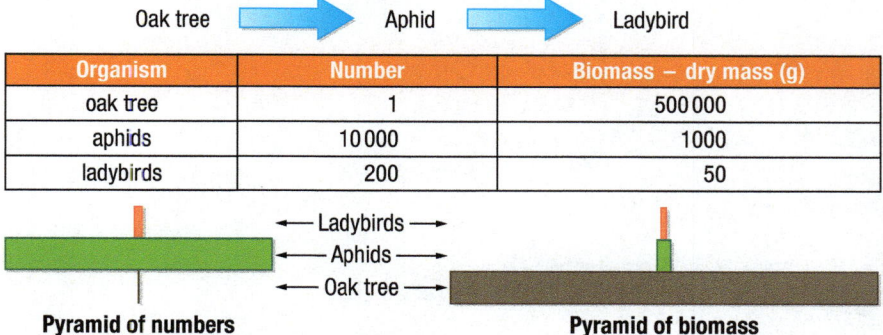

Organism	Number	Biomass – dry mass (g)
oak tree	1	500 000
aphids	10 000	1000
ladybirds	200	50

Pyramid of numbers **Pyramid of biomass**

Figure 4 *A pyramid of biomass is drawn to scale to represent the biomass of the organisms at each level of a food chain*

Figure 3 *Any food chain can be turned into a pyramid of biomass such as this. The blocks should always be drawn to scale*

Biomass of tertiary consumer (carnivore)
Biomass of secondary consumer (carnivore)
Biomass of primary consumer (herbivore)
Biomass of plants (producers)

Study tip

Remember that a pyramid of biomass gets smaller as you go up and that the plants go on the bottom step.

Summary questions

1 a What is biomass?
   b Why is a pyramid of biomass always drawn to scale?

2

Organism	Biomass – dry mass (g)
grass	100 000
sheep	5000
sheep ticks	30

   a Draw a pyramid of biomass for this grassland food chain.
   b Explain why the sheep ticks have so much less biomass than the grass cropped by the sheep.

3 Using the data in Figure 4, calculate the percentage biomass passed on from:
   a the producers to the primary consumers
   b the primary consumers to the secondary consumers.

Key points

- Radiation from the Sun is the source of energy for most living communities. Plants and algae transfer about 1% of the incident energy from light for photosynthesis. This energy is stored in the substances that make up the cells of the plants.

- The biomass at each stage can be drawn to scale and shown as a pyramid of biomass.

13.2 Energy transfers

After this topic, you should know:

- what happens to the material and energy in the biomass of organisms at each stage of a food chain

- how some of the energy is transferred to the environment.

The amounts of biomass and energy contained in living things get less as you progress up a food chain. As you have seen, only about 10% of the biomass from each trophic level is transferred to the level above. What happens to the rest?

Figure 1 *The amount of biomass in a lion is a lot less than the amount of biomass in the grass that feeds the zebra it preys on. But where does all the biomass go?*

Energy transfer in waste

The biomass that an animal eats is a source of energy, but not all of the energy can be used.

Firstly, herbivores cannot digest all of the plant material they eat. The material they can't digest is passed out of the body in faeces (Figure 2).

Figure 2 *Animals such as camels and horses produce a lot of dung, which is made up of all the biomass they can't digest*

The meat that carnivores eat is easier to digest than plants. This means that carnivores need to eat less often and produce less waste. However, as with herbivores, most carnivores cannot digest all of their prey. Hooves, claws, bones, and teeth are often indigestible. Therefore, some of the biomass that they eat is lost in their faeces.

When an animal eats more protein than it needs, the excess is broken down. It gets passed out as urea in the urine. This is another way in which biomass – and energy – are transferred from the body to the surroundings.

Energy transfer due to movement

Part of the biomass eaten by an animal is used for respiration in its cells. This supplies all the energy needs for the living processes taking place within the body, including movement.

Figure 3 *Sea anemones are animals that don't move much, so they don't need much to eat*

Movement needs a great deal of energy. The muscles use energy to contract and also get hot. So the more an animal moves about, the more energy (and biomass) it uses from its food.

Keeping a constant body temperature

Respiration supplies all the energy needed for living processes, including movement. Much of this energy is eventually transferred to the surroundings, warming them.

Energy transfers to the surroundings are particularly large in mammals and birds. This is because they use energy all the time to keep their bodies at a constant temperature (i.e., to keep warm when it's cold or to cool down when it's hot). So mammals and birds need to eat far more food than animals such as fish and amphibians to achieve the same increase in biomass.

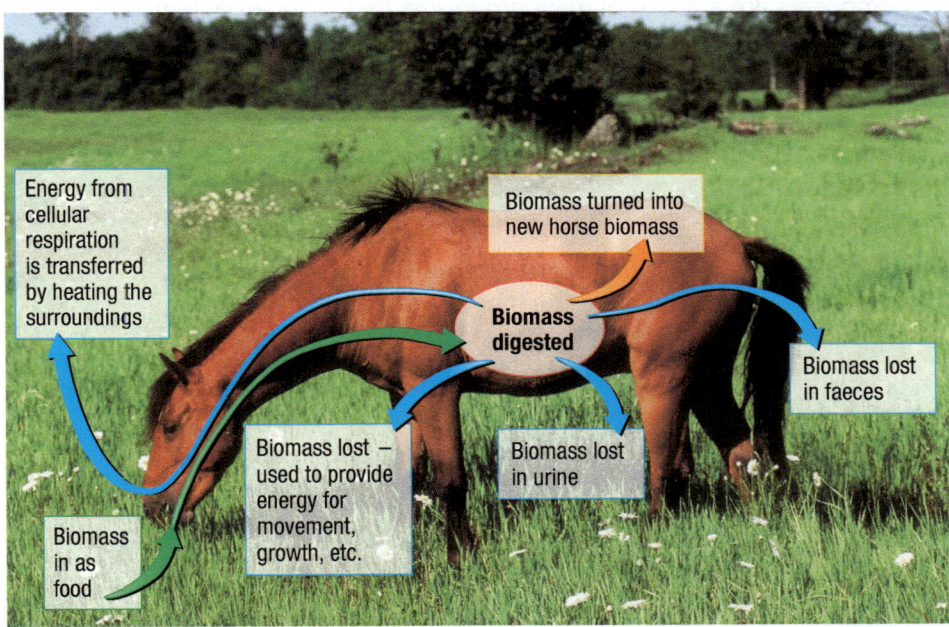

Figure 4 *Only between 2% and 10% of the biomass eaten by an animal such as this horse will get turned into new horse. The rest of the stored energy will be used for movement, transferred by heating the surroundings, or stored in waste materials*

Summary questions

1 **a** Why is biomass lost in faeces?

   b Why do animals that move around a lot use up more of the biomass they eat than animals that don't move much?

2 Explain why so much of the energy from the Sun that lands on the surface of the Earth is not turned into biomass in animals.

Key points

- Only about 10% of the biomass from each trophic level is transferred to the level above it because materials are lost and energy is transferred to the environment in the organism's waste material. Much of the energy from respiration is eventually transferred by heating the surroundings.

13.3 Decay processes

Living organisms remove materials from the environment for growth and other processes. For example, plants take nutrients from the soil all the time. These nutrients are passed on into animals through food chains and food webs. If this was a one-way process, the resources of the Earth would have been exhausted long ago.

Fortunately, all these materials are returned to the environment and recycled. For example, many trees shed their leaves each year, and most animals produce droppings at least once a day. Animals and plants eventually die as well. A group of organisms known as the **decomposers** then break down the waste and the dead animals and plants. In this process, decomposers return the nutrients and other materials to the environment. The same material is recycled over and over again. This process often leads to very stable communities of organisms.

The decay process

Decomposers are a group of microorganisms that include bacteria and fungi. They feed on waste droppings and dead organisms.

Detritus feeders, or **detritivores**, such as maggots and some types of worms, often start the process of decay. They eat dead animals and produce waste material. The bacteria and fungi then digest everything – dead animals, plants, and detritus feeders plus their waste. They use some of the nutrients to grow and reproduce. They also release waste products.

The waste products of decomposers are carbon dioxide, water, and nutrients that plants can use. When you say that things decay, they are actually being broken down and digested by microorganisms.

The decay process releases substances that plants need to grow. It makes sure that the soil contains the mineral ions that plants take up through their roots and use to make proteins and other chemicals in their cells. The decomposers also clean up the environment, removing the bodies of all the dead organisms.

Figure 1 *This orange is slowly being broken down by the action of decomposers. You can see the fungi clearly, but the bacteria are too small to be seen*

Conditions for decay

The speed at which things decay depends partly on the temperature. Chemical reactions in microorganisms, like those in most living things, work faster in warm conditions. They slow down and might even stop if conditions are too cold. Decay also stops if it gets too hot. The enzymes in the decomposers are denatured (change shape and stop working).

Most microorganisms also grow better in moist conditions. The moisture makes it easier for them to digest their food and also prevents them from drying out. So the decay of dead plants and animals – as well as leaves and dung – takes place far more rapidly in warm, moist conditions than it does in cold, dry conditions.

Although some microorganisms are anaerobic (can survive without oxygen), most decomposers respire aerobically. This means that they need oxygen to release energy, grow, and reproduce. This is why decay takes place more rapidly in aerobic conditions when there is plenty of oxygen available.

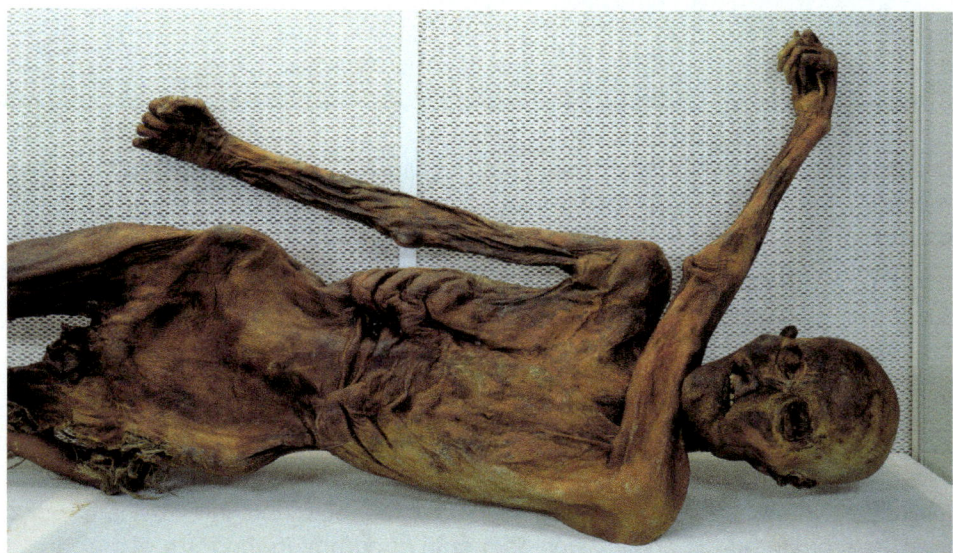

Figure 2 *Decomposers cannot function at low temperatures, so if an organism – such as this 4000-year-old man – is frozen as it dies, it will be preserved with very little decay*

The importance of decay in recycling

Decomposers are vital for recycling resources in the natural world. What's more, people can take advantage of the process of decay to help them recycle their waste.

In **sewage treatment plants**, microorganisms are used to break down the bodily waste humans produce. This makes the waste safe to release into rivers or the sea. These sewage works have been designed to provide the bacteria and other microorganisms with the conditions they need. This includes a good supply of oxygen.

Another place where the decomposers are useful is in the garden. Many gardeners have a **compost heap**. Grass cuttings, vegetable peelings, and weeds are put onto the compost heap. It is then left, to allow decomposing microorganisms to break all the plant material down. It forms a brown, crumbly substance known as compost, which can be used as a fertiliser. This may take weeks or months, depending on the temperature.

Practical

Investigating decay

Plan an investigation into the effect of temperature on how quickly things decay.

- Write a question that can be used as the title of this investigation.
- Identify the independent variable in the investigation.

Study tip

Decomposing microorganisms recycle all the molecules of life. Carbon goes into the atmosphere as carbon dioxide and mineral ions go into the soil to be used again by growing plants.

Key points

- Living things remove materials from the environment as they grow. These materials are returned to the environment either in waste materials or when living things die and decay.

- Materials decay because they are broken down (digested) by microorganisms. Microorganisms are more active and digest materials faster in warm, moist, aerobic conditions.

- The decay process releases substances that plants need to grow.

Summary questions

1. a What types of organisms are involved in the processes of decay?
   b Why are the processes of decay so important in keeping the soil fertile?

2. a Garden and kitchen waste added to a compost bin rots down and becomes compost much more rapidly in summer than in winter. Why is this?
   b During a particularly hot, dry summer, compost formation may slow down. Give a possible explanation for this.
   c Turning over the contents of a compost bin every so often can increase the rate at which decomposition takes place. Why is this?

13.4　The carbon cycle

Learning objectives

After this topic, you should know:

- what the carbon cycle is

- the processes that remove carbon dioxide from the atmosphere and return it again.

Figure 1 *Within the natural cycle of life and death in the living world, mineral nutrients are cycled between living organisms and the physical environment*

Imagine a stable community of plants and animals. The processes that remove materials from the environment are balanced by processes that return materials. Materials are constantly cycled through the environment. One of the most important of these materials is carbon.

All of the main **molecules** that make up our bodies (carbohydrates, proteins, fats, and DNA) are based on carbon atoms combined with other **elements**.

The amount of carbon on the Earth is fixed. Some of the carbon is 'locked up' in **fossil fuels** such as coal, oil, and gas. They are known as **carbon sinks**. The carbon is only released when you burn the fossil fuels.

Huge amounts of carbon are combined with other elements in carbonate rocks such as limestone and chalk. There is a pool of carbon in the form of carbon dioxide in the air. Carbon dioxide is also found dissolved in the water of rivers, lakes, and oceans. These things all act as carbon sinks. This stored carbon is described as 'sequestered'.

All the time, a relatively small amount of available carbon is cycled between living things and the environment. This constant cycling of carbon is called the **carbon cycle**.

Photosynthesis

Green plants and algae remove carbon dioxide from the atmosphere for photosynthesis. They use the carbon from carbon dioxide to make carbohydrates, proteins, and fats. These make up the biomass of the plants and algae. The carbon is passed on to animals that eat the green plants and algae. The carbon goes on to become part of the carbohydrates, proteins, and fats in these animal bodies. When these animals are eaten by other animals, some of the carbon becomes in turn the carbohydrates, fats, and proteins that make up their bodies.

This is how carbon is taken out of the environment. But how is it returned?

Respiration

Living organisms respire all the time. Plants, algae, and animals all use oxygen to break down glucose, providing energy for their cells. Carbon dioxide is produced as a waste product. This is how carbon is returned to the atmosphere.

When plants, algae, and animals die, their bodies are broken down by detritus feeders and decomposers. The animals which feed on dead bodies and waste are called detritus feeders. They include animals such as worms, centipedes, and many insects. The decomposers are the bacteria and fungi which complete the breakdown process

Carbon is released into the atmosphere as carbon dioxide when these organisms and microorganisms respire. All of the carbon (in the form of carbon dioxide) released by the various living organisms is then available again. It is ready to be taken up by plants and algae in photosynthesis.

Combustion

Wood from trees contains lots of carbon, locked into the molecules of the plant during photosynthesis over many years. Fossil fuels also contain lots of carbon, which was locked away by photosynthesising organisms millions of years ago.

When you burn wood or fossil fuels, carbon dioxide is produced, so you release some of that carbon back into the atmosphere. Huge quantities of fossil fuels are burnt worldwide to power our vehicles and to make electricity, whilst wood is burnt to heat homes and (in many countries) to cook food.

Photosynthesis: carbon dioxide + water **(+ energy)** → glucose + oxygen
Respiration: glucose + oxygen → carbon dioxide + water **(+ energy)**
Combustion: fossil fuel or wood + oxygen → carbon dioxide + water **(+ energy)**

The constant cycling of carbon in the carbon cycle is summarised in Figure 2.

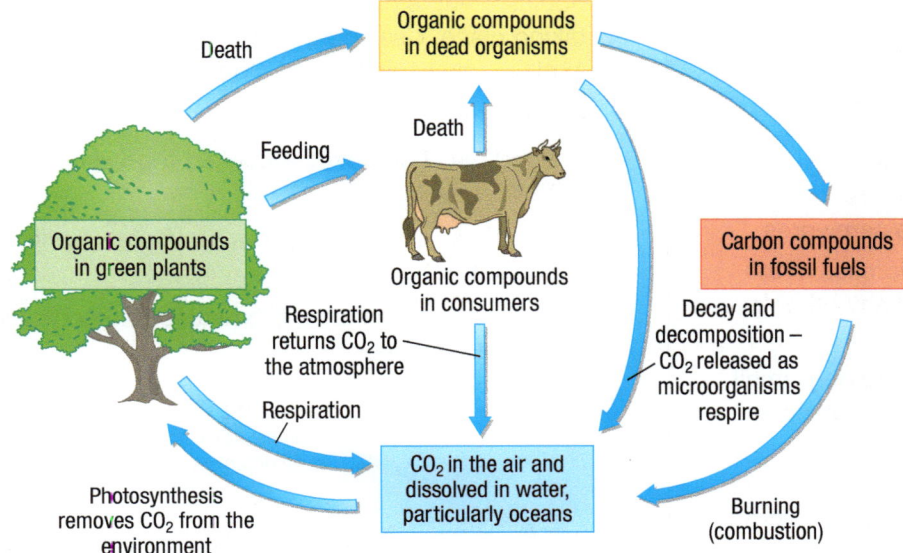

Figure 2 *The carbon cycle in nature*

Figure 3 *Burning wood and fossil fuels to keep us warm, power our cars, or make our electricity, all releases locked-in carbon in the form of carbon dioxide*

Energy transfers

When plants and algae photosynthesise, they transfer energy into the food they make. This chemical energy is transferred from one organism to another through the carbon cycle. Some of the energy can be used for movement or transferred by heating the organisms and their surroundings at each stage. The decomposers break down all the waste and dead organisms and cycle the materials as plant nutrients. By this time, all of the energy originally absorbed by green plants and algae during photosynthesis has been transferred elsewhere.

For millions of years, the carbon cycle has regulated itself. However, as people burn more fossil fuels they are pouring increasing amounts of carbon dioxide into the atmosphere. Scientists fear that the carbon cycle may not cope as the levels of carbon dioxide in our atmosphere increase, and it may lead to climate change.

Summary questions

1 a What is the carbon cycle?
  b What are the main processes involved in the carbon cycle?
  c Why is the carbon cycle so important for life on Earth?

2 a Where does the carbon come from that is used in photosynthesis?
  b Explain carefully how carbon is transferred through an ecosystem.

3 Explain the links between the processes of photosynthesis, respiration, and combustion, and describe the role of each process in the carbon cycle.

Key points

- The constant cycling of carbon in nature is called the carbon cycle.

- Carbon dioxide is removed from the atmosphere by photosynthesis in plants and algae. It is returned to the atmosphere through respiration of all living organisms, including the decomposers, and through combustion of wood and fossil fuels.

Summary questions

1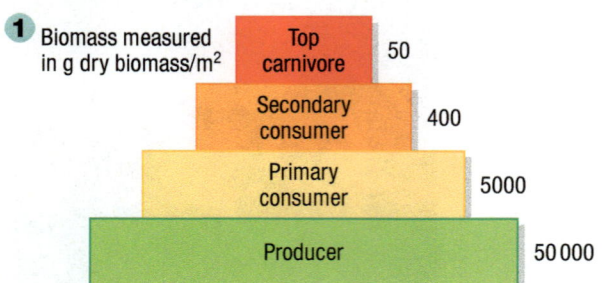

Biomass measured in g dry biomass/m^2

Top carnivore 50
Secondary consumer 400
Primary consumer 5000
Producer 50 000

Figure 1

a Use the information in Figure 1 to calculate the percentage biomass passed on:

   i from producers to primary consumers

   ii from primary to secondary consumers

   iii from secondary consumers to top carnivores.

b In any food chain or food web, the biomass of the producers is much larger than that of any other level of the pyramid. Why is this?

c In any food chain or food web, there are only small numbers of top carnivores. Use your calculations from **a** to help you explain why.

d All of the animals in the pyramid of biomass shown here are unable to maintain a constant warm body temperature. What difference would it have made to the average percentage of biomass passed on between the levels if mammals and birds had been involved? Explain the difference.

2 The world population is increasing and there are food shortages in many parts of the world. Explain, using pyramids of biomass to help you, why there would be more efficient use of resources if people everywhere consumed much less meat and more plant material.

3 Chickens are often farmed intensively to provide meat as cheaply as possible. The birds arrive in the shed where they will be reared as 1-day-old chicks. They are slaughtered at 42 days of age when they weigh about 2 kg. The temperature, amount of food and water, and light levels are carefully controlled. About 20 000 chickens are reared together in one house. The table below shows their weight gain.

Age (days)	1	7	14	21	28	35	42
Mass (g)	36	141	404	795	1180	1657	1998

a Draw a graph to show the growth rate of these chickens.

b The chickens are fed much more than 2 kg of food during their lives. Explain why the body mass they gain is less than the mass of the food they take in.

c Suggest reasons why the amount of food, water given to the hens and the temperature of the sheds is so carefully controlled.

d Why are birds for eating reared like this?

e Draw a second line on the graph drawn in **a** to show how you would expect a chicken reared outside in a free-range system to gain in mass, and explain the difference.

4 Microorganisms decompose organic waste and dead bodies. People preserve food to stop this decomposition taking place. Use your knowledge of decomposition to explain how each method stops food going bad:

a Food may be frozen.

b Food may be cooked – cooked food keeps longer than fresh food.

c Food may be stored in a vacuum pack – with all the air sucked out.

d Food may be tinned – it is heated and sealed in an airtight container.

5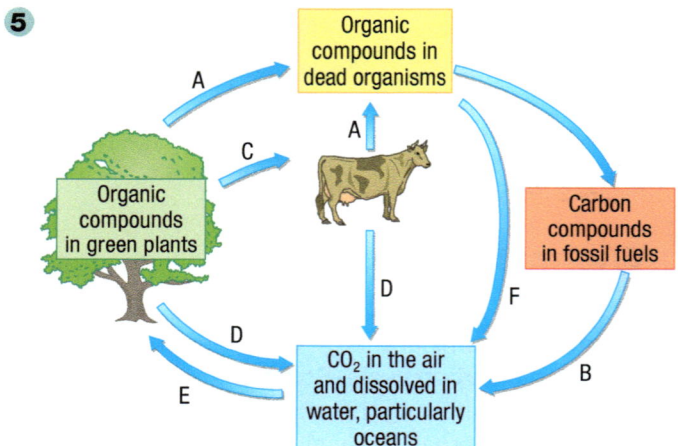

Figure 2

a How is carbon dioxide removed from the atmosphere in the carbon cycle?

b How does carbon dioxide get into the atmosphere?

c Where is most of the carbon stored?

d Why is the carbon cycle so important, and what could happen if the balance of the reactions was disturbed?

e List each of the processes labelled A–F in Figure 2.

6 a The temperature in the middle of a compost heap will be quite warm. Energy is transferred as microbes respire. How does this help the compost to be broken down more quickly?

b In sewage works, oxygen is bubbled through the tanks containing sewage and microorganisms. How does this help to ensure that human waste is broken down completely?

Practice questions

1 A woodland habitat contained the following:

- 40 trees
- 10 000 caterpillars, eating the leaves
- 350 birds, eating the caterpillars.

a Draw and label a pyramid of biomass for this woodland habitat. (3)

b Name the source of energy for the habitat. (1)

c Explain how this energy is captured, converted into chemical energy, and transferred to chemical components in the bodies of the caterpillars. (6)

d Scientists estimated the amount of energy contained in each layer of the pyramid.

The results are shown in the following table.

Organism	Energy in MJ
Trees	5 000 000
Caterpillars	20 000
Birds	1000

i Calculate the percentage of energy present in caterpillars that is transferred to the birds. (2)

ii Suggest **two** reasons why not all the energy present in the caterpillars is transferred to the birds. (2)

e *In this question you will be assessed on using good English, organising information clearly, and using specialist terms where appropriate.*

In autumn, the leaves fall from the trees.

Describe how the carbon in the dead leaves is recycled so that the trees can use it again. (6)

2 Figure 1 shows what happens to the energy in the food a calf eats.

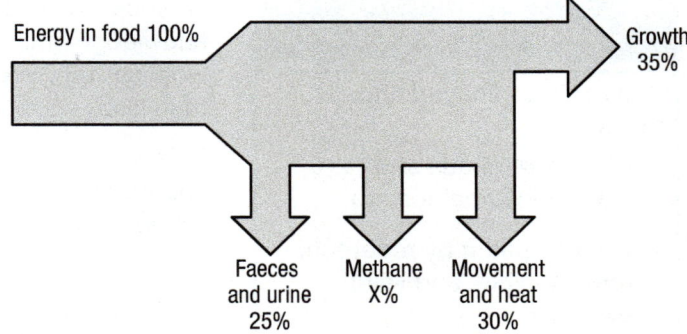

Energy in food 100%

Growth 35%

Faeces and urine 25%

Methane X%

Movement and heat 30%

Figure 1

In the calculations for the questions below, show clearly how you work out your answer.

a Calculate the percentage of energy transferred to the environment in methane (X). (2)

b The energy in the food the calf eats in one day is 10 megajoules.

Calculate the amount of this energy that would be transferred to the environment in faeces and urine. (2)

c Name the process that transfers the energy from the food into movement. (1)

d The farmer decides to move his calf indoors so that it will grow more quickly.

Suggest **two** reasons why. (2)

e The farmer's wife says she does not think that this is a good idea. Suggest a reason why. (1)

Investigations

Science works for us all day, every day. Working as a scientist you will have knowledge of the world around you and particularly about the subject you are working with. You will observe the world around you. An enquiring mind will then lead you to start asking questions about what you have observed.

Science usually moves forward by slow, steady steps. Each small step is important in its own way. It builds on the body of knowledge that scientists already have.

Thinking scientifically

Deciding on what to measure

Variables can be one of two different types:

- A **categoric variable** is one that is best described by a label (usually a word). The colour of eyes is a categoric variable (e.g., blue or brown eyes).
- A **continuous variable** is one that you measure, so its value could be any number. Temperature (as measured by a thermometer or temperature sensor) is a continuous variable (e.g., 37.6 °C, 45.2 °C). Continuous variables can have values (called a quantity) that can be given by any measurements made (e.g., light intensity, flow rate, etc.).

When designing an investigation you should always try to measure continuous **data** whenever you can. If this is not always possible, you should then try to use ordered data. If there is no other way to measure your variable then you have to use a label (categoric variable).

Making your investigation repeatable, reproducible, and valid

When you are designing an investigation you must make sure that others can repeat any results you get – this makes it **reproducible**. You should also plan to make each result **repeatable**. You can do this by getting consistent sets of repeat measurements.

You must also make sure that you are measuring the actual thing you want to measure. If you don't, your data can't be used to answer your original question. This seems very obvious, but it is not always quite so easy. You need to make sure that you have controlled as many other variables as you can, so that no one can say that your investigation is not **valid**.

How might an independent variable be linked to a dependent variable?

The **independent variable** is the one you choose to vary in your investigation.

The **dependent variable** is used to judge the effect of varying the independent variable.

These variables may be linked together. If there is a pattern to be seen (e.g., as one thing gets bigger the other also gets bigger), it may be that:

- changing one has caused the other to change
- the two are related, but one is not necessarily the cause of the other.

Learning objectives

After this topic, you should know:

- what continuous and categoric variables are
- what is meant by repeatable, reproducible, and valid evidence
- what the link is between the independent and dependent variables
- what a hypothesis and a prediction are
- about risks in hazardous situations.

Starting an investigation

Observation

As a scientist you use observations to ask questions. You can only ask useful questions if you know something about the observed event. You will not have all of the answers, but you know enough to start asking the correct questions.

When you are designing an investigation you have to observe carefully which variables are likely to have an effect.

What is a hypothesis?

A **hypothesis** is an idea based on observation that has some really good science to try to explain it.

When making hypotheses you can be very imaginative with your ideas. However, you should have some scientific reasoning behind those ideas so that they are not totally bizarre.

Remember, your explanation might not be correct, but you think it is. The only way you can check out your hypothesis is to make it into a prediction and then test it by carrying out an investigation:

observation + knowledge → hypothesis → prediction → investigation

Starting to design an investigation

An investigation starts with a question, followed by a **prediction**. You, as the scientist, predict that there is a relationship between two variables.

You should think about a preliminary investigation to find the most suitable range and interval for the independent variable.

Making your investigation safe

Remember that when you design your investigation, you must:
- look for any potential **hazards**
- decide how you will reduce any **risk**.

You will need to write these down in your plan:
- write down your plan
- make a risk assessment
- make a prediction
- draw a blank table ready for the results.

Study tip

Observations, backed up by creative thinking and good scientific knowledge, can lead to a hypothesis.

Key points

- Continuous data can give you more information than other types of data.
- You must design investigations that produce repeatable, reproducible, and valid results if you are to be believed.
- Be aware that just because two variables are related, this does not mean that there is a causal link.
- Hypotheses can lead to predictions and investigations.
- You must make a risk assessment, make a prediction, and write a plan.

Setting up investigations

Learning objectives

After this topic, you should know:

- what a fair test is
- how a survey is set up
- what a control group is
- how to decide on variables, range, and intervals
- how to ensure accuracy and precision
- the causes of error and anomalies.

Study tip

Trial runs will tell you a lot about how your investigation might work out. They should get you to ask yourself:

- Do you have the correct conditions?
- Have you chosen a sensible range?
- Have you got enough readings that are close together?
- Will you need to repeat your readings?

Study tip

A word of caution!

Just because your results show precision does not mean that your results are accurate.

Imagine that you carry out an investigation into the energy value of a type of crisp. You get readings of the amount of energy released that are all about the same. This means that your data will have precision, but it doesn't mean that they are necessarily accurate.

Fair testing

A **fair test** is one in which only the independent variable affects the dependent variable. All other variables are controlled.

This is easy to set up in the laboratory, but almost impossible in fieldwork. Plants and animals do not live in environments that are simple and easy to control. They live complex lives with variables changing constantly.

So how can you set up the fieldwork investigations? The best you can do is to make sure that all of the many variables change in much the same way, except for the one that you are investigating. Then at least the plants get the same weather, for example, even if it is constantly changing.

If you are investigating two variables in a large population then you will need to do a survey. Again, it is impossible to control all of the variables. Imagine if scientists were investigating the effect of diet on diabetes. They would have to choose people of the same age and family history to test. The larger the **sample size** tested, the more valid the results would be.

Control groups are used in these investigations to try to make sure that you are measuring the variable that you intend to measure. For example, when investigating the effects of a new drug, the control group will be given a **placebo**. The control group think they are taking the test drug, but the placebo does not contain the drug. In this way you can control the variable of 'thinking that the drug is working' and separate out the actual effect of the drug.

Designing an investigation

Accuracy

Your investigation must provide **accurate** data. Accurate data is essential if your results are going to have any meaning.

How do you know if you have accurate data?

It is very difficult to be certain. Accurate results are very close to the true value. It is not always possible to know what that true value is.

- Sometimes you can calculate a theoretical value and check it against the experimental evidence. Close agreement between these two values could indicate accurate data.
- You can draw a graph of your results and see how close each result is to the line of best fit.
- Try repeating your measurements with a different instrument and see if you get the same readings.

How do you get accurate data?

- Using instruments that measure accurately will help.
- The more carefully you use the measuring instruments, the more accuracy you will get.

Precision

Your investigation must provide data with sufficient precision. If it doesn't, you will not be able to make a valid conclusion.

How do you get precise and repeatable data?

- You have to repeat your tests as often as necessary to improve repeatability.
- You have to repeat your tests in exactly the same way each time.
- Use measuring instruments that have the appropriate scale divisions needed for a particular investigation. Smaller scale divisions have better resolution.

Making measurements

Using instruments

You cannot expect perfect results. When you choose an instrument, you need to know that it will give you the accuracy that you want (i.e., that it will give you a true reading).

When you choose an instrument, you need to decide how precise you need to be. Some instruments have smaller scale divisions than others. Instruments that measure the same thing can have different sensitivities. The resolution of an instrument refers to the smallest change in a value that can be detected. Choosing the wrong scale can cause you to miss important data or make silly conclusions.

You also need to be able to use an instrument properly.

Errors

Even when an instrument is used correctly, the results can still show differences. Results may differ because of a **random error**. This is most likely to be due to a poor measurement being made. It could be due to not carrying out the method consistently.

The error may be a **systematic error**. This means that the method was carried out consistently, but an error was being repeated.

Anomalies

Anomalies are results that are clearly out of line. They are not those that are due to the natural variation that you get from any measurement. These should be looked at carefully. There might be a very interesting reason why they are so different. If they are simply due to a random error then they should be ignored.

If anomalies can be identified whilst you are doing an investigation, then it is best to repeat that part of the investigation. If you find anomalies after you have finished collecting the data for an investigation, the anomalous results must be discarded.

Did you know ... ?

Imagine measuring the temperature after a set time when a sugar is used to heat a fixed volume of water.

Two students repeated this experiment, four times each. Their results are marked on Figure 1 below:

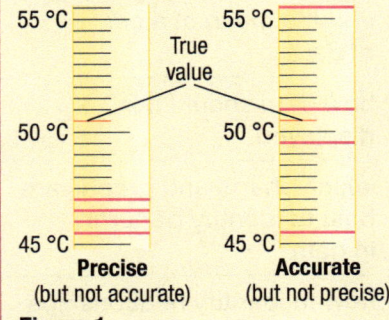

Figure 1

- A **precise** set of results is grouped closely together.
- An accurate set of results will have a mean (average) close to the true value.

Key points

- Care must be taken to ensure fair testing.
- You can use a trial run to make sure that you choose the best values for your variables.
- Careful use of the correct equipment can improve accuracy.
- If you repeat your results carefully you can improve precision.
- Results will nearly always vary. Better instruments give more accurate results.
- Resolution in an instrument is the smallest change that it can detect.
- Human error can produce random and systematic errors.
- You must examine anomalies.

Using data

After this topic, you should know:

- what is meant by the range and the mean of a set of data

- how data should be displayed

- which charts and graphs are best to identify patterns in data

- how to identify relationships within data

- how scientists draw valid conclusions from relationships

- how to evaluate the reproducibility of an investigation.

Presenting data

Tables

Tables are really good for getting your results down quickly and clearly. You should design your table before you start your investigation.

The range of the data

Pick out the maximum and the minimum values and you have the range. You should always quote these two numbers when asked for a range. For example, the range is between … (the lowest value) and … (the highest value) Don't forget to include the units!

The mean of the data

Add up all of the measurements and divide by how many there are.

Bar charts

If you have a categoric independent variable and a continuous dependent variable you should use a **bar chart**.

Line graphs

If you have a continuous independent and a continuous dependent variable you should use a **line graph**.

Scatter graphs

These are used in much the same way as a line graph, but you might not expect to be able to draw such a clear line of best fit. For example, if you want to see if lung capacity is related to how long people can hold their breath, you might draw a scatter graph of your results.

Using data to draw conclusions

Identifying patterns and relationships

Now that you have a bar chart or a graph of your results, you can begin looking for patterns in your results. You must have an open mind at this point.

Firstly, there could still be some anomalous results. You might not have picked these out earlier. How do you spot an anomaly? It must be a significant distance away from the pattern, not just within normal variation.

A line of best fit will help to identify any anomalies at this stage. Ask yourself – do the anomalies represent something important, or were they just a mistake?

Secondly, remember that a line of best fit can be a straight line or it can be a curve – you have to decide based on your results.

The line of best fit will also lead you to consider what the relationship is between your two variables. You need to consider whether your graph shows a **linear relationship**. This simply means asking yourself whether you can you be confident about drawing a straight line of best fit on your graph. If the answer is yes, is this line positive or negative?

A **directly proportional** relationship is shown by a positive straight line that goes through the origin (0, 0).

Your results might also show a curved line of best fit. These can be predictable, complex, or very complex!

Drawing conclusions

Your graphs are designed to show the relationship between your two chosen variables. You need to consider what that relationship means for your conclusion. You must also take into account the repeatability and the validity of the data you are considering.

You must continue to have an open mind about your conclusion.

You will have made a prediction. This could be supported by your results, it might not be supported, or it could be partly supported. It might suggest some other hypothesis to you.

You must be willing to think carefully about your results. Remember that it is quite rare for a set of results to completely support a prediction and to be completely repeatable.

Look for possible links between variables. It may be that:
- changing one has caused the other to change
- the two are related, but one is not necessarily the cause of the other.

You must decide which is the most likely. Remember that a positive relationship does not always mean a causal link between the two variables.

Your conclusion must go no further than the **evidence** that you have. Any patterns you spot are only strictly valid with in the range of values that you tested. Further tests are needed to check whether the pattern continues beyond this range.

The purpose of the prediction was to test a hypothesis. The hypothesis can:
- be supported
- be refuted, or
- lead to another hypothesis.

You have to decide which it is using the evidence available.

Evaluation

If you are still uncertain about a conclusion, it might be down to the repeatability, reproducibility, and validity of the results. You could check reproducibility by:
- looking for other similar work on the Internet or from others in your class
- getting somebody else to redo your investigation
- trying an alternative method to see whether you get the same results.

Key points

- The range states the maximum and the minimum value.
- The mean is the sum of the values divided by how many values there are.
- Tables are best used during an investigation to record results.
- Bar charts are used when you have a categoric independent variable and a continuous dependent variable.
- Line graphs are used to display data that are continuous.
- Drawing lines of best fit help us to study the relationship between variables. The possible relationships are: linear, positive, and negative; directly proportional; predictable; and complex curves.
- Conclusions must go no further than the data available.
- The reproducibility of data can be checked by looking at other similar work done by others, perhaps on the Internet. It can also be checked by using a different method or by others checking your method.

In fish and chip shops, potatoes are cut into chips several hours before they are cooked.

The mass of water in the chips must be kept constant during this time.

To keep the water in the chips constant, the chips are kept in sodium chloride solution.

1 Figure 1 shows some apparatus and materials.

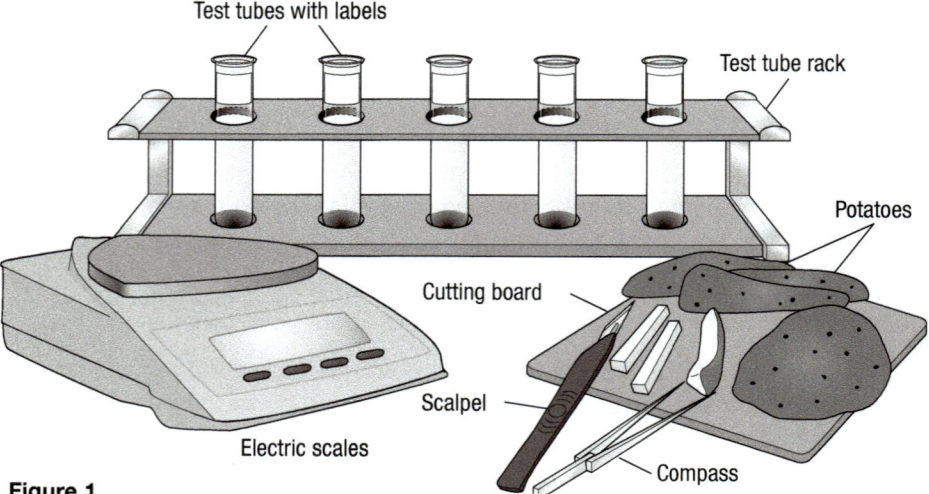

Figure 1

In this question you will be assessed on using good English, organising information clearly, and using scientific terms where appropriate.

Describe how you would use the apparatus and materials shown in Figure 1 to find the concentration of sodium chloride in which to keep the chips so that the mass of water in the chips remains constant.

You should include:
– the measurements you would make
– how you would make the investigation a fair test.

First, I would cut up the potato using the scalpel so that I had chips exactly the same cross-section (say 5 mm × 5 mm) and length (say 5 cm). I would then dilute some 1.0 mol/dm³ sodium chloride solution with distilled water to give 0.1, 0.2, 0.3, 0.4, 0.5, 0.6, 0.7, 0.8 mol/dm³ solutions in separate test tubes and label them. I would use pure water as well. I would use forceps to blot each chip dry, find its mass on the balance, and record the mass in a table. I would put one chip in each test tube, making sure I had the correct mass opposite each solution in my table.
I would leave them all for 24 hours. Next day, I would remove each one with forceps, blot it dry as before and find its mass. I would record the new mass in the next column of my table.
The concentration of sodium chloride where there was least change in mass would be the best one to use.

(6)

This answer would score 6 marks. The candidate has included all the necessary science points, has used appropriate scientific terms for the apparatus and materials, and has described a logical and detailed method.

In a similar investigation, a student investigated the effect of the concentration of sodium chloride solution on standard-sized cylinders cut from a potato.

Table 1 shows the student's results.

	Concentration of sodium chloride solution (mol/dm³)					
	0	0.2	0.4	0.6	0.8	1.0
Change in length of cylinders (mm)	+4.1	+1.5	−1.4	−3.6	−4.6	−5.2

2   On the graph paper below, draw a graph to display the student's results.
   – Add a suitable scale and label to the *y*-axis.
   – Plot the student's results.
   – Draw a line of best fit.

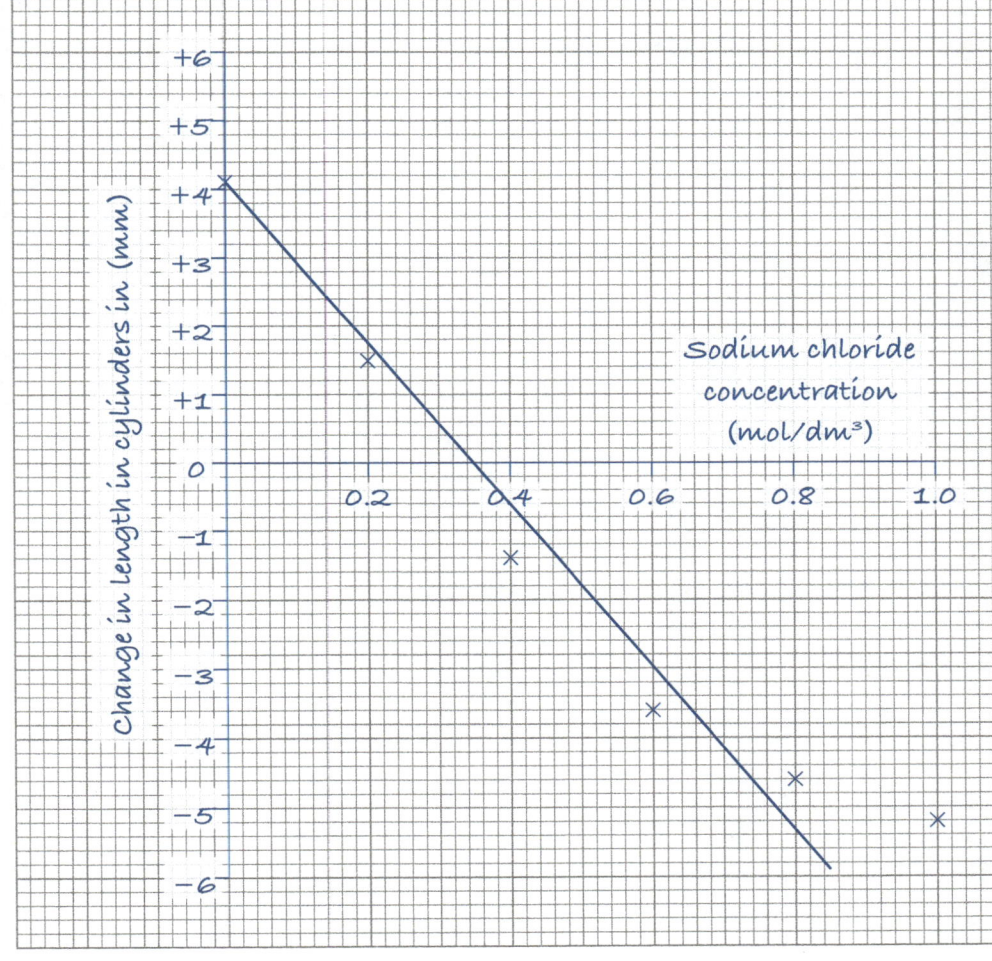

The candidate would score 3 marks, 1 for correct scale and label on the *y*-axis, and 2 marks for correct plots. However, the line of best fit is not close enough, as it should show a smooth curve.

(4)

3 In which concentration of sodium chloride would the chips *not* change mass?

Concentration 0.35 mol/dm³

This answer gains 1 mark as it is correct for the candidate's line.

(1)

4 Explain the changes in length of potato cylinders that were placed in the 1 mol/dm³ sodium chloride solution.

The cylinders shrank by 5.2 mm. This is because the liquid outside the cells had a higher concentration of salt than the liquid inside the cells and so water moved out of the cells by osmosis. This made the cells and the cylinders shrink in length.

(3)

The candidate would gain 2 marks for correctly identifying the concentration gradient and for knowing that therefore water moves out of the cells into the liquid outside. However, they did not gain the final point, which would note that this is able to happen because the cell membranes are partially permeable.

Glossary

A

Abdomen The lower region of the body. In humans it contains the digestive organs, kidneys, etc.

Accurate Describes a measurement judged to be close to the true value.

Acid rain Rain that is acidic due to dissolved gases, such as sulfur dioxide, produced by the burning of fossil fuels.

Active site Site on an enzyme where the reactants bind.

Active transport Movement of substances against a concentration gradient and/or across a cell membrane, using energy.

Adaptation Special feature that makes an organism particularly well-suited to the environment where it lives.

ADH Anti-diuretic hormone secreted by the pituitary gland in the brain that affects the amount of water lost through the kidneys in the urine.

Adult cell cloning Process in which the nucleus of an adult cell of one animal is fused with an empty egg from another animal. The resulting embryo is placed inside the uterus of a third animal to develop.

Adult stem cell Stem cell (cells with the potential to differentiate and form a variety of other cell types) that is found in small quantities in adult tissues.

Aerobic respiration Breaking down food using oxygen to release energy for the cells.

Agar Nutrient jelly on which many microorganisms are cultured.

Agglutinate Stick together.

Agriculture Growing plants or other organisms on farms to supply human needs (e.g., for food, clothing, etc).

Algae Single-celled or simple multicellular organisms that can photosynthesise but are not plants.

Algal cell The cells of algae, single-celled, or simple multicellular organisms, which can photosynthesise but are not plants.

Allele Version of a particular gene.

Alveoli Tiny air sacs in the lungs which increase the surface area for gaseous exchange.

Amino acid Building block of protein. Protease enzymes break down proteins into amino acids.

Amylase Enzyme made in the salivary glands, pancreas, and small intestine, which speeds up the breakdown of starch into simple sugars.

Anaerobic respiration Breaking down food without oxygen to release energy for the cells.

Anomalous result Result that does not match the pattern seen in the other data collected or is well outside the range of other repeat readings. It should be retested and, if necessary, discarded.

Antibiotic Drug that destroys bacteria inside the body without damaging human cells.

Antibodies Proteins made by white blood cells which bind to specific antigens.

Antigen Unique protein on the surface of a cell. It is recognised by the immune system as 'self' or 'non-self'.

Aorta Main artery leaving the left ventricle carrying oxygenated blood to the body.

Artery Blood vessel which carries blood away from the heart. It usually carries oxygenated blood and it has a pulse.

Artificial pacemaker Electrical device which can be implanted to act as pacemaker for the heart when the natural pacemaker region fails.

Asexual reproduction Reproduction that involves only one individual with no fusing of gametes to produce the offspring. The offspring are genetically identical to the parent.

Atrium Small upper chambers of the heart. The right atrium receives blood from the body and the left atrium receives blood from the lungs.

B

Bacteria Single-celled microorganisms that can reproduce very rapidly. Many bacteria are useful (e.g., gut bacteria and decomposing bacteria), but some cause disease.

Bacterial colony Population of billions of bacteria grown in culture.

Bar chart Chart with rectangular bars with lengths proportional to the values that they represent. The bars should be of equal width and are usually plotted horizontally or vertically. Also called a bar graph.

Bases Nitrogenous compounds that make up part of the structure of DNA.

Benign tumour Tumour that grows in one location and does not invade other tissues.

Biconcave disc The shape of red blood cells – a disc which is dimpled inwards on both sides.

Bile Yellowy-green liquid made in the liver and stored in the gall bladder. It is released into the small intestine and emulsifies fats.

Biodiversity The number and variety of different organisms found in a specified area.

Biofuel Fuel produced from biological material which is renewable and sustainable.

Biological detergent Washing detergent that contains enzymes.

Biomass Biological material from living or recently living organisms.

Bladder Organ where urine is stored until it is released from the body.

Blood circulation system System by which blood is pumped around the body.

Blood vessel Tube which carries blood around the body (i.e., arteries, veins, and capillaries).

Blood Liquid which is pumped around the body by the heart. It contains blood cells, dissolved food, oxygen, waste products, mineral ions, hormones, and other substances needed in the body or needing to be removed from the body.

Breathing The physical movement of air into and out of the lungs. In humans this is brought about by the action of the intercostal muscles on the ribs and the diaphragm.

C

Callus Mass of unspecialised plant tissue.

Cancer Common name for a malignant tumour.

Capillary Smallest type of blood vessels which run between individual cells. Capillaries have a wall which is only one cell thick.

Carbohydrate Molecules which provide us with energy. Carbohydrates contain the chemical elements carbon, hydrogen, and oxygen and are made up of single sugar units.

Carbon cycle The cycling of carbon through the living and non-living world.

Carbon sink Something that takes up more carbon dioxide than it produces (e.g., plants, the oceans).

Carcinogen Chemical that can cause mutations in cells and so trigger the formation of malignant tumours.

Carnivore Animal that eats other animals.

Carrier Individual that is heterozygous for a faulty allele that causes a genetic disease in the homozygous form.

Catalyst Substance which speeds up a chemical reaction. At the end of the reaction the catalyst remains chemically unchanged.

Categoric variable See Variable – categoric.

Cell cycle The sequence of events by which cells grow and divide.

Cell membrane Membrane around the contents of a cell which controls what moves in and out of the cell.

Cell wall Rigid structure which surrounds the cells of living organisms apart from animals.

Cellulose Large carbohydrate molecule which makes up plant and algal cell walls.

Central nervous system (CNS) Made up of the brain and spinal cord where information is processed.

Charles Darwin Victorian scientist who developed the theory of evolution by a process of natural selection.

Chemotherapy Treatment in which chemicals are used to either stop cancer cells dividing or to make them self-destruct.

Chlorophyll Green pigment contained in the chloroplasts.

Chloroplast Organelle in which photosynthesis takes place.

Chromosome Thread-like structure carrying the genetic information found in the nucleus of a cell.

Classification The organisation of living things into groups according to their similarities.

Clone Offspring produced by asexual reproduction which is genetically identical to its parent organism.

Cloning The production of offspring which are genetically identical to the parent organism.

Combustion The process of burning.

Competition The process by which living organisms compete with each other for limited resources such as food, light, or reproductive partners.

Complex carbohydrate Carbohydrate made up of long chains of single sugar units (e.g., starch, cellulose).

Compost heap A site where garden rubbish and kitchen waste are decomposed by microorganisms.

Concentration gradient Gradient between an area where a substance is at a high concentration and an area where it is at a low concentration.

Continuous variable See Variable – continuous.

Control group If an experiment is to determine the effect of changing a single variable, a control is often set up in which the independent variable is not changed, therefore enabling a comparison to be made. If the investigation is of the survey type, a control group is usually established to serve the same purpose.

Control variable See Variable – control.

Core body temperature Internal temperature of the body.

Coronary artery Artery which carries oxygenated blood to the muscle of the heart.

Coronary heart disease Heart disease caused by problems with the coronary arteries that supply the heart muscle with oxygenated blood.

Culture medium Substance containing the nutrients needed for microorganisms to grow.

Cuticle Waxy covering of a leaf (or an insect) which reduces water loss from the surface.

Cystic fibrosis Genetic disease that affects the lungs, digestive, and reproductive systems. It is inherited through a recessive allele.

Cytoplasm Water-based gel in which the organelles of all living cells are suspended.

D

Data Information, either qualitative or quantitative, that has been collected.

Decomposer Microorganism that breaks down waste products and dead bodies.

Deforestation Removal of forests by felling, burning, etc.

Denatured Shape of an enzyme has been changed so that it can no longer speed up a reaction.

Deoxygenated Lacking in oxygen.

Dependent variable See Variable – dependent.

Detritivore Organism that feeds on organic waste from animals and the dead bodies of animals and plants.

Detritus feeder A detritus feeder is the same as a detritivore. Detritivores are larger animals that feed on dead plants, dead animals, and their wastes.

Dialysis machine Machine used to remove urea and excess mineral ions from the blood when the kidneys fail.

Dialysis The process of cleansing the blood through a dialysis machine when the kidneys have failed.

Diaphragm A strong sheet of muscle that separates the thorax from the digestive organs, used to change the volume of the chest during ventilation of the lungs.

Differentiated Specialised for a particular function.

Diffusion The net movement of particles of a gas or a solute from an area of high concentration to an area of low concentration (down a concentration gradient).

Digested Broken down into small molecules by the digestive enzymes.

Digestive juices The mixture of enzymes and other chemicals produced by the digestive system.

Digestive system Organ system where food is digested, running from the mouth to the anus.

Direct contact Means of spreading infectious diseases by skin contact between two people.

Directly proportional Relationship that, when drawn on a line graph, shows a positive linear relationship that crosses through the origin.

DNA fingerprint Pattern produced by analysing DNA which can be used to identify an individual.

DNA Deoxyribonucleic acid, the material of inheritance.

Dominant Describes a characteristic that will show up in the offspring even if only one of the alleles is inherited.

Donor Person who gives material from their body to another person who needs healthy tissues or organs (e.g., blood, kidneys). Donors may be alive or dead.

Double circulatory system The separate circulation of the blood from the heart to the lungs and then back to the heart and on to the body.

Droplet infection Means of spreading infectious diseases through tiny droplets full of pathogens, which are expelled from your body when you cough, sneeze, or talk.

Drug Chemical which causes changes in the body. Medical drugs cure disease or relieve symptoms. Recreational drugs alter the state of your mind and/or body.

E

Ecology The scientific study of the relationships between living organisms and their environment.

Effector organ Muscles or glands which responds to impulses from the nervous system.

Electron microscope Instrument used to magnify specimens using a beam of electrons.

Element Substance made up of only one type of atom. An element cannot be broken down chemically into any simpler substance.

Embryonic stem cell Stem cell with the potential to form a number of different specialised cell types, which is taken from an early embryo.

Emulsifies Breaks down into tiny droplets which will form an emulsion.

Endemic When a species evolves in isolation and is found in only one place in the world– it is said to be endemic (particular) to that area.

Environmental cause External, not inherited, condition that affects the way in which characteristics of organisms develop.

Environmental isolation This occurs when the climate changes in one area where an organism lives but not in others.

Enzyme Protein molecule which acts as a biological catalyst. It changes the rate of chemical reactions without being affected itself at the end of the reaction.

Epidermal tissue Tissue of the epidermis – the outer layer of an organism.

Epithelial tissue Tissue made up of relatively unspecialised cells which covers and lines some parts of the body.

Error – human Often present in the collection of data, may be random or systematic. For example, the effect of human reaction time when recording short time intervals with a stopwatch.

Error – random Causes readings to be spread about the true value, due to results varying in an unpredictable way from one measurement to the next. Random errors are present when any measurement is made and cannot be corrected. The effect of random errors can be reduced by making more measurements and by calculating a new mean.

Error – systematic Causes readings to be spread about some value other than the true value, due to results differing from the true value by a consistent amount each time a measurement is made. Sources of systematic error can include the environment, methods of observation, or instruments used. Systematic errors cannot be dealt with by simple repeats. If a systematic error is suspected, the data collection should be repeated using a different technique or a different set of equipment, and the results compared.

Eutrophication Process by which excessive nutrients in water lead to very fast plant growth. When the plants die they are decomposed, which uses up a lot of oxygen so the water can no longer sustain animal life.

Evaporation The change of a liquid to a vapour at a temperature below its boiling point.

Evidence Data which has been shown to be valid.

Evolution Process of slow change in living organisms over long periods of time as those best-adapted to survive breed successfully.

Exchange surface Surface where materials are exchanged.

Extinction The permanent loss of all the members of a species.

Extremophile Organism which lives in environments that are very extreme (e.g., very high or very low temperatures, high salt levels, or high pressures).

F

Fair test A test in which only the independent variable has been allowed to affect the dependent variable.

False negative A test that shows that a specific problem is not present when in fact it is.

False positive A test that shows that a specific problem is present when in fact it is not.

Fatty acid Building block of lipids.

Fermentation Reaction in which the enzymes in yeast turn glucose into ethanol and carbon dioxide.

Fertile Describes soil that contains enough minerals (e.g., nitrates) to supply crop plants with all the nutrients needed for healthy growth.

Fertiliser Substance provided for plants that supplies them with essential nutrients for healthy growth.

Fossil fuel Fuel obtained from long-dead biological material.

Fossil Remains of an organism from many thousands or millions of years ago that have been preserved in rock, ice, amber, peat, etc.

G

Gamete Sex cell which has half the chromosome number of an ordinary cell.

Gametocytes The stage in the lifecycle of the malaria parasite Plasmodium that reproduces sexually and infects female mosquitos.

Gaseous exchange Exchange of gases (e.g., the exchange of oxygen and carbon dioxide which occurs between the air in the lungs and the blood).

Gene Short section of DNA carrying genetic information.

Genetic cause The alleles inherited by an organism that determine its characteristics directly.

Genetic disorder Disease which is inherited.

Genetic engineering Technique for changing the genetic information of a cell.

Genetic material DNA which carries the instructions for making a new cell or a new individual.

Genetically modified Describes an organism that has had its genetic material modified, usually by the addition of at least one new gene.

Genetically modified crop (GM crop) Crop that has had its genes modified by genetic engineering techniques.

Genotype The genetic make-up of an individual regarding a particular characteristic.

Geographical isolation This is when two populations become physically isolated by a geographical feature.

Glandular tissue Tissue which makes up the glands and secretes chemicals (e.g., enzymes, hormones).

Global warming Warming of the Earth due to greenhouse gases in the atmosphere trapping infrared radiation from the surface.

Glucagon Hormone involved in the control of blood sugar levels.

Glucose A simple sugar.

Glycerol Building block of lipids.

Glycogen Carbohydrate store in animals, including the muscles and liver of the human body.

Greenhouse effect The trapping of infrared radiation from the Sun as a result of greenhouse gases (e.g., carbon dioxide and methane) in the Earth's atmosphere. The greenhouse effect maintains the surface of the Earth at a temperature suitable for life.

Greenhouse gas Gases (e.g., carbon dioxide and methane), which absorb energy radiated from the Earth, and result in warming up the atmosphere.

Guard cell Cells which surround stomata in the leaves of plants and control their opening and closing.

H

Haemoglobin Red pigment which carries oxygen around the body.

Hazard Something (e.g., an object, a property of a substance, or an activity) that can cause harm.

Heart Muscular organ which pumps blood around the body.

Herbicide Chemical which kills plants.

Herbivore Animal which feeds on plants.

Herd immunity The target of vaccination programmes – when a large percentage of the population are immune to a disease, the spread of the pathogen is greatly reduced and it may disappear completely from a population.

Heterozygous An individual with different alleles for a characteristic.

Homeostasis Maintenance of constant internal body conditions.

Homozygous An individual with two identical alleles for a characteristic.

Hormone Chemical produced in glands which carries chemical messages around the body.

Horticulture Growing plants for food and for pleasure in gardens.

Hypertonic Solution with a higher concentration of solute molecules than another solution.

Hypothermia The state which occurs when the core body temperature falls below the normal range.

Hypothesis Proposal intended to explain certain facts or observations.

Hypotonic Solution with a lower concentration of solute molecules than another solution.

I

Immune response Response of the immune system to cells carrying foreign antigens. It results in the production of antibodies against the foreign cells and the destruction of those cells.

Immune system Body system which recognises and destroys foreign cells or proteins (e.g., invading pathogens).

Immunisation Giving a vaccine that allows immunity to develop without exposure to the disease itself.

Immunosuppressant drug Drug which suppresses the immune system of the recipient of a transplanted organ to prevent rejection.

Impulse Electrical signal carried along the neurones.

Independent variable See Variable – independent.

Industrial waste Waste produced by industrial processes.

Infectious disease Disease which can be passed from one individual to another.

Infectious Capable of causing infection.

Inheritance of acquired characteristics Jean-Baptiste Lamarck's theory of how evolution took place.

Inherited disorder Passed on from parents to their offspring through genes.

Inoculate To make someone immune to a disease by injecting them with a vaccine which stimulates the immune system to make antibodies against the disease.

Insoluble molecule Molecule which will not dissolve in a particular solvent such as water.

Insulin Hormone involved in the control of blood sugar levels.

Intercostal muscles Muscles between the ribs which raise and lower them during breathing movements.

Internal environment Conditions inside the body.

Inter-specific competition Competition for resources between members of different species.

Intra-specific competition Competition for resources between members of the same species.

Ion Charged particle produced by the loss or gain of electrons.

Ionising radiation Radiation made of particles which produce ions in the materials that they pass through, which in turn can make them biologically active and may result in mutation and cancer.

Isotonic Having the same concentration of solutes as another solution.

J

Jean-Baptiste Lamarck French biologist who developed a theory of evolution based on the inheritance of acquired characteristics.

K

Kidney Organ which filters the blood and removes urea, excess salts, and water.

Kidney transplant Replacement of failed kidneys with a healthy kidney from a donor.

Kidney tubule Structure in the kidney where substances are reabsorbed back into the blood.

L

Lactic acid One product of anaerobic respiration. It builds up in muscles with exercise. Important in yoghurt- and cheese-making processes.

Light microscope Instrument used to magnify specimens using lenses and light.

Limiting factor Factor which limits the rate of a reaction (e.g., temperature, pH, and light levels limit photosynthesis).

Line graph Used when both variables are continuous. The line should normally be a line of best fit, and may be straight or a smooth curve. (Exceptionally, in some investigations, the line may be a point-to-point line.)

Linear relationship Relationship between two continuous variables that can be represented by a straight line on a graph.

Lipase Enzyme which breaks down fats and oils into fatty acids and glycerol.

Lipid Oil or fat.

Liver Large organ in the abdomen which carries out a wide range of functions in the body.

M

Malignant tumour Tumour that can spread around the body, invading healthy tissues as well as splitting and forming secondary tumours.

Mean The arithmetical average of a series of numbers.

Median The middle value in a list of data.

Meiosis Two-stage process of cell division which reduces the chromosome number of the daughter cells. It is involved in making the gametes for sexual reproduction.

Metastase The way in which malignant tumours spread around the body.

Methane Hydrocarbon gas that makes up the main flammable component of biogas.

Microorganism Bacteria, viruses, and other organisms which can only be seen using a microscope.

Mineral ion Chemical needed in small amounts as part of a balanced diet to keep the body healthy.

Mitochondria The site of aerobic cellular respiration in a cell.

Mitosis Asexual cell division where two identical cells are formed.

Molecule Particle made up of two or more atoms bonded together.

Monitor Make observations over a period of time.

Monohybrid cross Genetic cross involving the inheritance of a single gene.

Motor neurone Neurone that carries impulses from the central nervous system to the effector organs.

MRSA Methicillin-resistant *Staphylococcus aureus*. An antibiotic-resistant bacterium.

Multicellular organism Organism which is made up of many different cells which work together. Some of the cells are specialised for different functions in the organism.

Muscular tissue Tissue which makes up the muscles. It can contract and relax.

Mutation Change in the genetic material of an organism.

N

Natural selection Process by which evolution takes place. Organisms produce more offspring than the environment can support, so only those which are most suited to their environment – the 'fittest' – will survive to breed and pass on their useful alleles.

Negative feedback system System of control based on an increase in one substance triggering the release of another substance which brings about a reduction in levels of the initial stimulus.

Negative pressure Describes a system in which the external pressure is lower than the internal pressure.

Nerve Bundle of hundreds or even thousands of neurones.

Nervous system See Central nervous system.

Net movement Overall movement of a substance.

Neurone Basic cell of the nervous system which carries minute electrical impulses around the body.

Nitrate ion Ion needed by plants to make proteins.

Non-renewable Something which cannot be replaced once it is used up.

Nucleus (of a cell) Organelle found in many living cells, containing the genetic information.

O

Obese Very overweight, with a BMI of over 30.

Optic nerve Nerve carrying impulses from the retina of the eye to the brain.

Organ Group of different tissues working together to carry out a particular function.

Organ system Group of organs working together to carry out a particular function.

Osmosis Diffusion of water from a dilute to a more concentrated solution through a partially permeable membrane that allows the passage of water molecules.

Ova Female sex cells (gametes or egg cells) in animals.

Ovary Female sex organ which contains the eggs and produces sex hormones during the menstrual cycle.

Overweight Describes a person whose body carries excess fat and if whose BMI is between 25 and 30.

Oxygen debt Extra oxygen that must be taken into the body after exercise has stopped to complete the aerobic respiration of lactic acid.

Oxygenated Containing oxygen.

Oxyhaemoglobin Molecule formed when haemoglobin binds to oxygen molecules.

P

Pacemaker Something biological or artificial that sets the basic rhythm of the heart.

Palisade mesophyll Upper layer of mesophyll tissue in plant leaves that contains many chloroplasts for photosynthesis.

Pancreas Organ that produces the hormone insulin and many digestive enzymes.

Parasite Organism which lives in or on other living organisms and gets some or all of its nourishment from this host organism.

Partially permeable membrane Allows only certain substances to pass through.

Pathogen Microorganism which causes disease.

Permanent vacuole Space in the cytoplasm filled with cell sap which is there all the time.

Pesticide Chemical that kills animals.

Petal Feature of a plant adapted to contain the sex organs. May be brightly coloured or patterned to attract insects and other pollinators.

Phenotype The physical appearance/biochemistry of an individual regarding a particular characteristic.

Phloem tissue Living transport tissue in plants which carries sugars around the plant.

Photosynthesis Process by which plants make food using carbon dioxide, water, and light energy.

Pigment Coloured molecule.

Pituitary gland Small gland in the brain which produces a range of hormones controlling body functions.

Placebo Substance used in clinical trials which does not contain any drug at all.

Plasma Clear, yellow, liquid part of the blood which carries dissolved substances and blood cells around the body.

Plasmid Extra circle of DNA found in bacterial cytoplasm.

Plasmolysis The state of a plant cell when large amounts of water have moved out by osmosis and the protoplasm shrinks, pulling the cell membrane away from the cell wall.

Platelet Fragment of cell in the blood which is vital for the clotting mechanism to work.

Polydactyly Genetic condition inherited through a dominant allele which results in extra fingers and toes.

Positive pressure Describes a system where the external pressure is higher than the internal pressure.

Precise Describes a measurement in which there is very little spread about the mean value. Precision depends only on the extent of random errors – it gives no indication of how close results are to the true value.

Predator Animal which preys on other animals for food.

Prediction Forecast or statement about the way in which something will happen in the future. In science it is not just a simple guess, because it is based on some prior knowledge or on a hypothesis.

Protease Enzyme which breaks down proteins.

Protein synthesis Process by which proteins are made on the ribosomes based on information from the genes in the nucleus.

Puberty Stage of development when the sexual organs and the body become adult.

Pulmonary artery Large blood vessel taking deoxygenated blood from the right ventricle of the heart to the lungs.

Pulmonary vein Large blood vessel bringing blood into the left atrium of the heart from the lungs.

Pyramid of biomass A model of the mass of biological material in the organisms at each level of a food chain.

R

Radiotherapy Cancer treatment in which cells are destroyed by targeted doses of radiation.

Random error See Error – random.

Range The maximum and minimum values of the independent or dependent variables, important in ensuring that any pattern is detected.

Receptor Special sensory cell that detects changes in the environment.

Recessive Describes a characteristic that will show up in offspring only if both of the alleles are inherited.

Recipient Person who receives a donor organ.

Red blood cell Blood cell which contains the red pigment haemoglobin. It is a biconcave disc shape and gives the blood its red colour.

Reflex arc Describes the sense organ, sensory neurone, relay neuron, motor neuron, and effector organ which bring about a reflex action.

Reflex Rapid automatic response of the nervous system that does not involve conscious thought.

Relay neurone Neurone located in the central nervous system (spinal cord or brain) which links a sensory neurone and a motor neurone in a reflex response.

Repeatable Describes a measurement for which the original experimenter can repeat the investigation using same method and equipment and obtain the same results.

Reproducible A measurement is reproducible if the investigation is repeated by another person, or by using different equipment or techniques, and the same results are obtained.

Respiration Process by which food molecules are broken down to release energy for the cells.

Respiratory (breathing) system System of the body including the airways and lungs that is specially adapted for the exchange of gases between the air and the blood.

Ribosome Site of protein synthesis in a cell.

Risk Likelihood that a hazard will actually cause harm. You can reduce risk by identifying the hazard and doing something to protect against that hazard.

Root hair cell Cell on the root of a plant with microscopic hairs which increases the surface area for the absorption of water from the soil.

S

Salivary gland Gland in the mouth which produces saliva containing the enzyme amylase.

Sample size Size of a sample in an investigation.

Secrete Release chemicals (e.g., hormones or enzymes).

Selective reabsorption Absorption of varying amounts of water and dissolved mineral ions back into the blood in the kidney, depending on what is needed by the body.

Sense organ Collection of special cells known as receptors which respond to changes in the surroundings (e.g., eye, ear).

Sensory neurone Neurone which carries impulses from the sensory organs to the central nervous system.

Sequestered Describes the storage of carbon dioxide directly or indirectly in plant material and water.

Sewage treatment plant Site where human waste is broken down using microorganisms.

Sewage A combination of bodily waste, waste water from homes, and rainfall overflow from street drains.

Sex chromosome Chromosome which carries the information about the sex of an individual.

Sexual reproduction Reproduction which involves the joining (fusion) of male and female gametes, producing genetic variety in the offspring.

Sickle-cell anaemia Genetic disorder affecting the structure of haemoglobin, which in turn affects the shape of red blood cells, making them sickle-shaped so they don't carry oxygen efficiently.

Simple sugar Small carbohydrate molecule made up of single sugar units or two sugar units joined together.

Small intestine Region of the digestive system where most of the digestion of food takes place.

Smog Haze of small particles and acidic gases which forms in the air over major cities as a result of the burning of fossil fuels in vehicles and pollution from industrial processes.

Solute Solid which dissolves in a solvent to form a solution.

Specialised Adapted for a particular function.

Speciation Formation of a new species.

Species Group of organisms with many features in common which can breed successfully, producing fertile offspring.

Sperm Male sex cell (gamete) in animals.

Spongy mesophyll Lower layer of mesophyll tissue in plant leaves that contains some chloroplasts and also has big air spaces to give a large surface area for the diffusion of gases.

Stem cell Undifferentiated cell with the potential to form a wide variety of different cell types.

Stent Metal mesh placed in an artery which is used to open up the blood vessel by the inflation of a tiny balloon.

Stimulus Change in the environment that is detected by sensory receptors.

Stomata Openings in the leaves of plants (particularly the underside) which allow gases to enter and leave the leaf. They are opened and closed by guard cells.

Sulfur dioxide Polluting gas formed when fossil fuels containing sulfur impurities are burnt.

Sustainable food production Methods of producing food which can be sustained over time without destroying the fertility of the land or oceans.

Synapse Gap between neurones where the transmission of information is chemical rather than electrical.

Systematic error See Error – systematic.

T

Territory Area where an animal lives and feeds, which it may mark out or defend against other animals.

Therapeutic cloning Cloning by transferring the nucleus of an adult cell to an empty egg to produce tissues or organs which could be used in medicine.

Thermoregulatory centre Area of the brain which is sensitive to blood temperature.

Thorax The upper (chest) region of the body. In humans it includes the ribcage, heart, and lungs.

Tissue culture Using small groups of cells from a plant to make new plants.

Tissue Group of specialised cells all carrying out the same function.

Trachea Main tube lined with cartilage rings which carries air from the nose and mouth down towards the lungs.

Translocation Movement of sugars from the leaves to the rest of a plant.

Transpiration stream Movement of water through a plant from the roots to the leaves as a result of the loss of water by evaporation from the surface of the leaves.

Transpiration Loss of water vapour from the leaves of plants through the stomata when they are opened to allow gas exchange for photosynthesis.

Transport system System for transporting substances around a multicellular living organism.

Tuber Modified part of a plant which is used to store food in the form of starch.

Tumour Mass of abnormally growing cells that forms when cells do not respond to the normal mechanisms which control the cell cycle.

Turgor The state of a plant cell when the pressure of the cell wall on the cytoplasm cancels out the tendency for water to move in by osmosis, so the cell is rigid.

Type 1 diabetes Form of diabetes caused when the pancreas cannot make insulin. It usually occurs in children and young adults and can be treated by regular insulin injections.

Type 2 diabetes Form of diabetes linked to obesity, diet, and exercise levels as well as genetics. The pancreas still makes insulin (although the levels may reduce) but the body cells stop responding to insulin.

U

Urea Waste product formed by the breakdown of excess amino acids in the liver.

Urine Liquid produced by the kidneys containing the metabolic waste product urea along with excess water and salts from the body.

Urobilin Yellow pigment that comes from the breakdown of haemoglobin in the liver.

V

Vaccination Introducing small quantities of dead or inactive pathogens into the body to stimulate the white blood cells to produce antibodies that destroy the pathogens. This makes the person immune to future infection.

Vaccine Dead or inactive pathogen material used in vaccination.

Vacuum An area with little or no gas pressure.

Valid Describes whether an investigative procedure is suitable to answer the question being asked.

Valve Structure which prevents the backflow of liquid (e.g., the valves of the heart or the veins).

Variable Physical, chemical, or biological quantity or characteristic.

Vein Blood vessel which carries blood away from the heart. It usually carries deoxygenated blood and has valves to prevent the backflow of blood.

Vena cava Large vein going into the right atrium of the heart carrying deoxygenated blood from the body.

Ventricles Large chambers at the bottom of the heart. The right ventricle pumps blood to the lungs, the left ventricle pumps blood around the body.

Villi Finger-like projections from the lining of the small intestine which increase the surface area for the absorption of digested food into the blood.

Virus Microorganism which takes over body cells and reproduces rapidly, causing disease.

W

White blood cell Blood cell which is involved in the immune system of the body, engulfing bacteria, making antibodies, and making antitoxins.

Wilting Process by which plants droop when they are short of water or too hot. This reduces further water loss and prevents cell damage.

X

Xylem tissue Non-living transport tissue in plants, which transports water around the plant.

Y

Yeast Single-celled fungi which produce ethanol when they respire carbohydrates anaerobically.

Z

Zygote Cell formed when the male and female gametes fuse at fertilisation.

Answers

1 Cell structure and organisation

1.1
1. **a** Nucleus, cytoplasm, cell membrane, mitochondria, ribosomes.
   b Cell wall, chloroplasts, permanent vacuole.
   c Cell wall provides support and strengthening for the cell and the plant; chloroplasts for photosynthesis; permanent vacuole keeps the cells rigid to support the plant.
2. The nucleus controls all the activities of the cell and contains the instructions for making new cells or new organisms. Mitochondria are the site of aerobic respiration, so they produce energy for the cell.
3. Root cells in a plant do not have chloroplasts because they don't carry out photosynthesis – they are underground so have no light.

1.2
1. **a** It isn't contained in a nucleus and there are extra genes known as plasmids separate from the main genetic material.
   b Bacteria cells.
   c Plant cell walls are made of cellulose, bacteria cell walls are not.

2.

	Useful	Damaging
Bacteria	Food production, e.g., yoghurt and cheese Sewage treatment Making medicines Decomposers in natural cycles, e.g., carbon and nitrogen cycle Healthy gut	Diseases Decay
Fungi	Food production – bread Alcoholic drinks Antibiotics Decomposers in food chains and webs	Diseases Decay of food stuffs

3.

Feature	Animal cell	Plant or algal cell	Bacterial cell	Yeast cell
Cell membrane	Yes	Yes	Yes	Yes
Nucleus	Yes	Yes	No	Yes
Plasmids	No	No	Yes	No
Chloroplasts	No	Yes	No	No
Cell wall	No	Yes	Yes	Yes
Cytoplasm	Yes	Yes	Yes	Yes

1.3
1. Fat cells: not much cytoplasm so room for fat storage; ability to expand to store fat; few mitochondria as they do not need much energy, so do not waste space. Cone cells from human eye: outer segment containing visual pigment; middle segment packed full of mitochondria; specialised nerve ending. Root hair cells: no chloroplasts so no photosynthesis; root hair increases surface area for water uptake; vacuole to facilitate water movement; close to xylem tissue. Sperm cells: tail for movement to egg; mitochondria to provide energy for movement; acrosome full of digestive enzymes to break down the outside layers of the egg cell; large nucleus full of genetic material.
2. **a** It takes lots of energy for muscle cells to contract and move things around and this is supplied by the mitochrondria, which are the cell organelles where energy is released.
   b Chloroplasts make food for the plant by photosynthesis. They need light energy for this. So having the cells near the top of the leaf packed with chloroplasts means they can make the best possible use of the light available.
3. Mitochondria – the number of mitochondria give an idea of how much energy the cell uses.
   Flagella or cilia – used to move the cell around or to move substances past the cell.
   Nucleus – tells you if the cell is capable of reproduction. Storage materials such as fat or starch.
   Cellulose cell walls – suggest plant cell.
   Chloroplasts – show photosynthesis takes place.
   Any other valid point.

1.4
1. **a** A tissue is a group of cells with similar structure and function.
   b Organs are made up of tissues. One organ can contain several tissues, all working together.
2. **a** Specialised cell – single cell adapted to function as a gamete
   b Organ – the kidney is a group of tissues with a collective function
   c Organ – the stomach is a group of tissues with one function
3. **a** Many chloroplasts in palisade mesophyll layer for photosynthesis; vascular bundle to maintain concentration gradients of gases, air spaces for gas exchange; stomata in underside of leaf to allow for gas exchange.
   b Folded lining increases surface area for absorption of products; glandular tissue for production of digestive juices; epithelial tissue for a good blood supply; muscular tissue churns food and digestive juices to encourage digestion and move food through digestive system.

1.5
1. A 3, B 4, C 1, D 2
2. Organs are made up of tissues. One organ can contain several tissues, all working together eg the stomach. Organ systems are groups of organs that perform a particular function eg the digestive system.
3. In the digestive system the salivary glands produce enzymes to start digestion, then the stomach churns up the food to a pulp. In the small intestine the gall bladder produces bile which helps in the digestion of fats in the small intestine The pancreas produces enzymes which chemically break down food molecules in the small intestine. Each organ is dependent on others in the system to function at its best.

1.6
1. Particles in a liquid or a gas move randomly. As the particles move they bump into each other and this makes them move apart. Diffusion is the spreading of the particles of a gas or of a substance in solution along a concentration gradient and this happens as a result of the random collisions. When the particles are concentrated, there are more collisions. Many particles will move randomly towards the area of low concentration. Only a few will move randomly in the other direction.
2. **a** Heating the gas or solution will speed up diffusion as the particles are moving faster.
   b Folded membranes provide an increased surface area so diffusion can take place more quickly.
3. **a** Concentration gradient between the gut (high concentration of digested food) and blood stream (low concentration of digested food) so digested food molecules move from the gut into the blood stream by diffusion. The large surface area of the lining of the small intestine gives a big area for diffusion to take place over and increases the rate at which it occurs.
   b There is a concentration gradient between the carbon dioxide concentration in the blood (high) and the air in the lungs (relatively low) so carbon dioxide moves from the blood into the air in the alveoli of the lungs along a concentration gradient by diffusion. Again, a large surface area and rich blood supply speed up the process.
   c The female moth produces chemicals which spread out into the air around her by diffusion. The further the distance from the female moth, the fewer molecules of the attractive chemical there will be. The male moth is sensitive to the chemical and flies up the concentration gradient, following the chemical as it gets stronger until it brings him to the female moth.

1.7
1. **a** In diffusion all the particles move freely along concentration gradients. In osmosis only water molecules move across a partially permeable membrane from an area of high concentration of water to an area of low concentration of water.
   b If the cell makes water during chemical reactions and the cytoplasm becomes too dilute, water moves out of the cell by osmosis. If the cell uses up water in chemical reactions and the cytoplasm becomes too concentrated, water moves in by osmosis to restore the balance.
2. **a** Isotonic solution – a solution with the same concentration of solutes as the inside of a cell. Hypotonic solution – a solution which has a lower concentration of solutes than the cytoplasm of a cell. Hypertonic solution – a solution which has a higher concentration of solutes than the inside of a cell.

b If the solute concentration of the fluid surrounding the cells of the body is lower than the cell contents, water will move into the cells by osmosis, they will swell and may burst. If the solute concentration is higher than the cells then water will leave the cells by osmosis. The cells will shrink and stop working properly. So it is important for solute concentration of the fluid surrounding the body cells to be as constant as possible to minimise the changes in the size and shape of the cells of the body and to keep them working normally.

3 Plants rely on osmosis to support their stems and leaves. Water moves into plant cells by osmosis from the xylem. This causes the vacuole to swell and press the cytoplasm against the plant cell walls. The pressure builds up until turgor is reached when the pressure is so great that no more water can physically enter the cell. Turgor pressure makes the cells hard and rigid, which in turn keeps the leaves and stems of the plant rigid and firm.

4 The cytoplasm of *Amoeba* contains a lower concentration of water particles than the water in which the organism lives. The cell membrane is partially permeable, so water constantly moves into *Amoeba* from its surroundings by osmosis. If this continued without stopping, the organism would burst. Water can be moved into the vacuole by active transport, and then the vacuole moved to the outside of the cell using energy as well.

1.8

1 Transport protein or system in the membrane is usually used. The substrate molecule binds to the transport protein in the membrane. This moves across the membrane carrying the substance to the other side. The substrate is released and the carrier molecule returns to its original position. This all uses energy.

2 **a** In active transport substances are moved along a concentration gradient or across a partially permeable membrane which they cannot cross by diffusion. The process uses energy. Osmosis and diffusion both involve the movement of substances down a concentration gradient or across a partially permeable membrane and they do not use energy produced by the cell.

b Cellular respiration releases the energy needed for active transport. Cellular respiration takes place in the mitochondria so cells which carry out lots of active transport often have lots of mitochondria to provide the energy they need.

3 **a** They need to get rid of the excess salt from the salt water in the sea and they use active transport to secrete the salt against a concentration gradient into special salt glands that remove it from the body.

b Plants need to move mineral ions from the soil into their roots. Mineral ions are much more concentrated in the cytoplasm of plant cells than in soil water, so they have to be moved against a concentration gradient. This involves active transport and the use of energy from cellular respiration.

Answers to end of chapter summary questions

1 **a** A genetic material, B cytoplasm, C cell membrane, D cell wall, E Plasmids.

b F cell membrane
G Golgi body
H Cell wall
I Ribosomes
J Cytoplasm
K Vacuole
L Chloroplasts
M Nucleus

c Similarities: Both have cell walls, cell membrane and cytoplasm. Both have genetic material but that of the plant involves chromosomes contained in a nucleus and that of a bacterial cell is a long circular strand of DNA found free in the cytoplasm with additional small loops of DNA known as plasmids.
Differences: Bacterial cells are much smaller than plant cells. Plant cells contain chloroplasts which can carry out photosynthesis, bacteria do not. Plant cells have permanent vacuoles, bacterial cells do not, bacterial cells may have slime capsules, plant cells do not. Bacterial cells may have flagella to move them about, plant cells do not.

d Similarities: Both have cell membrane, cytoplasm and nucleus.
Differences: Bacterial cells are much smaller than animal cells. Bacterial cells have food storage granules which most animal cells do not. Bacterial cells have cell walls, animal cells don't, Bacterial cells have permanent vacuoles, animal cells do not.

2 **a**

Diffusion	Osmosis	Active transport
The net movement of particles from an area of high concentration to an area of lower concentration.	The net movement of water from a high concentration of water molecules to a lower concentration (dilute to more concentrated solution) across a partially permeable membrane.	The movement of a substance from a low concentration to a higher concentration, or across a partially permeable membrane.
Takes place because of the random movements of the particles of a gas or of a substance in solution in water.	Although all the particles are moving randomly, only the water molecules can pass through the partially permeable membrane.	Involves transport or carrier proteins which carry specific substances across a membrane.
Takes place along a concentration gradient.	Takes place along a concentration gradient of water molecules.	Takes place against a concentration gradient.
No energy from the cell is involved.	No energy from the cell is involved.	Uses energy from cellular respiration.

b Expect water to move into both A and B by osmosis as the inside of the bag is hypertonic to the outside. Expect water to move into bag B faster than bag A because at a higher temperature. Increase in temperature gives increased rate of random movements of the particles and so would increase the rate at which water particles would pass through the partially permeable membrane, so the rate of osmosis would increase.

3 **a** Xylem; phloem; epithelial tissue;

b Leaves; for photosynthesis; stem; for structural support; roots; to absorb water and nutrients and to anchor into ground;

c Xylem; for transport of water; phloem; for transport of mineral ions;

4 **a** Organ systems are groups of organs that all work together to perform a particular function.

b

A	Mouth – chewing food
B	Oesophagus – passing food from mouth to stomach
C	Stomach – digesting protein
D	Pancreas – producing digestive juices that chemically break down food molecules
E	Large intestine – absorbing water from undigested food, producing faeces
F	Anus – passing faeces
G	Small intestine – digesting and absorbing soluble food into bloodstream
H	Duodenum – chemical digesting
I	Liver – producing bile

Answers to end of chapter practice questions

1 **a** A = nucleus; B = Golgi apparatus; C = Plasmid

b Controls all the activities of the cell (1); contains the genetic material (1)

c Plasmids (1); and loop of genetic material (1)

d **i** 0.1 mm;
ii 0.04 mm;
iii 0.002 mm (3)

e Two from: enzymes or named enzyme; to make cell parts or named cell parts; antibodies; antigens; hormones (2)

f Bacterial cells are too small to contain them (1)

2 **a** Root hair cells absorb water from the soil (1) by osmosis (1), for photosynthesis. The mesophyll leaf cell absorbs light energy from the sun (1) which it uses in photosynthesis (1) to make glucose for the plant. (1)

b Compare: both have cell wall for support; cell membrane for exchange of materials; mitochondria to provide energy by cell respiration and ribosomes to make proteins. *(Maximum of 3 marks)*
Contrast: leaf cell has chloroplasts with chlorophyll to capture (absorb) light energy but root hair cells have none as they are underground in dark; root hair cell has long, thin projection to give maximum surface area for water absorption/active transport of minerals; root hair cell has larger permanent vacuole to store more water and pass it on to adjacent cells. *(Maximum of 3 marks)*

3 a Pot A = turgid cell (1) – cell membrane in contact with cell wall and large amount of cell sap (1); Pot B = plasmolysed cell (1) – cell membrane coming away from cell wall and small amount of cell sap (1)
b Water from roots passed up via xylem to stem and leaf cells (1) which keeps them turgid (1) so the contents press out against cell membrane and cell wall giving strength (1). (Answer could be reverse for wilting plant.)

4 a Isotonic (1)
b 0.28 mol/dm³ (1)
c There is a higher concentration (of salts/sugars) inside the cell/higher concentration of water outside the cell (1), so the solution is hypotonic. (1) This causes water to diffuse into the cell (1) by osmosis (1), across the partially/semi-permeable cell membrane. (1) The cell fills up with water, swells and eventually bursts, (1) as it does not have a rigid cell wall like a plant cell to contain it. (1) *(Maximum of 6 marks)*

2 Cell division and differentiation

2.1

1 a Chromosomes are structures made of DNA found in pairs in the nucleus of the cells which contain the inherited material.
b A gene is a small packet of information that controls a characteristic, or part of a characteristic, of your body. It is a section of DNA.
c An allele is a particular form of a gene which codes for a particular characteristic.

2 a Cells need to be replaced with identical cells to do the same job.
b In mitosis a cell divides to form two identical daughter cells. Copies are made of the chromosomes in the nucleus. Then the cell divides in two which each get a copy of the full set of chromosomes.
Mitosis is important for forming the identical cells needed for growth or the replacement of tissues.

3 a Differentiation is the process by which cells become specialised and adapted to carry out a particular function in the body of a living organism. It is important because all of the cells of an early embryo are the same but organisms need different cells to carry out different roles in the body, e.g., muscle cells, sperm cells and gut lining cells.
b In animals, it occurs during embryo development and is permanent. In plants, it occurs throughout life and can be reversed or changed.
c Plants can be cloned relatively easily. Differentiation can be reversed, mitosis is induced, conditions can be changed and more mitosis induced. The cells redifferentiate into new plant tissues. In animals, differentiation cannot be reversed, so clones cannot be made easily. In order to make clones, embryos have to be made.

2.2

1 a 23 (pairs)
b 23
c 46

2 Sexual reproduction involves the fusing of two special sex cells or gametes so it brings genetic information from two individuals together, which introduces variety. The gametes are made during the process of meiosis. During meiosis each gamete receives a random mixture of the original chromosome pairs so the gametes are all different from the original cells as well, which also introduces variety.

3 a Meiosis. After the chromosomes are copied, the cell divides twice quickly resulting in sex cells each with half the number of chromosomes.
b In the reproductive organs/in the ovary or the testes.
c Meiosis is important because it halves the chromosome number in the gametes, Then when two gametes fuse during sexual reproduction, the new individual has the correct normal number of chromosomes in pairs in the cells, e.g., in humans, 23 chromosomes in the gametes, 46 chromosomes in 23 pairs in a normal somatic cell.
Meiosis is also important because it introduces variety as each gamete receives a random mixture of the original chromosomes.

2.3

1 a A stem cell is undifferentiated cell which has the potential to differentiate and form different specialised cells in the body. A normal body cell is specialised for a specific function and if divided by mitosis can only form cells with the same specialisation.
b Embryonic stem cells, adult stem cells.

2 a They can be used to make any type of adult cell to repair or replace damaged tissues, with no rejection issues.
b Treating paralysis, treating degenerative diseases of the brain, growing new organs for transplant surgery, treating blindness, any other sensible suggestion.
c Changing all the time – so far stem cells used successfully to treat cancers of the bone marrow, beginning to be used for restoring eyesight, some success in healing hearts after heart attack, any valid ideas as this will change from year to year.

3 a There are ethical objections and concerns over possible side effects.
b By using stem cells from umbilical blood, adult stem cells and therapeutic cloning.

Answers to end of chapter summary questions

1 a Mitosis is cell division that takes place in the normal body cells and produces genetically identical daughter cells.
b [Marks awarded for correct sequence of diagrams with suitable annotations.]
c All the divisions from the fertilised egg to the baby are mitosis. After birth, all the divisions for growth are mitosis, together with all the divisions involved in repair and replacement of damaged tissues.

2 Meiosis is a special form of cell division to produce gametes in which the chromosome number is reduced by half. It takes place in the reproductive organs (the ovaries and testes).

3 a Meiosis is important because it halves the chromosome number of the cells, so that when two gametes fuse at fertilisation, the normal chromosome number is restored. It also allows variety to be introduced.
b [Marks awarded for correct sequence of diagrams with appropriate annotations.]

4 a Stem cells are unspecialised cells which can differentiate (divide and change into many different types of cell) when they are needed.
b They may be used to repair damaged body parts, e.g., grow new spinal nerves to cure paralysis; grow new organs for transplants; repair brains in demented patients. [Accept any other sensible suggestions.]
c For: They offer tremendous hope of new treatments; they remove the need for donors in transplants; they could cure paralysis, heart disease, dementia etc; can grow tissues to order.
Against: Many embryonic stem cells come from aborted embryos or from fertility treatment and so this raises ethical issues about the use of unborn humans in research. Some people are also concerned about whether using stem cells may trigger cancer. There are also issues surrounding the cost and amount of time stem cell research takes.

5 a Mitosis
b Meristem
c In mitosis in the meristems identical daughter cells are formed containing the same number of chromosomes as the parent cells. In meiosis in the sex organs the number of chromosomes is halved to make non-identical gametes

Answers to end of chapter practice questions

1 One mark for each correctly matched term and definition, as follows:
Chromosome – a structure carrying a large number of genes.
Allele – a different form of one gene.
Gene – a section of genetic material coding for one characteristic.
Nucleus – the part of a cell that contains the genetic material.
Gamete – a cell with a single set of chromosomes.

2 a B with chromosomes all doubled to make 12 in total (1); C with 2 cells each containing 3 pairs (1); D with all cells containing 3 different chromosomes (1).
b Gametes, ova, eggs, sperm. (1)
c Identical to cell at A. (1)

3 a One mark for each correct row   *(Maximum of 5 marks).*

Differentiated cells	v
Cells with a single set of chromosomes	i
Undifferentiated cells	ii or iii or iv
Cells dividing rapidly by mitosis	iii or iv
An embryo	ii

b CELL → TISSUE → ORGAN → ORGAN SYSTEM → ORGANISM
(2 marks all correct; 1 mark if 1 error)
c Phloem = tissue (1); stem = organ (1); root hair = cell (1); water transport system = organ system (1); sunflower plant = organism (1)

3 Human biology – Breathing

3.1

1 As living organisms get bigger and more complex, their surface area to volume ratio gets smaller (diagrams to show this are useful here). As a result it is increasingly difficult to exchange materials quickly enough with the outside world. Gases and food molecules can no longer reach every cell inside the organism by simple diffusion and metabolic waste cannot be removed fast enough to avoid poisoning the cells. So in many larger organisms, there are special surfaces with very large surface areas where the exchange of materials takes place.
2 a It is covered in tiny buds with a very good blood supply.
   b The tiny buds give it a large surface area so the turtle can use them for breathing – they absorb oxygen from the water which diffuses into the blood. It is so effective that the turtles can stay underwater for months at a time.
3 a Having a large surface area that provides a big area over which exchange can take place; being thin, which provides a short diffusion path having an efficient blood supply (in animals); this moves the diffusing substances away and maintains a concentration (diffusion) gradient; being ventilated (in animals) to make gaseous exchange more efficient by maintaining steep concentration gradients.
   b Any three suitable exchange surfaces, e.g., alveoli of lungs, villi of small intestine, gills of fish, roots and leaves of plants with correct adaptations, e.g., examples of large surface area, thin, etc.

3.2

1 Intercostal muscles contract to move your ribs up and out and diaphragm muscles contract to flatten the diaphragm, so the volume of your thorax increases, the pressure decreases and air moves in. Intercostal muscles relax and the ribs move down and in and the diaphragm relaxes and domes up so the volume of your thorax decreases. The pressure increases and air is forced out.
2 Gaseous exchange is the exchange of the gases oxygen and carbon dioxide in the lungs. This is vital because oxygen is needed by the cells for cellular respiration to provide energy, whilst carbon dioxide is a poisonous waste product which must not be allowed to build up.
3 a [Marks awarded for well-drawn bar chart correctly labelled.]
   b Bar chart shows that you breathe in air which is mainly nitrogen with oxygen and a tiny bit of carbon dioxide. The air you breathe out has less oxygen and more carbon dioxide. So you take oxygen out of the air into the blood and pass carbon dioxide out of the blood into the air and change the composition of the air. BUT you only breathe in oxygen and only breathe out carbon dioxide.
   c Good ventilation system – breathing – to maintain a good concentration gradient; large surface area; good blood supply; small diffusion distances – alveoli.

3.3

1 a Glucose + oxygen → carbon dioxide + water (+ energy)
   b $C_6H_{12}O_6 + 6O_2 → 6CO_2 + 6H_2O$ (+ energy)
   c Muscle cells are very active and need a lot of energy so they need large numbers of mitochondria to supply the energy. Fat cells use very little energy so need very few mitochondria.
2 a The main uses of energy in the body are for movement, building new molecules and heat generation.
   b The symptoms of starvation are: people become very thin; stored energy is used up and growth stops; new proteins are not made and result in a lack of energy or raw materials; people lack energy, as there is a lack of fuel for the mitochondria, people feel cold, as there is not enough fuel for the mitochondria; to produce heat energy.
3 [See Practical box 'Investigating respiration' on page 32 of the Student Book. Any sensible suggestions for practical investigations.]

3.4

1 a Glycogen is a complex carbohydrate stored in the muscles.
   b Glycogen can be converted rapidly to glucose to provide fuel for aerobic respiration, which provides the body cells with energy. Muscle tissue often needs sudden supplies of energy to rapid contraction in a way that most other tissues do not, so muscle needs a glycogen store. Other tissues don't need the energy in the same way so have not evolved to have glycogen stores.
2 Heart rate: increases before exercise starts as a result of anticipation. It rises rapidly, followed by a steady rise and then falls quite sharply as the exercise finishes. Increased heart rate supplies muscles with the extra blood they need to bring glucose/sugar and oxygen to the muscle fibres, and to remove the carbon dioxide that rapidly builds up.
Breathing rate: increases more slowly and evenly than the heart rate, but remains high for some time after exercise. To begin with, increased heart rate supplies enough oxygen, then the breathing rate needs to increase to meet demand. When exercise stops, breathing rate remains high until the oxygen debt is paid off.
3 [Award marks based on ideas presented when predicting results. Look for clear, sensible ideas, safe investigation, realistic expectations, appropriate methods of recording and analysing, awareness of weakness in investigation. Look also for clear understanding of independent, dependent and control variables.]

3.5

1 The muscles become fatigued. After a long period of exercise, your muscles become short of oxygen and switch from aerobic to anaerobic respiration, which is less efficient. The glucose molecules are not broken down completely, so less energy is released than during aerobic respiration. The end products of anaerobic respiration are lactic acid and a small amount of energy.
2 The waste lactic acid you produce during exercise as a result of anaerobic respiration has to be broken down to produce carbon dioxide and water. This needs oxygen, and the amount of oxygen needed to break down the lactic acid is known as the oxygen debt. Even though your leg muscles have stopped, your heart rate and breathing rate stay high to supply extra oxygen until you have broken down all the lactic acid and paid off the oxygen debt.
3 a Cellular respiration which takes place without oxygen.
   b Animals – anaerobic respiration takes place in the muscles when there is not enough oxygen and the waste product is lactic acid and a relatively small amount of energy. This allows animals to continue running, etc., even when they cannot breathe fast enough to supply the oxygen they need.
Glucose → lactic acid (+ energy)
$C_6H_{12}O_6 → 2C_3H_6O_3$ (+ energy)
Plants and yeast – when they respire anaerobically they form ethanol and carbon dioxide. This allows them to continue to respire in low oxygen atmospheres. Not common in plants as they form oxygen during photosynthesis. Quite common in yeasts. People make use of it and deprive yeasts of oxygen to make alcoholic drinks.
Glucose → ethanol + carbon dioxide (+ energy)
$C_6H_{12}O_6 → 2C_2H_5OH + 2CO_2$ (+ energy)

Answers to end of chapter summary questions

1 a The alveoli provide a very large surface area with thin walls and a rich blood supply.
   b Air is moved in and out of the lungs by movements of the ribcage and diaphragm. Breathing in: intercostal muscles contract, pulling ribs upwards and outwards. Diaphragm muscles contract flattening the diaphragm. These things increase the volume of the thorax, which lowers the air pressure so it is lower than the outside air. This is then pushed into the lungs by atmospheric pressure.
Breathing out: intercostal muscles relax so ribs drop down and in. Diaphragm muscle relaxes so diaphragm domes up. These things reduce the volume of the thorax and increase the pressure, so air is forced out of the lungs.
Constantly refreshing the air in the lungs maintains the best possible concentration gradients between the air and the blood for the movement of oxygen from the air into the blood and carbon dioxide out of the blood into the air in the lungs. This makes gaseous exchange more efficient.
2 a A system which forces a carefully measured unit or breath of air into the lungs under positive pressure, rather like blowing up a balloon. Once the lungs are inflated the positive pressure stops and the lungs deflate as the ribs move down, forcing the air back out of the lungs.
   b In normal breathing, the increase in volume of the chest creates a negative pressure – so that air is drawn into the lungs by the force of the atmospheric air. In positive pressure ventilation, the pressure in the chest does not change and air is forced in under pressure from the outside.

c Doesn't involve the patient being encased in an artificial lung or shell, so much easier for them to be mobile.

3 a [Award marks for standard of graphs, axes, etc.]

b As the peas start to grow, they began to respire aerobically. As a result, a small amount of heat energy is produced so the temperature increased.

c Because the seeds were dry and not growing, so there was no respiration or heat produced.

d As a control level.

e i Any reasonable explanation, e.g., the important thing about flask C is that the peas are dead so the temperature for the first five days remains at 20 °C as they are not respiring.

ii Peas had gone mouldy and mould respiring so temperature goes up. Anomaly, e.g., sun on thermometer, poor reading, etc.

4 a [Credit will be given in the subsequent answers for extracting and using the information on the bar charts.]

b i Increased fitness means that the heart has a greater volume and pumps more blood at each beat. The heart therefore beats more slowly at rest.

ii Increased fitness affects the lungs by lowering the breathing rate.

5 a The breakdown of glucose in a cell using oxygen to release energy that can be used by the cell. Carbon dioxide and water are waste products of the reaction.
Glucose + oxygen → carbon dioxide + water (+ energy)
$C_6H_{12}O_6 + 6O_2 \rightarrow 6CO_2 + 6H_2O$ (+ energy)

b The breakdown of glucose in the cell in the absence of oxygen to release a small amount of energy to be used by the cell.

c In a human being the waste product is lactic acid.
Glucose → lactic acid (+ energy)
$C_6H_{12}O_6 \rightarrow 2C_3H_6O_3$ (+ energy)
In a yeast cell the waste products are ethanol and carbon dioxide.
Glucose → ethanol + carbon dioxide (+ energy)
$C_6H_{12}O_6 \rightarrow 2C_2H_5OH + 2CO_2$ (+ energy)

d This is the amount of oxygen needed to convert the lactic acid produced during a period of anaerobic exercise in the muscles to carbon dioxide and water with the release of energy.

e When exercise begins the heart and breathing increase to bring more oxygen into the body. The capacity of the heart and lungs will be bigger in a fit individual than in an unfit person, and so the breathing and heart rate will not increase as much as a fit person can bring more air into the body and pump more oxygenated blood around the body with each breath or heartbeat than an unfit individual. The muscles of a fit individual will be bigger with a better blood supply than the muscles of an unfit individual, so they will contract more efficiently and use aerobic respiration for longer. So a fit individual will build up a smaller oxygen debt than an unfit individual for the same amount of exercise, and will be able to convert the lactic acid to carbon dioxide and water faster as they bring more oxygen into their body.

6 a Aerobic respiration produces more energy to allow the muscles to contract more efficiently, so athletes want it to continue as long as possible before changing to less efficient anaerobic respiration.

b Red blood cells carry oxygen to the tissues, so if you have more red blood cells, you have more oxygen so aerobic respiration continues longer and muscles work more effectively.

c It increases the red blood cells in the body just before a performance and so allows more oxygen to be carried to the working muscles.

d They start anaerobic respiration where glucose is incompletely broken down to form lactic acid. Less energy is produced and the lactic acid can cause muscle fatigue.

Answers to end of chapter practice questions

1 a A = intercostal muscles (1); B = bronchi (1); C = diaphragm (1)

b Any five from the following (must be in correct order):
The intercostal muscles contract (1) and the ribcage is pulled up and out (1). The diaphragm contracts (1), which causes it to flatten (1). This increases the volume in the thorax (1), so the pressure decreases (1), pulling air in/air pushed in from the atmosphere where the pressure is now greater/there is greater atmospheric pressure (1).

2 a Rounded/bubble shape (1) gives maximum surface area (1). Because walls are only one-cell thick/there are thin cells lining alveolus and/or capillary (1) this gives shorter diffusion path/there is a short distance for gases to diffuse (1). Allow – layer of water lining alveolus (1) to dissolve oxygen molecules (1). *(Maximum of 4 marks)*

b Fresh air with more oxygen is continually brought into alveolus by ventilation/breathing in (1). Blood in capillary is continually moving away taking oxygen with it (1).

c i Glucose (1), water (1), C6H12O6 (1), 6 H2O (1) *(Ignore energy)*

ii Mitochondria (1)

iii Muscle cells require a lot of energy to contract (1), and mitochondria are the cell parts where energy is released in aerobic respiration. (1)

3 • There is a clear description of most of the features of a normal lung which must be copied and at least two advantages of the artificial lung. The answer shows almost faultless spelling, punctuation and grammar. It is coherent and in an organised, logical sequence. It contains a range of appropriate or relevant specialist terms used accurately. *(5–6 marks)*

• There is a description of at least three features of a normal lung which must be copied and at least one advantage of the artificial lung. There are some errors in spelling, punctuation and grammar. The answer has some structure and organisation. The use of specialist terms has been attempted, but not always accurately. *(3–4 marks)*

• There is a description of at least two features of the lung which must be copied and at least one advantage of the artificial lung. The spelling, punctuation and grammar are very weak. The answer is poorly organised with almost no specialist terms and/or their use demonstrating a general lack of understanding of their meaning. *(1–2 marks)*

• No relevant content. *(0 marks)*

Examples of biology points made in the response:
- large surface area
- method of removing carbon dioxide
- thin membrane
- method of filtering the air going in/ventilation described
- no need for tissue matching
- no operation needed
- few lungs become available
- no need for (immunosuppressant) drugs
- reference to ethics involved with transplants.

4 a Glucose (1) → carbon dioxide (1) + ethanol (1) *(Ignore energy)*

b Carbon dioxide (1) as it is a gas and will form bubbles (1).

c i 40 °C (1)

ii Anaerobic respiration in yeasts is controlled by enzyme(s) (1). At 0 °C enzymes are inactive (1) so no carbon dioxide is produced and the volume does not increase (1). 40° C is the optimum temperature so there is most activity and volume increase (1). By 60° C/80° C/ higher temperatures, the enzyme(s) have been denatured (1), so there is no volume increase (1). *(Maximum of 5 marks)*

4 Human biology – Circulation

4.1

1 Most of the cells of the body are too far away from the air or even from the lungs to be able to get oxygen and get rid of their waste carbon dioxide by diffusion. They are also too far from the digestive system to be able to get the food they need by diffusion, and from the excretory organs to get rid of waste products. This is why people need a circulatory system. The blood carries everything that is needed by the cells and is carried close to every cell in the body in the capillary network, so food and oxygen can pass into the cells by diffusion along a concentration gradient. Waste products diffuse from the cells into the blood along a concentration gradient. The circulation of the blood by the pumping of the heart means substances are constantly renewed or removed which maintains the steep concentration gradients into and out of the cells. Any other sensible points.

2 Blood carried from heart to lungs is deoxygenated blood from the body, so it is dark (purply) red until it picks up oxygen again in the lungs. It is called an artery because it carries blood leaving the heart.

3 a Blood enters the atria of the heart (the top chambers). Deoxygenated blood comes into the right atrium from the body through the vena cava. Oxygenated blood from the lungs comes into the left atrium through the pulmonary vein. The atria contract together and force blood down into the larger lower chambers, the ventricles. Valves close to stop the blood flowing backwards as the ventricles contract. The right ventricle sends deoxygenated blood to the lungs in the pulmonary artery. The left ventricle pumps oxygenated blood around the body in the aorta. As the blood leaves the heart, more valves close to prevent it flowing backwards.

b i The heart valves prevent the blood flowing backwards into the chambers they have just left, which makes the heart more efficient.

ii The coronary arteries supply the heart muscle with oxygenated blood so that they can respire aerobically and contract efficiently.

iii The thickened muscle of the left ventricle wall allows the heart to pump the blood all around the body very efficiently as it can pump harder than the right side, which only has to send blood to the lungs.

4.2

1 a Artery – blood vessel that carries blood away from the heart; has pulse from blood forced through them from the heart beat; have a small lumen and thick walls of muscle and elastic fibres.

b Vein – blood vessel that carries blood towards the heart; no pulse; valves to keep blood flowing in the right direction; large lumen; relatively thin walls.

c Capillary – very tiny vessel with narrow lumen and walls one-cell thick, so ideal for diffusion of substances in and out.

2 a [Make sure students' diagrams show the capillary network between arteries and veins and link the arteries and veins to the heart.]

b [These should be labelled: heart, lungs, artery to lungs, capillaries in lungs, vein to heart, artery to body, capillaries in organs of the body, vein to heart.]

4.3

1 Any three from: transporting oxygen from the lungs to the cells of the body; transporting carbon dioxide from the cells of the body to the lungs; transporting digested food molecules from the gut to the cells; transporting urea from the liver to the kidneys; transporting the white blood cells of the immune system around the body; any other sensible point.

2 a Blood plasma is a yellow liquid with cells suspended in it.

b Red blood cells.

c Any three from: transports waste products, digested food, carbon dioxide, blood cells, hormones.

3 White blood cells form antibodies and actively digest microorganisms. Platelets help with clotting, which keeps the microorganisms out.

Answers to end of chapter summary questions

1 a A Vena cava
B Right atrium
C Aorta
D Left ventricle
E Heart valves
F Pulmonary artery
G Pulmonary vein

b Vena cava, right atrium, atrium contracts, blood through valve, right ventricle, ventricle contracts, blood out through valves into pulmonary artery, to lungs where blood is oxygenated, back to heart through pulmonary vein, through valve into left atrium, left atrium contracts, blood through valve into left ventricle, left ventricle contracts, blood through valve into aorta, round body.

c Artery – blood vessel that carries blood away from the heart; has pulse from blood forced through them from the heart beat; have a small lumen and thick walls of muscle and elastic fibres.
Vein – blood vessel that carries blood towards the heart; no pulse; valves to keep blood flowing in the right direction; large lumen; relatively thin walls.
Capillary – very tiny vessel with narrow lumen and walls one-cell thick, so ideal for diffusion of substances in and out.

2 a Not enough blood is pumped out of the heart into the circulation so the patient will suffer from a lack of oxygen. Because the heart never empties properly it will not pump as efficiently.

b The gap in the centre of the heart allows the oxygenated blood on the left of the heart to become mixed with the deoxygenated blood on the right side of the heart. This means that the blood pumped around the body is not fully oxygenated so the baby will suffer the symptoms of lack of oxygen – lack of energy, blue colour, etc.

c The coronary arteries supply the oxygen needed by the heart muscle to beat and pump blood around the body. If they are narrowed or blocked then not enough oxygen reaches the heart muscle so it cannot contract properly or may die. This is particularly noticeable when a person exercises and their heart needs more oxygen as it needs to beat harder and faster. If the heart cannot pump properly, not enough oxygen gets to the body either.

3 Large surface area – gives a bigger area over which exchange can take place – eg villi of small intestine, alveoli of lungs, gills of fish
Thin – to provide a short diffusion path eg single cell thick walls of alveoli and blood capillaries
Efficient blood supply in animals to maintain steep concentration gradients by bring diffusiong substances in and removing others eg rich blood supply to alveoli, villi etc
Being ventilated (in animals) to maintain steep concentration gradients by bring substances and removing substances all the time

4 a Blood has three main functions: transport, protection and regulation.

b

	Plasma	Red blood cells	White blood cells	Platelets
Structure	Yellow liquid	No nucleus Packed with haemoglobin	Large cells with a nucleus	Small fragments of cells

No nucleus |
| **Function/ functions** | Carries red blood cells, white blood cells and platelets around the body

Transports carbon dioxide from the organs to the lungs

Transports the soluble products of digestion from the small intestine to the other organs

Transports urea from the liver to the kidneys | Transport oxygen from lungs to organs | Part of the defence system of the body against microorganisms | Help blood to clot at the site of a wound |

5 If the body is wounded, blood escapes. Platelets collect at the site of a wound and help cause blood clotting. This is a series of enzyme controlled reactions which result in the change of fibrinogen in the plasma into fibrin. This forms a net of fibres. Red blood cells and more platelets get caught in the fibrin net to form a clot. This dries and hardens to form a scab. Blood clotting is very important because if the blood did not clot, you could bleed to death from a simple cut. The scab also protects the new skin as it grows and prevents bacteria getting into a wound.

Answers to end of chapter practice questions

1 a i Right atrium (1)
ii pulmonary artery (1)
iii pulmonary vein (1)
iv aorta (1)

b To prevent the blood flowing back (into the left atrium) (1)

2 a Because the blood cells are microscopic and are in suspension in the plasma so the blood looks like a liquid (1)
There are more red blood cells than anything else so their red colour is what we see. (1)

b Helps defend the body against pathogens/engulfs pathogens/produces antibodies against pathogens (1).

c Starts the clotting mechanism at a wound site (1).

d Absorbed by diffusion in lungs (1); binds to haemoglobin in red cells to form oxyhaemoglobin (1); splits from haemoglobin in tissues and diffuses into cells (1).

e Any two from: urea (1); waste product from liver to kidneys for removal (1); carbon dioxide (1); waste product from cells to lungs for removal (1); soluble food molecules/glucose/amino acids/fatty acids/glycerol(1); from gut to cells/liver (1).

3 a Without clotting if you cut yourself you will bleed to death. (1) The clot also prevents the entry of microorganisms into the wound. (1) The clot dries to form a scab which protects the new skin as it forms underneath (1)

b A = red blood cell (1) B = fibrin (protein) strand (1) C = platelet (1)

c Blood clotting is a series of enzyme-controlled reactions that result in the change of fibrinogen into fibrin. (1) This produces a network (or web) of protein fibres. (1) The fibres then capture lots of red blood cells and more platelets to form a jelly-like clot. (1)

4 a A = artery (1) B = vein (1) C = capillary (1)

b artery: thick walls made of elastic fibres and muscle with a small lumen. (1) Can withstand the pressure of the blood pumped out from the heart and return to normal size once pulse moves through (1)
Vein: Thin walls with large lumen (1) because the blood in veins is under lower pressure and there is no pulse. Often have valves to prevent the backflow of blood. (1)
Capillary: vessels with single cell thin walls (1) - form a huge network between cells - blood at very low pressure, thin walls allow diffusion of substances into and out of the blood.(1)

5 Human biology – Digestion

5.1

1 a A molecule made up of long chains of amino acids.

b As structural components, as hormones, as antibodies and as catalysts (enzymes).

2 Similarities: all vital components of a balanced diet; all contain carbon, hydrogen and oxygen; all large molecules made up of smaller molecules joined together; any other sensible point.
Differences: Carbohydrates are made of sugar units. Complex carbohydrates are long chains of sugar units joined together by condensation reactions. They are broken down into glucose in the body, which is used to provide energy in cellular respiration. Carbohydrate-rich foods include bread, potatoes, rice and pasta. [Any other sensible point.]
Proteins are made up of single units called amino acids joined together. They contain nitrogen as well as carbon, hydrogen and oxygen. They are joined by peptide links. Molecules have complex 3-D shapes. Protein-rich foods include meat, fish, pulses and cheese. [Any other sensible point.]
Lipids may be solids (fats) or liquids (oils). They are made up of three fatty acid molecules joined to a molecule of glycerol. They are the most energy-rich food. They are insoluble in water. Lipid-rich foods include olive oil, butter, cream. [Any other sensible point.]

3 a Iodine test – iodine turns from yellowy-red to blue-black in the presence of starch.

b Ethanol test – cloudy white layer forms at the boundary if lipid is present.

4 a It depends on the fatty acids which are joined to the glycerol.

b Complex carbohydrates are made up of long chains of simple sugars joined together by condensation reactions when a molecule of water is released each time two simple sugars are joined.

5.2

1 a Catalyst: a substance that speeds up a chemical reaction but is not used up or involved in the reaction and can be used many times over.

b Enzyme: large protein molecules which act as biological catalysts.

c Active site: an area in the structure of the enzyme that is a specific shape and which enables the substrate of the catalysed reaction to fit into the enzyme protein. This allows the enzyme to catalyse the reaction.

2 a Protein

b The substrate (reactant) of the reaction to be catalysed fits into the active site of the enzyme like a lock and key. Once it is in place, the enzyme and the substrate bind together. The reaction takes place rapidly and the products are released from the surface of the enzyme, which is then ready to catalyse another reaction. [A diagram can help with this explanation and students should be given credit for a well labelled diagram used as part or all of their explanation.]

3 a Building up large molecules from smaller ones; breaking down large insoluble molecules into smaller soluble ones; changing one molecule into another; any other sensible suggestion.

b The reactions needed for life to continue could not take place fast enough without enzymes to speed them up. Enzymes also control the many reactions so that they can take place in the same small area without interfering with each other.

5.3

1 To begin with, enzyme controlled reactions go faster as the temperature increases – increase in the speed of random movements of particles mean collisions between enzyme and substrate are more likely. However, once the temperature reaches around 40 °C the structure of the protein making up the enzyme starts to be affected. The bonds holding the protein in its complex 3-D shape start to break down and the shape of the active site changes. The substrate can no longer bind to the active site and so the enzyme cannot catalyse the reaction. [Students should include a diagram to show the effect of temperature on enzyme action and could draw a diagram to show the change in the shape of the active site.]

2 a About pH 2.

b About pH 8.

c The activity levels fall fast.

d The increase in pH affects the shape of the active site of the enzyme, so it no longer bonds to the substrate. It is denatured and no longer catalyses the reaction.

5.4

1

Enzyme	Where it is made	Reaction catalysed	Where it works
Amylase	Salivary glands, pancreas, small intestine	Starch → sugars/ glucose	Mouth, small intestine
Protease	Stomach, pancreas, small intestine	Proteins → amino acids	Stomach, small intestine
Lipase	Pancreas, small intestine	Lipids → fatty acids and glycerol	Small intestine

2 Large insoluble molecules in food cannot be absorbed into the blood so have to be digested to form small insoluble molecules that can be absorbed.

5.5

1 a Acidic conditions

b Hydrochloric acid is made in glands in the stomach.

c Alkaline/alkali

d The liver produces bile that is stored in the gall bladder and released when food comes into the small intestine.

2 [Marks for a good diagram showing a large fat droplet coated in bile splitting into many small fat droplets.] This produces a larger surface area so enzymes can get to more fat molecules and break them down quickly.

3 Bread is mainly carbohydrate, butter is a lipid and egg has protein.
All food taken into the mouth and chewed (physically broken up) and coated in saliva to make it easier to swallow and to start the digestion of starch using amylase from the salivary glands.
Food swallowed down the oesophagus.
Stomach muscles churn the food and mix it with digestive juices – protease enzymes (pepsin) to break down proteins to amino acids and hydrochloric acid to give the acid pH needed for pepsin to work.
Food squirted out of the stomach into the first part of the small intestine (duodenum). Bile added from gall bladder to emulsify fats and give a bigger surface area for digestive enzymes to work on. It also neutralises the acid from the stomach and gives an alkaline pH, which is needed for the enzymes from the pancreas to work at their best.
In the duodenum (first part of the small intestine), digestive enzymes are added from the pancreas – amylase breaks down starch (carbohydrates) to glucose, proteases break down protein to amino acids and lipases break down lipids to give fatty acids and glycerol.
Semi-digested food squeezed on by peristalsis to the small intestine. Amylase, proteases and lipases are made in the wall of the small intestine. The lining is covered with villi giving a large surface area so the products of digestion can be absorbed into the blood stream efficiently.
The remains of the sandwich, which cannot be digested, pass into the large intestine. Here water is removed. What is left is the faeces and these are passed out of the body through the anus.
Any other sensible points.

5.6

1 A 3, B 4, C 1, D 2
2 A folded gut wall has a much larger surface area over which nutrients can be absorbed.
3 a Because they have flattened villi, so much smaller surface area is available for absorption of digested food; so much less food is absorbed; they don't get enough glucose and other nutrients and so lose weight and tend to be thin.
   b Someone with coeliac disease is affected by gluten. The villi are flattened and surface area for absorption of digested food products is lost. Without gluten in the diet, the gut can recover and the villi will reappear. Then the body can absorb nutrients properly from the gut and so gain weight, etc.

Answers to end of chapter summary questions

1 salivary amylase and/or pancreatic amylase and/or small intestine amylase break down insoluble starch into soluble sugars/glucose
   • stomach/pancreatic/small intestine protease breaks down insoluble proteins into soluble amino acids
   • pancreatic/small intestine lipase breaks down insoluble fats/lipids into soluble fatty acids and glycerol
   • the small soluble molecules can be absorbed into the blood from the small intestine
2 a Acid in the stomach creates optimal conditions for protease enzymes to catalyse the breakdown of food and kills bacteria.
     The alkaline pH of the small intestine creates the optimal conditions for the enzymes of the small intestine to work.
   b Bile neutralises the acid of the stomach to make the alkaline conditions of the small intestine.
   c Villi greatly increase the uptake of digested food by diffusion. Only a certain number of digested food molecules can diffuse over a given surface area of gut lining at any one time. Increasing the surface area means that there is more room for diffusion to take place.
3 a A molecule made up of long chains of amino acids.
   b

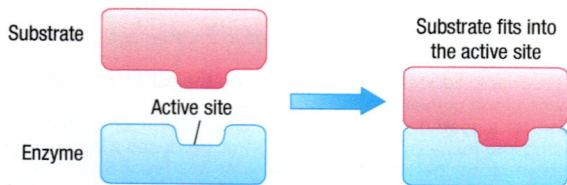

Substrate

Substrate fits into the active site

Active site

Enzyme

   c Enzyme reactions speed up with the increase in temperature as the enzymes and substrates are moving faster with more energy, so more likely to collide and react. An enzyme can only work efficiently at the correct pH for it because the pH affects the forces that form the specific active site.
4 a [Award marks for clearly labelled axes, correct axes, accurate plotting of points on graph etc.]
   b The reaction speeds up with the increase in temperature.
     Particles move faster with more energy, so more likely to collide react.
   c [Award marks for clearly labelled axes, correct axes, accurate plotting of points on graph etc.]
   d That it increases the rate up to about 40 °C and after that, the rate of the reaction decreases and eventually stops.
   e Manganese(IV) oxide is a chemical and not adversely affected by temperature. Catalase is an enzyme made of protein – as temperature goes up, the enzyme is denatured, the shape of the active site is lost and it can no longer catalyse the reaction.
5 a Any two from: temperature of test tubes; volume of solution; volume of A and B protease; concentration of A and B protease.
   b Repeated the investigation/repeat each test three times and take a mean/repeat and discard any anomalous results.
   c Optimum activity of enzyme A is pH 10. optimum pH of B is 2.
   d Enzyme B must be pepsin because stomach is acid/has a pH of about 2. Enzyme A must be trypsin as pancreatic enzymes work best at alkaline pH/pH of 8–9.

Answers to end of chapter practice questions

1 a i A = Large intestine (1), B = pancreas (1), C = gall bladder (1), D = stomach (1)
     ii D (1); C (1); B (1) and D (1); A (1).
   b Scientific points:
     • salivary amylase and/or pancreatic amylase and/or small intestine amylase break down insoluble starch into soluble sugars/glucose
     • stomach/pancreatic/small intestine protease breaks down insoluble proteins into soluble amino acids
     • pancreatic/small intestine lipase breaks down insoluble fats/lipids into soluble fatty acids and glycerol
     • the small soluble molecules can be absorbed into the blood from the small intestine. *(6 marks maximum)*
2 a i monosaccharides (1)   ii Glycerol (1)
     iii amino acids (1)   iv fatty acids (1)
   b i Biological catalysts, they increase the rate of chemical reactions without being used up. (1)
     ii the active site is complementary to the substrate, like a lock and key, to allow the enzyme and substrate to bind together. (3)
     iii Denatured (1)
     iv High temperature (1); incorrect pH (1)
   c Muscles (1) - structural and locomotory (1)
     Antibodies (1) - defence, destroy pathogens (1)
     Hormones (1) - regulatory, homeostasis (1)
3 a Microorganisms (1)
   b i Proteases (1) ii Amino acids (both words) (1)
   c i 14 minutes (1)
     ii Enzyme Z (1) – it takes the least time (to pre-digest protein)/works fastest (1) *(Allow only 7 minutes/less time/faster; do not allow 'works best'.)*
     iii Temperature (1); pH (1)

6 Nervous coordination and behaviour

6.1

1 a To take in information from the environment around you and coordinate the response of the body so you can react to your surroundings.
   b A neurone is a single nerve cell, a nerve is a bundle of hundreds or thousands of neurones.
   c A sensory neurone carries impulses from your sense organs to the central nervous system, a motor neurone carries information from the CNS to the effector organs – muscles and glands – of your body.
2 [Marks awarded for table showing receptors for light, sound, position, smell (could also have temperature, pain, pressure) with student example of a stimulus for each one.]
3 Light from the fruit is detected by the sensory receptors in the eyes, an impulse travels along the sensory neurone to the brain, information is processed in the brain and an impulse is sent along a motor neurone to the muscles of the arm and hand so you pick up the fruit and put it in your mouth. [Give credit if students add anything further, e.g., sensory impulses from mouth/nose to brain with information about taste, smell of fruit, touch sensors send impulses about presence of fruit, motor impulses to muscles for chewing, etc.]

6.2

1 a They enable you to avoid damage and danger because they happen very fast. They control many of the vital functions of the body, such as breathing, without the need for conscious thought.
   b It would slow the process down so would not be so effective at preventing damage. It would be very difficult to consciously control breathing, heart rate, digestion, etc., and still be able to do anything else.
2 Reflex actions to operate automatically, even when you are asleep, so cannot rely on conscious thought processes, unlike speaking and eating, which you choose when to do.
3 Stimulus → receptor → sensory neurone → synapse → chemical message → relay neurone → synapse → chemical message → motor neurone → muscles in leg lift the foot.

6.3

1 An action made in response to a stimulus that modifies the relationship between the organism and the environment.
2 Make a table to compare the five types of behaviour.
3 Research and write a report.

6.4

1 Sound signals – making, hearing and interpreting sounds.
Visual signals – actions, ornaments and colours
Chemical signals – pheromones, used to identify other individuals
2 a Loud singing from birds indicates where their territory is to other birds and other species.
  b Dogs bark to show aggression towards another animal
  c Male cicadas make sounds by rubbing parts of their body to attract a female
3 Discuss the statement.

6.5

1 Mating for life, having several mates over a lifetime, having one mate for a breeding season, and having several mates in a single breeding season
2 a To display how strong and fit they are without the physical risk of fighting another male
  b Fighting another male to compete for the female - advantageous for strong animals but a risk to life.
   Bringing gifts of food – no risk to life but requires energy to find the food Having extravagant plumage – no risk to life by displaying it but can have a detrimental effect on the animal the rest of the time,for example impeding flight speed so easier target for predators
3 a Leaving offspring after birth to look after themselves, advantage is it allows parents to not use more energy on them so they can go on to produce more offspring, but the risk is that fewer will survive without parental protection.
   Parents that look after offspring can suffer from starvation and are therefore easily predated, and harder for them to produce more offspring quickly. However the offspring are more likely to survive.
  b Discuss the statement

6.6

1 By giving a positive reward in response to the desired behaviour and not to the undesirable one.
2 Dogs trained by operant conditioning to respond by sitting and staring or barking at people trapped in the rubble. Difficulties involve differentiating between people trapped and people not trapped - dog must only be searching for trapped ones, trained as above. Also dangerous and busy scenario so dog must be trained by rewarding him for staying calm and focused in these situations.
3 Discuss the purposes, changed behaviours and ways that this has been done.

Answers to end of chapter summary questions

1 a F   d B
  b C   e A
  c D   f E
2 a It enables you to react to your surroundings and to coordinate your behaviour.
  b i Eye   iii Skin
    ii Ear   iv Skin
  c Diagram of reflex arc. The explanation needs to include the following points: reference to three types of neurone: a sensory neurone; a motor neurone; a relay neurone. The relay neurone is found in the CNS, often in the spinal cord. An electrical impulse passes from the sensory receptor, along the sensory neurone to the CNS. It then passes to a relay neurone and straight back along a motor neurone to the effector organ (usually a muscle in a reflex). This is known as the 'reflex arc'. The junction between one neurone and the next is known as a 'synapse'. The time between the stimulus and the reflex action is as short as possible. This allows you to react to danger without thinking about it.
3 a Habituation takes place when a stimulus is repeated many times and has no effect on an organism, either good or bad. Eventually the organism stops responding to the stimulus and the response is lost permanently.
   Importance: enables animals to stop reacting to neutral features of the natural world such as the wind, water movements, rain etc so their nervous system is not constantly firing false alarms

  b i Imprinting
    ii Allows young animals to identify and attach to their parents when young and to recognise and respond to other organisms of the same species when adult.
    iii Scientists presented newly hatched birds with a microlight lane so they imprinted on that. They could then use the plane as the parent, feeding them from it, use it to teach them to fly and mimic the parents in teaching them the migration route, increasing the chances of the birds surviving and going on the meet and bred with their own kind.
4 a Operant conditioning
  b Pavlov's dogs showed classical conditioning – they linked normal behaviour eg the response to food, to a neutral signal – the ringing of a bell. In operant conditioning the animal carries out a random behaviour that delivers a reward or punishment. This reinforces the use of the behaviour (or prevents it) so the animal learns that carrying out a particular behaviour has a particular consequence and repeats it
  c Humans present animals with the opportunity to carry out a random behaviour which is what they want them to do. They then reward that behaviour with attention, treats etc When the animal does not carry out the desired behaviour it is 'punished' by lack of attention, no treats. The animal learns that a particular behaviour gets reward and repeats it. Two extra marks for specific example.
5 a Visual signals, sound, chemical signals 3 marks
  b Sound – eg can be very specific, lots of variations eg human speech, primate communication; can carry a long way eg howler monkeys, whales; any other sensible point; any damage to sound producing mechanism prevents communication, loss of ability to hear prevents communication, any other sensible point
   Visual signals – can be very species specific eg peacock tail, can be very dramatic or very subtle eg facial expressions, tail movements, any other sensible point; any damage to the ability to see eg blindness prevents communication, loss of signalling parts eg tail can reduce communication
   Smell: Very species specific eg moths, can be used to communicate sexual state eg most mammals, any other sensible point; Can easily be lost in wind or water; anything that reduces ability to smell eg infection of nose reduces ability to communicate.
6 a A set of behaviours that help determine which animals will mate and develop bonds between the pair. It advertises the quality of a male as a potential mate.
  b Impala – size, strength, size of horns, ability to fight other males
   Bower bird – ability to build a nest, supply decoration and food for female and her offspring
  c Impala adv: winner of fights usually gets mating rights over a group of females, females can see that the winning male is big and strong and has good genes to pass on to their offspring: disadv – males may be injured during fights, effort taken to keep fighting may mean males cannot survive after the breeding season – any other sensible suggestions
   Bower bird: Av – male bird does not risk harm in his display, male can demonstrate ability to provide for the female and her off-spring, if one female doesn't like the bower another may do; disadv – it can be exhausting, takes a lot of effort and no females may choose – any other sensible points
  d Benefits – huge numbers of offspring so some likely to survive even if most die, relatively little drain on parental resources; limitations are mainly that the parent has no way of increasing the survival chances of the offspring and they could all be wiped out

Answers to end of chapter practice questions

1 a Receptor(s) (1), effector(s) (1)
   b i Nose/tongue (1)   iii Eye (1)
      ii Ear (1)   iv Skin (1)
   c By a nerve impulse to a muscle to contract (1); by causing a gland to secrete a hormone into the blood (1).
2 a Receptor (1), relay neurone (1), motor neurone (1), response (1)
   b Stimulus = hot saucepan/heat/high temperature (1); Effector = (arm) muscle (1)
   c i When the (electrical) impulse reaches the synapse it causes a chemical substance to be released (1), which diffuses across the gap and causes an impulse to be initiated/triggered/started in the next neurone (1).
      ii Central nervous system (allow brain, spinal cord) (1).
   d 1.2/0.02 = 60 metres per second (*2 marks for correct answer, even if no working shown. If answer is incorrect allow 1 mark for 1.2/0.02*)
3 a Independent variable = number of frames it took the hammer to move to the knee/speed of hammer (1); dependent variable = distance moved by lower leg (1); control variable = distance hammer moved (1)
   b Can measure very short times/fast speeds (1); keeps a permanent record (1).
   c Repeating the trial and calculating means (1).
   d The faster the speed of the hammer the greater the movement (1), up to a maximum of 10 cm movement (1).

7 Homeostasis

7.1

1 a Chemical substances that coordinate many body processes. Made by endocrine glands and secreted directly into the blood stream. They are carried to their target organs in the blood.
   b Hormones: chemical, made by glands and carried in the blood stream, often relatively slow response, often act over a long period of time. Nerves: electrical, impulses travel in neurones, very fast response, act instantly.
2 a To stop too much water moving in or out of cells, damaging and destroying them.
   b Because the enzymes work best at 37 °C.
   c Because blood sugar that is too high or too low causes problems in the body.
3 a i Losing water through sweating.
      ii Losing salt through sweating.
      iii Temperature going up with exercising.
   b Sweating cools you down and helps to keep the body temperature constant – a costume makes you sweat more (as you get hotter), which means you lose more water – but also makes it harder for sweat to evaporate (so you don't cool so effectively). Also, a costume is heavy so it's harder work to run.

7.2

1 a This is the temperature at which enzymes work best.
   b Above around 40 °C the enzymes which are made of protein start to denature. This means the shape of the active sites change and the enzymes can no longer catalyse the reactions of the cell so the cells die. This quickly results in death of the whole organism and is irreversible. Below 35 °C the reactions of the body slow down even with enzymes catalysing them and they cannot take place fast enough to maintain life.
2 a The thermoregulatory centre in the brain is sensitive to the temperature of the blood flowing through it. It also receives information about the skin temperature from receptors in the skin and coordinates the body responses to keep the core temperature at 37 °C.
   b Temperature sensors in the skin send impulses to the thermoregulatory centre in the brain giving information about the temperature of the skin and the things it touches. This is important for maintaining the core temperature because if the external surroundings and the skin are cold, the body will tend to conserve heat to keep the core temperature up, and vice versa.
3 If core temperature increases, to lower body temperature: blood vessels supplying capillaries in skin dilate; more blood in capillaries so more heat is lost, more sweat produced by sweat glands, which cools the body as it evaporates. If core temperature decreases, to raise body temperature: blood vessels supplying blood to skin capillaries constrict; less blood transported to surface of skin so less heat is lost, shivering occurs by rapid muscle movement, which needs respiration – releasing heat energy.

7.3

1 a Hormone: a chemical message carried in the blood which causes a change in the body.
   b Insulin: a hormone made in the pancreas which causes glucose to pass from the blood into the cells where it is needed for energy.
   c Diabetes: a condition when the pancreas cannot make enough insulin to control the blood sugar.
   d Glycogen: an insoluble carbohydrate stored in the liver.
2 a Blood glucose levels go up above the ideal range. This is detected by the pancreas, which then secretes insulin. Insulin causes the liver to convert glucose to glycogen. This causes glucose to move out of the blood into the cells of the body, therefore lowering blood glucose levels. When the blood sugar level falls, glucose is released back into the blood.
      If the blood glucose level drops below the ideal range, this is detected by the pancreas. The pancreas secretes glucagon, which causes the liver to convert glycogen into glucose, which increases the blood glucose level.
   b Glucose is needed for cellular respiration, which releases energy for everything. Too much or too little glucose in the blood causes problems.
3 Type 1 diabetes is a condition where the pancreas does not make enough or any insulin. It is treated by injections of insulin to help control blood glucose levels. Your diet needs to be carefully controlled with regular meals and the intake of carbohydrate carefully monitored.
   Type 2 diabetes is a condition where either the pancreas does not make enough insulin or your body does not react properly to the insulin than is produced. Type 2 diabetes is linked to obesity, old age and a lack of exercise. It can be treated by improving the diet, the amount of exercise and losing weight as well as insulin injections.

7.4

1 a People with type 1 diabetes need insulin injections but they also need to monitor and control their food intake carefully.
   b Type 2 diabetes can often be treated and controlled or even cured by eating a carefully controlled balanced diet, losing weight and taking regular exercise. Because it can be caused by too little insulin being produced or by the cells becoming insensitive to insulin, it can also sometimes be treated by drugs that help the insulin the body makes have a bigger effect on the body cells, help the pancreas make more insulin or reduce the amount of glucose absorbed from the gut. If none of these treatments work, the patient will have to use insulin.
2 a Original – from pancreases of cattle and pigs used for meat; no control over quantities as used what was available from slaughterhouses; not exactly the same chemically.
      Modern – produced by genetically modified bacteria; exact quantity and quality control; exactly the same as naturally occurring human insulin.
      Genetically modified insulin is better as it is a match for the natural hormone and both quantity and quality of the product can be controlled to give better glucose control.
   b Insulin treatment widely available; patient deals with it themselves; relatively cheap, etc.
      Pancreatic transplant is a good idea but complex surgery; high risk; expensive; patients have to be on immunosuppressant drugs for the rest of their lives; needs repeating eventually; not enough donors. These are the reasons why it is not more widely used.

Answers to end of chapter summary questions

1 a The maintenance of a constant internal environment, for example in terms of a constant core temperature, the water and ion content of the body and the blood glucose levels.
   b For cells to work properly they need to be at the right temperature (so enzymes work optimally); they need to be surrounded by the correct concentration of water and mineral ions in the blood so osmosis doesn't cause problems; they need glucose to provide energy and they need waste products to be removed as build up can change pH or poison systems. This is why the body systems must be controlled within fairly narrow limits.

2 a i

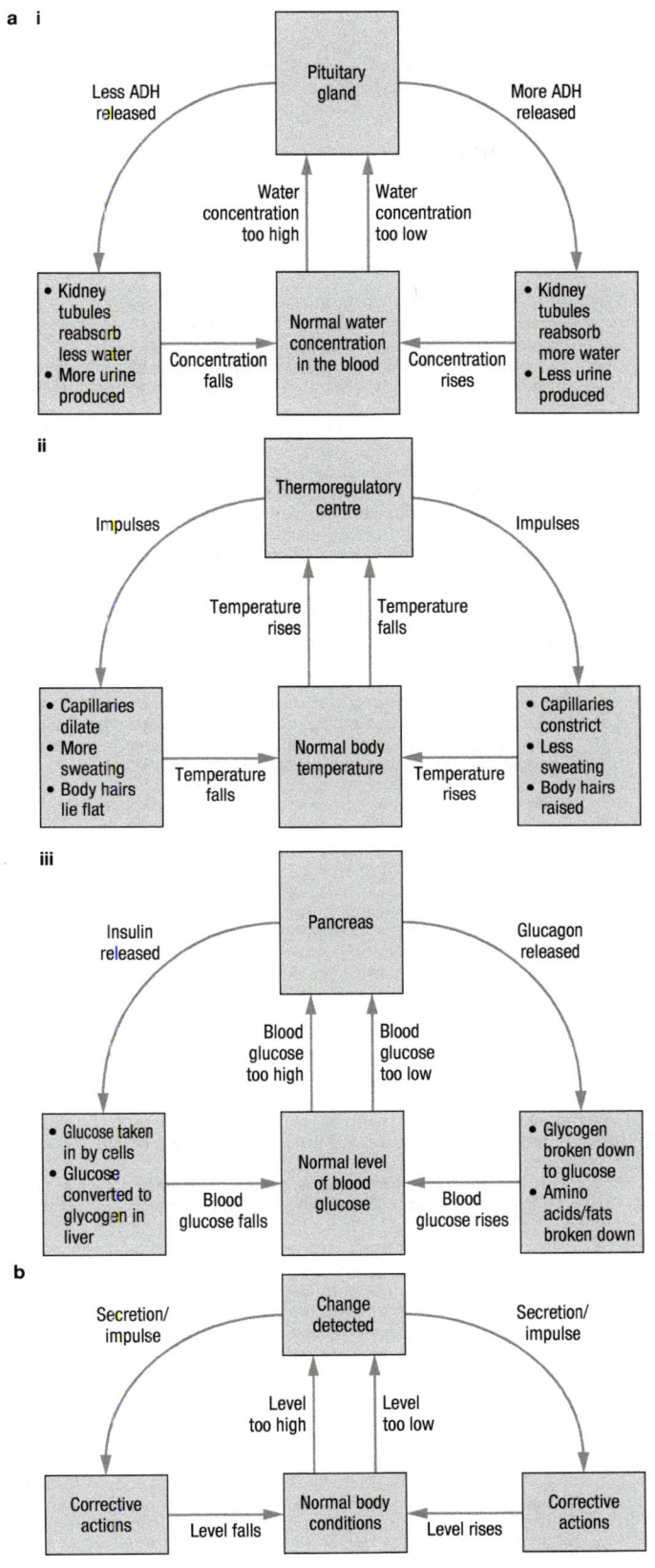

ii

iii

b

3 a Deaths increased – gradually at first and then by a lot.
   b Around 25 °C.
   c When very hot (often humid), sweat doesn't evaporate to cool people down; people lose a lot of water by sweating so become dehydrated and therefore can't sweat and cool down; exercise in heat generates heat in muscles; body can't get rid of it by sweating, etc.; any sensible points.
4 a They go up.
   b About 60–120 milligrams per litre.
   c About 50–310 milligrams per litre.
   d Insulin injections keep blood sugar levels within a reasonable range; prevent loss of blood sugar in the urine; allow cells to take up glucose etc. Limitations – can't keep blood sugar within the narrow range of natural insulin control.
   e Carbohydrates broken down into glucose (blood sugar), so the more carbohydrate-rich food eaten, the higher the blood sugar levels will climb and the harder it is for the insulin injections to maintain safe and healthy levels of blood sugar.

Answers to end of chapter practice questions

1 a Homeostasis (1)
   b Any three from: core body temperature, water content of body, ion content of body, blood glucose concentration (3).
   c i   Brain (1)
      ii Pancreas (1)
2 a i   Thermoregulatory centre (allow thermoregulation centre/ hypothalamus) (1).
      ii It has receptors (1) (ignore receptors in skin) that detect changes in the temperature of the blood/plasma (1).
   b Shivering causes the muscles to contract (1) resulting in increased respiration/more heat released/produced (1).
   c i   Blood vessels/arteries/arterioles dilate/widen (1).
      ii There will be more blood close to/near surface (1) so more heat is lost/heat is lost faster/the body cools faster (1).
3 a All small molecules (accept list if it contains both waste and useful molecules) pass into the filtrate (1), but large ones such as proteins and lipids do not (1). Useful molecules (accept glucose, amino acids, some ions, some water) are reabsorbed (1), but waste molecules (accept urea, water, ions, poisons) are passed into urine. (1)
   b i   Brain (1)
      ii ADH (anti-diuretic hormone) (1)
      iii Pituitary (1)
   c Scientific points:
      As it was hot the worker sweated a lot so the water content of his blood fell. This was detected in the brain and the pituitary gland released ADH. The ADH caused the kidneys to reabsorb more water, so his urine was low volume and concentrated. After he drank the water the water content of his blood went up, so the brain caused the pituitary to stop making ADH, so less water was reabsorbed by the kidney and the urine was higher in volume and dilute.
                                             (6 marks maximum)

8 Defending ourselves against disease

8.1

1 a Microorganisms known as pathogens. Pathogens include bacteria and viruses.
   b Bacteria may make you ill as a result of the toxins they produce as they divide rapidly or they may cause direct damage to your cells. Viruses take over the cells of your body as they reproduce, damaging and destroying cells, which is how they cause disease.
   c Pathogens make you ill as a result of the way your body reacts to the toxins they make or the cell damage they cause.
2 Any sensible suggestions should be accepted, such as: wiping work surfaces, cleaning toilets, using tissues to blow nose, washing hands before handling food, for example:
3 Any sensible points, e.g.,
   • Pathogens are very small so until the development of microscopes people had no way of seeing bacteria or viruses.
   • Because people couldn't see microorganisms it was very difficult to work out or understand how diseases spread.
   • Evidence, such as doctors handwashing reducing the deaths of women after giving birth, was seen as against the Bible, which was very powerful.
   • Still difficult to convince people if their ideas are entrenched, e.g., Barry Marshall attempting to show people that most stomach ulcers are the result of bacterial infections.

8.2

1 a Droplet infection, direct contact, contaminated food and drink, break in the skin.

b When you cough, sneeze or talk, droplets full of pathogens pass into the air to be breathed in by someone else.
Pathogens on skin passed to someone else's skin on contact.
Pathogens taken in on food or in drink.
Pathogens can get through the barrier of the skin to the tissue underneath.

2 a Pathogens cannot be stopped from getting into cuts.

b You have not got enough white blood cells to ingest pathogens or to produce antibodies/antitoxins, so pathogens are not destroyed.

3 a Prevents pathogens getting from your hands to the food.

b Removes pathogens from where they might come into contact with other people or get on your hands.

c Prevents pathogens from the gut being taken in with drinking water.

4 Explanation to include the ingestion of microorganisms, the production of antibodies and antitoxins.

8.3

1 a A unique protein found on the surface membrane of the cells of any organism.

b A protein made by white blood cells in response to specific antigens. The antibodies attach themselves to the antigens and destroy the cell/pathogen.

c Any sensible suggestions, for example: bacterial – TB, tetanus, diphtheria; viral – polio, measles, whooping cough.

2 a Every cell has unique proteins on its surface called 'antigens'. Your immune system recognises that the antigens on the microorganisms that get into your system are different from the ones on your own cells. Your white blood cells then make antibodies to destroy the antigens/pathogens. Once your white blood cells have learnt the right antibody needed to tackle a particular pathogen, they can make that antibody very quickly if the pathogen gets into your system again, and in this way you develop immunity to that disease.

b A small quantity of dead or inactive pathogen is introduced into your body. This gives your white blood cells the chance to develop the right antibodies against the pathogen without you getting ill. Then, if you meet the live pathogens, your body can respond rapidly, making the right antibodies just as if you had already had the disease.

3 Vaccines can be made using inactive viruses or bacteria so can stimulate antibody production against either type of pathogen, thereby developing immunity.

8.4

1 Paracetamol relieves symptoms/makes you feel better, whereas antibiotics kill the bacteria and actually make you better.

2 a He noticed a clear area around mould growing on bacterial plates.

b It was difficult to get much penicillin out of the mould and it does not keep easily.

c Florey and Chain.

3 Viral pathogens reproduce inside your cells, so it is very difficult to develop a drug that destroys them without destroying your cells as well.

8.5

1 Students should show clear understanding of the different stages involved in the development of antibiotic resistance. Colony of bacteria treated with antibiotic 1 → 5% have mutation and survive → the surviving bacteria are treated with antibiotic 2 → 5% have a mutation and are resistant to antibiotic 1 and 2 → etc.

2 a Bacterium

b MRSA has developed resistance to many antibiotics including methicillin as a result of them being used extensively in hospitals. Increasingly small colonies of antibiotic-resistant bacteria have survived and reproduced until now the majority of *Staphylococcus aureus* in hospitals are resistant to common antibiotics.

3 a Increased use of antibiotics leading to more resistant bacteria, lower hygiene standards in hospitals, people failing to wash their hands between patients, visitors, etc. Any other sensible point.

b Possible reasons include: men are more likely to pick up infections than women; men are less likely to wash their hands thoroughly or use the alcohol gel in hospitals as patients or visitors than women; more men die than women; more men go into hospital than women; men are less likely to complete a course of medicine than women and so develop a resistant strain; any other sensible point.

c A big effort has been made to reduce these infections including: reduction in prescription of antibiotics; treating conditions with very specific antibiotics for the pathogen rather than broad spectrum antibiotics; constant reminders to medical staff to wash hands/use alcohol gel between patients; constant reminders to patients and visitors to wash hands/use alcohol gels on entering and leaving hospitals, doctor's surgeries, wards, etc.; increasing hygiene standards in hospitals; nursing patients affected by antibiotic-resistant strains of bacteria in isolation; any other sensible point.

8.6

1 a To find out more about them. To find out which nutrients they need to grow and to investigate what will affect them and stop them growing.

b Agar jelly is a culture medium containing nutrients. It provides the carbohydrates and other nutrients needed by bacteria, which are grown on it.

2 Using up the available food and oxygen; build up of waste products such as carbon dioxide and other toxins.

3 a Because this reduces the chances of growing microorganisms which might be harmful to people.

b Because at higher temperatures the microorganisms grow much more rapidly.

Answers to end of chapter summary questions

1 a Breathed into the lungs; taken in through the mouth; through cuts and breaks in the skin; any other sensible suggestion.

b They reproduce rapidly and often produce toxins or may damage cells directly. The body's reaction to these situations causes the symptoms of disease.

c Viruses take over the cells of your body as they reproduce, damaging and destroying the cells. This, along with the reaction of the body to the damage, causes the symptoms of disease.

2 a Each time a colony of bacteria are exposed to an antibiotic, some individual bacteria may survive due to genetic mutations unique to them. The population of this resistant strain will steadily increase as the non-resistant strain are killed by the antibiotic. Resistance to vancomycin can develop through this process of natural selection. The pathogen can then spread quickly as patients will not have immunity against the new strain.

b Use antibiotics carefully – only when they are needed – and make sure people always finish the course.

3 a Use sterile Petri dish and agar.
Sterilise the inoculating loop by heating it to red hot in the flame of a Bunsen and then let it cool. Do not put the loop down as it cools.
Innoculate agar with zigzag streaks of bacteria using sterile loop.
Replace the lid on the dish quickly to avoid contamination.
Secure the lid with adhesive tape but do not seal.
Label the culture and incubate at no warmer than 25 °C.

b Include points such as: Inoculate agar plates with bacteria – ideally from school floor.
Add circles of filter paper soaked with different strengths of the disinfectant and incubate at no higher than 25 °C.
Look for areas of clear agar around the disinfectant-soaked disk.
Recommend lowest concentration that destroys the bacteria.

4 The skin covers the body and prevents entry of pathogens; if your skin is cut you bleed, washing pathogens out, and then the blood clots forming a seal over the healing skin and preventing the entry of any pathogens. The breathing system produces sticky mucus, which traps pathogens from the air, and cilia, which move the mucus away from the lungs out of the body or into the stomach. The stomach produces acid, which kills most of the pathogens taken in through. The mouth the white blood cells of the immune system form antibodies against the antigens on any pathogens that get into the body.
Some of the white blood cells of the immune system engulf and digest pathogens.
Some of the white blood cells of the immune system produce antitoxins against the toxins produced by some bacteria so they no longer make you ill.

5 a A vaccine contains a dead or weakened form of a disease-causing pathogen. It works by triggering the body's natural immune responses. A small amount of the vaccine is injected into the body. The white blood cells develop the antibodies needed to destroy the pathogen without you becoming ill. Then, if you take the live pathogens into your body, the immune system can provide the right antibodies very quickly (as if you had had the disease) and destroy the pathogen before it can make you ill.

b Some diseases are so dangerous that you can be dead or permanently damaged before the body has time to develop the right antibodies. These are the diseases you are usually vaccinated against. It is not worth the expense of vaccinating against less serious diseases.

6 a 1998

b It has varied but the general trend has been for it to increase, particularly in the late 2000s.

c It went up dramatically.

d Because initially the majority of children were still vaccinated against measles and mumps. As years passed there was a bigger and bigger population of unvaccinated children who were vulnerable to the diseases.

e i So that the population has herd immunity – when most people are immune to a disease, it cannot spread; this protects the small number of people not vaccinated.

ii Expect it to continue to rise for a few more years perhaps and then gradually fall again as the proportion of the population who are vaccinated continues to rise.

Answers to end of chapter practice questions

1 a Pathogens (1)
b Viruses (1)
c Toxins (1)
2 a i Lives inside cells (1)
ii Inactive (1)
iii Antibodies (1)
b i 1950 (1)
ii 8 (years) (1)
iii Any one from: disease could be reintroduced (from abroad); disease would spread if it came back; protection on holiday abroad; high proportion of immune people needed to prevent epidemic (1). *(1 mark maximum)*
3 a i Defence (accept specific functions of white cells) (1).
ii Forming clot at site of wound (1).
iii 100 ÷ 0.008 (1); equals 12 500 (1) (*Correct answer with or without working gains 2 marks; ignore any units*)
iv The size of red blood cell is approximately same size as capillary or red blood cell is too big (1); therefore there is no room for more than one cell or only one can fit (1). (*Allow use of numbers; do not accept capillaries are narrow*)
v In lungs, oxygen *diffuses* from the alveoli into the blood (1); in the red blood cell, oxygen combines with *haemoglobin*, forming *oxyhaemoglobin* (1); in tissues, *oxyhaemoglobin* splits up, releasing *oxygen*, which *diffuses* into the cells (1). (*For each mark, whole statement is required*)
b i (Student Y) because she had the lower resting heart rate (1); the lower heart rate increase (1) and the quicker recovery time (1). (*Accept converse for Student X*)
ii When exercising the *rate* of aerobic respiration in the muscles is higher (1); (the increased heart rate) increases *rate* of delivery of oxygen to the (respiring) muscles (1); and increases *rate* of delivery of glucose to the (respiring) muscles (1); and results in faster removal of carbon dioxide and lactic acid (1).

9 Plants as organisms

9.1

1 a CO_2 comes from the air; water from the soil; light energy from the Sun/electric light.
b Carbon dioxide and water from the water it lives in; light from Sun or electric light.
2 From the air into the air spaces in the leaf; into plant cells; into chloroplasts; joined with water to make glucose; converted to starch for storage.
3 When a plant is in the light it carries out photosynthesis. During photosynthesis, the plant makes glucose from carbon dioxide and water using energy from light. The plant in the light produces oxygen that will relight the glowing splint. The plant in the dark produces carbon dioxide which will extinguish it.

9.2

1 Carbon dioxide, light and temperature.
2 a i Light levels are low until sunrise, temperature falls overnight.
ii Carbon dioxide will limit photosynthesis.
iii Low light levels in winter, days are shorter, temperature colder.
iv Trees will limit the light, temperature will be warm so carbon dioxide will be limiting.
b Each case is within the natural environment and light, temperature and carbon dioxide levels can interact meaning that at any time, any one of them may be the limiting factor for photosynthesis.
3 a As light intensity increases, so does the rate of photosynthesis. This tells us that light intensity is a limiting factor.
b An increase in light intensity has no effect on the rate of photosynthesis, so it is no longer a limiting factor; something else probably is.
c Temperature acts as a normal limiting factor to begin with; increase in temperature increases the rate of photosynthesis. But above a certain temperature, the enzymes in the cells are destroyed so no photosynthesis can take place.

9.3

1 Respiration; energy for cell functions; growth; reproduction; building up smaller molecules into bigger molecules; converted into starch for storage; making cellulose; making amino acids; building up fats and oils for a food store in seeds.
2 a Glucose is soluble and would affect the movement of water into and out of the plant cells. Starch is insoluble and so does not disturb the water balance of the plant.
b Leaves, stems, roots and storage organs.
c [Any sensible suggestions involving a slice of potato and dilute iodine solution.]
3 Bogs are wet and peaty and the soil contains very few minerals, especially nitrates. Plants need nitrates form the soil to make amino acids and build them into proteins. So many plants cannot grow well on bogs. Carnivorous plants trap insects and digest their bodies, which provides a good supply of nitrates and other minerals. So they can grow and thrive on bogs as they do not rely on them for their minerals.

9.4

1 a To move food made in the leaves to the rest of the plant and to transport water and mineral ions taken from the soil to the rest of the plant.
b All the cells need dissolved sugar for cellular respiration and also as the basis for making new plant material. Water is needed for photosynthesis to make sugar, and also to keep the cells rigid to support the plant.

2

Xylem	Phloem
Mature cells are dead	Living cells
Transports water and minerals from the soils around the plant	Transports dissolved sugars from photosynthesis around the plant
Found on the inside of vascular bundles in the stem	Found on the outside of vascular bundles in the stem

3 The transport tissue of young trees is on the outside of the trunk under the bark. In young trees the bark is relatively soft and animals such as deer and even rabbits will eat it. If the bark is nibbled off all around the tree, water cannot move up from the roots and sugars cannot move down to the roots and the young tree will die. Plastic covers protect the young bark so it cannot be eaten. They can be removed once the trees are more mature. If this isn't done, most of the young trees are likely to be destroyed and the woodland will eventually die as the old, mature trees are not replaced.

Answers to end of chapter summary questions

1 a
carbon dioxide + water $\xrightarrow{\text{light energy}}$ glucose + oxygen

$6CO_2 + 6H_2O \xrightarrow{\text{light energy}} C_6H_{12}O_6 + 6O_2$

b Starch
c i Credit accurately drawn graphs, correctly labelled axes, etc.
ii Plants in higher light intensity photosynthesise faster and therefore produce more food and grow well. Light will not limit them – CO_2 or temperature might. For plants in lower light, the light is a limiting factor on their growth.
2 a Because conditions are good for photosynthesis – plenty of light, water and warm temperatures allow rapid photosynthesis, which gives rapid growth.
b Made from glucose made in photosynthesis.
c Energy store for the developing embryo as the seed germinates and grows but before it can photosynthesise.
d Energy for the plant cells for growth; to make complex carbohydrates such as cellulose for plant cell walls and starch for storage; combined with nitrates from the soil to make amino acids and proteins; any other sensible point.
3 Adaptations of leaves: thin and flat to give short diffusion distances; specialised cells with big air spaces and big surface area for gases to diffuse in and out of cells; xylem tissue bringing water into the leaf; stomata to allow carbon dioxide, oxygen and water vapour to diffuse in and out of leaves. All depends on passive processes of diffusion along concentration gradients.
Adaptations of roots: long and thin to give big surface area for the absorption of water and mineral ions; specialised root hair cells to increase the surface area for movement of water into the roots and active transport of mineral ions; movement of water passive along concentration gradients, movement of mineral ions active.

4 a A = xylem   B = phloem

b Students reproduce a diagram of tissue A with annotations to labels eg hollow tube of xylem allows free flow of water and minerals through plant; lignin spirals both withstand the pressure of the water moving through them and support and strengthen plant. Additional point: Xylem is a dead tissue and so not affected by heat, infection etc

c An annotated version of fig B phloem is a living tissue – needs energy transferred from respiration to move sugars into and out of cells; phloem vessels almost hollow – just sieve plates at ends of cells – so water containing dissolved sugars etc can move easily through the plant.; companion cells support the phloem vessels and contain mitochondria to provide the energy needed to move the dissolved sugars into and out of cells/up and down the plant.

5 a An increase in light level allows the maximum amount of photosynthesis to take place up to the point where other factors are limiting. It is important to provide enough light to allow maximum photosynthesis to take place but not to pay for more lighting than needed as that is a waste of money

b As you need carbon dioxide for photosynthesis an increase in carbon dioxide levels increases the rate of photosynthesis, but too much is toxic.

c Higher temperatures increase the rate of reaction but if the temperature is too high the enzymes will denature.

Answers to end of chapter practice questions

1 a Carbon dioxide (1), oxygen (1).

b The sun (1). The light energy is absorbed by chlorophyll/chloroplasts (1) in the leaf (mesophyll) cells (1).

c Temperature (1) may be limiting as the enzymes work slowly in cold temperatures(1), or light (1) may be limiting as December is a very dark month and it is early morning (1).

d Any **three** from: for storage as starch (1); for storage as fats/oils (1); to make proteins/enzymes/ named protein or enzyme (1); to make cellulose for cell walls.   *(3 marks maximum)*

2 a Translocation – The movement of sugars from the leaves to other tissues (1); Xylem – The cells which transport water in the plant (1); Stomata – Openings in the lower surface of the leaf for gas exchange (1); Phloem – The cells which transport sugars in the plant (1); Transpiration stream – The transport of water from roots to leaves (1).

b To provide turgor for cells to support leaves and stems (1); to dissolve gases/ions/small food molecules for transport (1).

c Any four from the following possible answers, but they must be in the correct order:
The transpiration stream is driven by the evaporation of water from the leaves (1). In hot weather, photosynthesis is faster and uses more water (1). Stomata will be open to take in more carbon dioxide (or at a faster rate), meaning that more water vapour can escape (1). Evaporation happens faster if it is hot, dry and windy, so more water is lost from leaves (1) and more water is drawn in by the roots to replace what is being used up and lost (1).   *(4 marks maximum)*

d Active transport (1), in which energy is required (1) (and a protein in the membrane transports the ion across against the concentration gradient).

e To make (amino acids for) proteins/for DNA synthesis (1).

3 a • There is a clear, balanced and detailed description referring to the data in the graph about light, temperature and carbon dioxide and how to set up a controlled experiment. The answer shows almost faultless spelling, punctuation and grammar. It is coherent and in an organised, logical sequence. It contains a range of appropriate or relevant specialist terms used accurately.   *(5–6 marks)*
• There is some description of setting up a controlled experiment, including at least two variables. There are some errors in spelling, punctuation and grammar. The answer has some structure and organisation. The use of specialist terms has been attempted, but not always accurately.   *(3–4 marks)*
• There is a brief description with reference to setting up several tunnels and mention of at least one variable, but little clarity and detail. The spelling, punctuation and grammar are very weak. The answer is poorly organised with almost no specialist terms and/or their use demonstrating a general lack of understanding of their meaning.   *(1–2 marks)*
• No relevant content.   *(0 marks)*
Examples of biology points made in the response:
• use of term limiting factors
• the more photosynthesis, the more growth
• carbon dioxide optimum around 4%
• plants need water
• control of light intensity
• types of light
• temperature control/25 °C
• idea that light changes with type of plastic/colour of plastic/ thickness of plastic

• idea that might need heating/ventilation to control/monitor temperature
• idea that need to contain the carbon dioxide/have a source of carbon dioxide gas
• reference to having different sets of conditions in each model tunnel to be able to determine optimum/idea that might try slightly lower/higher temperature/carbon dioxide level to check cost effectiveness.

b Any one from the following: possible to mimic large scale events/ idea of/on a small scale; can be used to predict changes/changes in variables; use of a description, e.g., to predict the spread of disease/ can predict the effect of a chemical on all bacteria using a safe organism/can use fast breeding organisms to mimic processes which occur slowly in others/can predict the effect of global warming on organisms in a locality.   *(1 mark maximum)*

4 When a plant is in the light it carries out photosynthesis. During photosynthesis, the plant makes glucose from carbon dioxide and water using energy from light.
The plant in the light produces oxygen that will relight the glowing splint. The plant in the dark is respiring and not photosynthesising and so produces carbon dioxide which will extinguish it.

10 Variation and inheritance

10.1

1 a The gene.

b Offspring inherit information from both parents and so end up with a combination of characteristics, some from the father and some from the mother.

2 a You inherit one set from each parent.

b Genes are carried on the chromosomes, so because chromosomes come in pairs, so do the genes – one from each parent.

c 20 000–25 000.

3 a This is because there are genetic differences between all the members of a species unless they are identical twins or produced by asexual reproduction.

b There are two sources of variety in individuals. One is the genetic information they inherit. The other is the variety in the environment in which they grow and develop. If both of these are identical – both the genetics and the environment – then the organisms should be identical too.

c It is almost impossible for two organisms to experience identical environmental conditions, so there will always be environmentally caused differences between them.

10.2

1 a No fusion of gametes, only one parent, no variety.

b Two parents, fusion of gametes, variety.

c Sex cell.

d The differences between individuals as a result of their genetic material.

2 Advantages of sexual reproduction: mixes genes, leads to variation, allows process of evolution, increases chances of a species surviving if environment changes.
Disadvantages of sexual reproduction: need to find a partner, risk of gametes not finding each other, generally slower.
Advantages of asexual reproduction: only involves one parent, simple and efficient, genes and characteristics conserved.
Disadvantages of sexual reproduction: no genetic variety – only produces clones, no evolution through natural selection.

3 a **i** By bulbs.   **ii** By flowers.

b They have the safety and reliability of asexual reproduction but the genetic variety which is introduced by sexual reproduction to help them survive changes in conditions.

c Asexually produced offspring will all be genetically identical to the single parent. Sexually produced offspring will be genetically different from both parents and from each other as each will be the result of the fusing of a different egg and sperm and each one will have a different genetic mixture.

10.3

1 Genetic variation, variation due to the environment of an organism, variation due to a combination of factors

2 In continuous variation there is a gradual transition between extremes and environment affects the basic genetic variation. The graph showing height and the differences in height of a population shows this continuous variation. In contrast blood groups are inherited as a single gene and you will have one specific blood group – there is no gradual transition between them and so this is an example of discontinuous variation

3 Recognition that as genetically identical (clones) if all other variables are kept the same except temperature, any differences should be the result of temperature on the growth of the seedlings Credit for any sensible suggestions along with recognition of the need to control variables, how to get the most reliable and valid data from the investigation, etc.

10.4
1 23; 22; sex chromosomes; XX; X; Y; male;
2 a He found that characteristics were inherited in clear and predictable patterns. He realised some characteristics were dominant over others and that they never mixed together.
  b No one could see the units of inheritance, so there was no proof of their existence. People were not used to studying careful records of results.
  c Once people could see chromosomes, a mechanism for Mendel's ideas of inheritance became possible.
3 50:50 chance each time of inheriting an X or Y from father, independent of previous children

10.5
1 a Dominant allele – an allele which controls the development of a characteristic even when it is present on only one of the chromosomes.
  b Recessive allele – an allele which only controls the development of a characteristic if it is present on both chromosomes.
  c Marks for each case where students identify correctly the single gene characteristic and the dominant and recessive alleles.
2 Marks awarded for drawing a Punnett square correctly with the appropriate gametes. DD, Dd, dD have dimples; dd has no dimples.
3 see question 3 in 10.6 below

10.6
1 a DNA is deoxyribonucleic acid. These are long strands that twist to form a double helix structure, making chromosomes. Genes are small sections of this DNA.
  b labelled diagram
2 1 Aa   11 aa
    2 aa   12 Aa or AA
    3 Aa   13 AA or Aa
    4 Aa   14 aa
    5 Aa   15 Aa
    6 Aa   16 Aa
    7 Aa   17 Aa
    8 Aa   18 AA or Aa
    9 aa   19 AA or Aa
    10 AA or Aa
3 a 1 and 2 Aa × aa

	A	a
a	Aa	aa
a	Aa	aa

    1Aa: 1aa
    50% or 1:1 chance of offspring being normal pigmentation
  b 7 and 8. Aa a Aa

	A	a
a	AA	Aa
a	Aa	aa

    1AA: 1Aa: 1aa
    3:1 or 75% chance of offspring having normal pigmentation

10.7
1 a A genetic disorder which causes extra fingers or toes.
  b The faulty allele is dominant, so only one parent needs to have the allele and pass it on for the offspring to be affected.
  c A Pp only – as produced a child that was unaffected.
    B Pp – because mother must pass on a recessive allele to produce two unaffected children.
    C–E – could be PP or Pp as each parent has the genotype Pp.
2 a Carriers have a normal dominant allele, so their body works normally.
  b CF (cystic fibrosis) is recessive – must inherit one from each parent to get the disease. But if parents had the disease themselves, they would almost certainly be infertile, so parents must be carriers.
3 [Marks awarded for genetic diagram based on Figure 3 in Student Book, showing how cc (cystic fibrosis) arises.]

10.8
1 a This is a disorder caused by a faulty recessive allele which can be passed on from one generation to another. Because it is recessive parents can carry the allele without realising it. Any affected offspring have to inherit a faulty allele from both parents.
  b The main symptoms are breathlessness, lack of energy and tiredness along with pain and tissue death and sometimes death of the individual. In sickle cell anaemia the red blood cells are sickle shaped instead of the normal biconcave discs. As a result they do not carry oxygen efficiently so the person does not get enough oxygen – they feel breathless and tired. The sickle cells can block small blood vessels and this can cause pain and tissue death. This leaves the person open to severe infections and even death.
2 a RR × Rr

	R	R
R	RR	RR
r	Rr	Rr

    None of the offspring have sickle cell anaemia. 50:50% (1 in 2) chance of a child being a carrier.
  b Rr × Rr

	R	r
R	RR	Rr
r	Rr	rr

    1 in 4 of the offspring have sickle cell anaemia. 50:50% (1 in 2) chance of a child being a carrier.
  c RR × rr

	R	R
r	Rr	Rr
r	Rr	Rr

    None of the offspring have sickle cell anaemia. All of the offspring are carriers.
3 a A whole chromosome disorder when an extra copy of chromosome 21 is inherited. It means the person affected has 47 chromosomes instead of 46.
  b Down's syndrome involves the whole chromosome not a single faulty allele.

Answers to end of chapter summary questions
1 a From a runner – a special stem from the parent plant with small new identical plant on the end.
  b Asexual
  c By sexual reproduction (flowers, pollination, etc.).
  d The new plants from the packet will be similar to, but not identical to their parents – each one will be genetically different. The plants produced by asexual reproduction will be identical to their parents.
2 a A sex cell, e.g., ovum and sperm in humans, pollen and ovule in plants.
  b There are half the number of chromosomes in a gamete than in a normal body cell.
3 a DNA
  b A section of DNA made up of hundreds and thousands of bases which code for a particular protein.
  c A DNA strand is made up of combinations of bases. They are grouped into threes, and three bases codes for a particular amino acid. The arrangement of the bases determines the string of amino acids which is joined together to make a protein. Proteins are important in the structure of cells and also form the enzymes that catalyse all the other reactions of a cell. So by determining the proteins which are made, the DNA strand determines how the organism is put together and what it looks like – its phenotype.
4 Mendel carried out large numbers of genetic crosses on plants, particularly peas, growing in the monastery garden where he lived and worked. He carried out specific breeding experiments and counted the different offspring carefully, kept careful records and analysed his results. This was very unusual for the time, when statistical analysis of data was very rare. Mendel realised that there were clear patterns emerging from his data and made the hypothesis that there were individual units of inheritance, that some characteristics were dominant over others and that the units did not mix. Chromosomes and genes had not been discovered, so people found it very hard to accept Mendel's ideas. Also, there was no model of how it worked. It was the development of the microscope that enabled people to see chromosomes and how they were passed on when cells divided; this meant people eventually recognised Mendel's work.

5 a Sami's alleles are ss. You know this because she has curved thumbs and the recessive allele is curved thumbs. She must have inherited two recessive alleles to have inherited the characteristic.

b If the baby has curved thumbs, then Josh is Ss. The baby has inherited a recessive allele from each parent, so Josh must have a recessive allele. You know he also has a dominant allele as he has straight thumbs.

	Sami	
	s	s
Josh S	Ss	Ss
s	ss	ss

c If the baby has straight thumbs, then Josh could be either Ss or SS. You know that the baby has inherited one recessive allele from the mother, and you know that Josh has one dominant allele, but you do not know if he has two dominant alleles.

	Sami	
	s	s
Josh S	Ss	Ss
S	Ss	Ss

6 a As an anomaly, the white-flowered plant should be investigated, e.g., to see if the colour was a result of a mutation or because of the particular conditions in which it was grown. He could breed from it, plant it in a different soil, etc.

b i To have white flowers, both of the parent plants must have contained a recessive white allele, so:

	P	P
P	PP	Pp
p	Pp	pp

ii Expect a 3 : 1 ratio; actual results 295 : 102 – very close.

c Suggest cross purple flowers with white flowers (pp). If purple flowers homozygous PP, all the offspring will be purple (Pp):

	P	P
p	Pp	Pp
p	Pp	Pp

If purple flowers heterozygous, half of the flowers will be purple (Pp) and half will be white (pp).

	P	p
p	Pp	pp
p	Pp	pp

7 a The inheritance of a single pair of genes which influence a particular phenotype characteristic.

b Dominant allele – controls the development of a phenotype characteristic even if only present on one of the chromosome pair. Recessive alleles – only control the development of a phenotype characteristic if they are present on both chromosomes of a pair.

c Any two correct answers; for example: dimples, dangly earlobes, straight or curved thumbs.

d There is a faulty recessive allele that causes the condition cystic fibrosis, but only if it is present on both chromosomes in a pair. An individual can be a carrier – have one recessive allele – but not be aware of it because they also have a healthy dominant allele and so have no symptoms of disease. For example, F dominant normal, f recessive cystic fibrosis, normal parents Ff – heterozygotes gives a 3:1 chance of having a child with cystic fibrosis.

	F	f
F	FF	Ff
f	Ff	ff

e H dominant Huntington's, h recessive normal. Affected Hh (50% or 1 in 2 chance of having affected children) or HH (all children will be affected as only affected alleles can be passed on by the homozygous parent, normal parent hh).

	H	h
h	Hh	hh
h	Hh	hh

Answers to end of chapter practice questions

1 a Genes (1); DNA (1).
b Chromosomes (1); gamete (1).
c Alleles (1).
d Heterozygous (1); homozygous (1).
2 a Asexual reproduction (1).
b Red (1) because the cutting plants have exactly the same genes/genetic material as the parent plant (1).
c Any **two** from: (different amounts of) water (1), light (1), fertiliser with mineral ions (1). *(2 marks maximum)*
d Having clones/all plants with same genes (1), will be a control variable (1).
3 DNA (1); protein (1); amino acids (1); DNA (1); double helix (1); three (1); amino acid (1).
4 • All, or nearly all, scientific points are included in a clear and logical order. There is almost faultless spelling, punctuation and grammar. *(5–6 marks)*
 • Many scientific points are correctly made, expressed clearly. There are very few mistakes in spelling, punctuation and grammar. *(3–4 marks)*
 • A few scientific points are made although they may not be clearly expressed and the order may be confused. There are many mistakes in spelling, punctuation and grammar. *(1–2 marks)*
Examples of scientific points are: identical twins have same genes/genetic material (1); girls have nearly the same height, so likely to be mainly genetic variation (1); girls have very different weights (1) and this must be due to the environment (1); probably different diets/amount of exercise (1); hair is very different now but would have been same when born (1); hair differences must be due to environment as girls have treated it differently (1); eyes are same colour as eye colour is controlled by genes and cannot change (1).
5 a Parental genotypes HbAHbs, HbAHbs (accept explanations in terms of the symbols A and s) (1); gamete genotypes HbAHbs, HbAHbs correctly derived (1); children's genotypes HbAHbA, HbAHbs, HbAHbs, HbsHbs correctly derived (1); HbsHbs clearly defined as having sickle-cell anaemia (1).
b i HbAHbA individuals are more likely to die of malaria (1); HbsHbs individuals are likely to die of the condition before maturity (1); but crosses between heterozygotes keeps frequency of Hbs allele high (1).
ii There is partial coincidence between distribution of malaria and sickle-cell allele (1), but there could be another factor that influences both distributions (1).

11 Genetic manipulation

11.1

1 a Embryo cloning: splitting cells apart from a developing embryo before they become specialised, to produce several identical embryos.
b Tissue cloning: getting a few cells from a desirable plant to make a big mass of identical cells, each of which can produce a tiny identical plant.
c Asexual reproduction: reproduction which involves only one parent; there is no joining of gametes and the offspring are genetically identical to the parent.
2 a It allows the production of far more calves from the best cows; can carry good breeding stock to poor areas of the world as frozen embryos; can replicate genetically engineered animals quickly.
b Either: cow given hormones to produce large numbers of eggs → then cow inseminated with sperm → embryos collected and taken to the lab → embryos split to make more identical embryos → cells grown on again to make more identical embryos → embryos transferred to host mothers.
Or: cow given hormones to produce large numbers of eggs → eggs collected and taken to lab → eggs and sperm mixed → embryos grown → embryos split up to make more identical embryos → cells grown on to make bigger embryos → embryos transferred to host mothers.
c Marks awarded for understanding of issues involved in embryo cloning. For example, economic issues – only wealthy farmers/wealthy countries can afford the technology, is it acceptable to produce large numbers of identical cattle, etc.

3

Tissue cloning in plants	Embryo cloning in cattle
Based on normal body cells	Based on cells from an embryo
Involves using hormones to make a mass of unspecialised tissue and then using different hormones to stimulate the production of many plantlets	Involves splitting an individual embryo into a number of cells and allowing them to grow into small balls of cells before implanting them in surrogate mothers prepared using hormones
Produces thousands of genetically identical offspring	Produces up to thirty genetically identical offspring a year

11.2

1 The nucleus is removed from an unfertilised egg cell → the nucleus is taken from an adult body cell → the nucleus from the adult cell is inserted (placed) in the empty egg cell → new cell is given a tiny electric shock → new cell fuses together → begins to divide to form embryo cells → ball of cells inserted into womb to continue its development.

2 Natural mammalian clones occur when an early embryo completely splits in two to form two identical embryos, which continue to develop into genetically identical twin individuals.
Embryo clones are made artificially when the cells of a very early embryo are split so the individual cells can all continue to grow and divide and form a number of genetically identical individuals.
Adult cell cloning involves taking the nucleus of an unfertilised egg cell and replacing it with the nucleus of a cell from an adult cell of another animal of the same species. A small electric shock is needed to start the new cell dividing and developing to form an embryo, which can be implanted into a surrogate mother and eventually forms a new individual that is a clone of the original source of the adult cell nucleus. Very few of these adult cell clones survive – it is still a very difficult and experimental technique.

3 Advantages: enables humans to clone adult animals so they can clone genetically engineered organisms, making it possible to clone new tissues and organs for people with diseases or needing transplants; could help infertile couples; could help conserve very endangered species. Any other valid points.
Disadvantages: people are concerned about human cloning; reduces variety in a population; objections to the formation of embryos that are then used to harvest tissues; people object to the cloning of endangered or extinct animals. Any other valid points. Students should also give an opinion as to the validity of the main concerns expressed. Award marks for reasoned consideration of the relevant scientific points.

11.3

1 Modifying the genome of different organisms enables us to use them to make useful substances, for example: bacteria making human insulin; modifying animals and plants to make human proteins in milk; modifying organisms to be resistant to disease, resistant to toxins or pesticides, to glow or change when attacked, to have an increased yield, shorter stems so less easily damaged, etc.; any other sensible points.

2 Cut a gene for a short stem from a plant closely related to the one you want to engineer using enzymes → take a plasmid out of a bacterium and cut it open using enzymes → insert the short stem gene into the bacterial plasmid, which acts as a vector → insert the vector into the cells of the plant you want to engineer → then clone those plant cells using tissue culture to make thousands of identical genetically engineered plants.

3 Any three sensible choices, e.g., golden rice, drought-resistant strains, plants that can withstand being under water, short-stemmed crops, pesticide resistance, etc. For whichever choices are made, award marks for sensible suggestions as to how the change has increased the crop yield and why it is important.

11.4

1 Advantages: improved growth rates of plants, bigger yields, plants that will grow in a range of conditions, plants that make their own pesticides or are resistant to herbicides. Disadvantages unknown but concerns about: insects may become pesticide-resistant if they eat pesticide-resistant plants repeatedly, GM organisms may affect human health, genes from GM plants may spread into wildlife and cross-breed with wild plants.

2 Credit for relevant comments backed by science. Examples of relevant points:
No: cloning has many potential benefits such as reproducing genetically engineered organisms, saving organisms from extinction, producing cheap plants. Some forms of cloning have been going on for centuries (cuttings) and these have been/may be used to produce medical treatments, etc.
Yes: most animals produced by adult cell cloning have problems, wasteful process, risk of human cloning for the wrong reasons, etc.

3 Relatively few successes so far; genetic engineering tried in cystic fibrosis so far with little success; some success in SCID; any other recent developments.

Answers to end of chapter summary questions

1 **a** Embryo cloning – flushing out early embryos and dividing them before replacing in surrogate mother cows.
b Both allow large numbers of genetically identical individuals to be produced from good parent stock much faster and more reliably than would be possible using traditional techniques.
c Cloning plants uses bits of the adult plant as the raw material for the cloning. Animal embryo cloning, as it is used at the moment, involves using embryos as the raw material for the cloning, although this may change in the future.
d There are more and more people in the world needing to be fed, so techniques for reproducing high yielding plants and animals are always helpful and are financially beneficial for farmers. Also, in developed countries people demand high quality but cheap food, so techniques that reproduce valuable animals and plants are valued.

2 **a** Clear description of adult cell cloning, e.g., the nucleus is removed from an unfertilised egg cell. At the same time the nucleus is taken from an adult body cell, e.g., a skin cell of another animal of the same species. The nucleus from the adult cell is inserted (placed) in the empty egg cell. The new cell is given a tiny electric shock, which fuses the new cells together and causes it to begin to divide to form embryo cells. These contain the same genetic information as the original adult cell and the original adult animal. When the embryo has developed into a ball of cells, it is inserted into the womb of an adult female to continue its development.
b Plant cloning has been accepted for a long time and doesn't threaten people – only advantages seen in general. Cloning animals is seen as worrying in itself but also raises concerns of human cloning. Cloning pets, etc., is seen as frivolous.

3 **a**

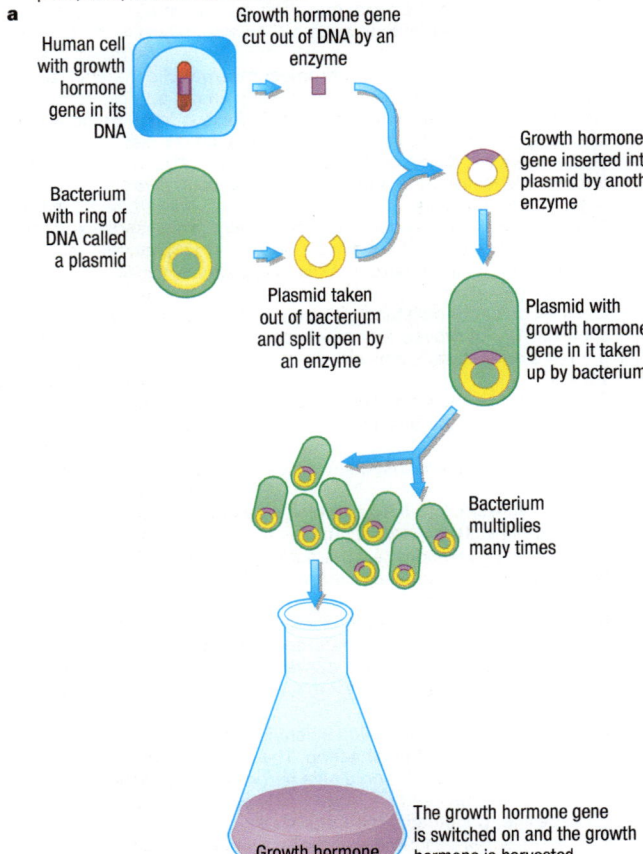

Human cell with growth hormone gene in its DNA

Growth hormone gene cut out of DNA by an enzyme

Growth hormone gene inserted into plasmid by another enzyme

Bacterium with ring of DNA called a plasmid

Plasmid taken out of bacterium and split open by an enzyme

Plasmid with growth hormone gene in it taken up by bacterium

Bacterium multiplies many times

Growth hormone

The growth hormone gene is switched on and the growth hormone is harvested

b It is pure – free from any contamination. It is the human version of a hormone. It can be produced in large amounts relatively easily and cheaply as and when it is needed.

4 **a** They have been raised and trained on different establishments so it is environmental factors which are influencing their racing ability.
b As a control – to see how the animal turns out if not trained up as a racing mule.
c Data on effect of diet, handling, intensity of training, etc., on the temperament, running speed, stamina, etc., of the mules.

5 a Crops that have been genetically modified by adding genetic material from another organism.

b People are concerned about effects on human health, possible spread to other wild organisms through cross-breeding, insects becoming pesticide resistant, etc.

c Increasing the yield by: getting bigger seed heads; shorter stems so less damage by wind; drought, temperature or flood resistant so get crops in difficult conditions; making pesticides so prevent crops from being destroyed by insects, etc.; any other sensible points.

d Any sensible suggestions – look for awareness of good experimental design.

Answers to end of chapter practice questions

1 a A clone is a group of organisms created by asexual reproduction from a single parent and has exactly the same genetic material as the parent. (1) (Credit answer in terms of mitosis, provided that identical genetic material is explained.)

b Tissue culture (1).

c An embryo is split (1) at a very early stage/when it is still a ball of cells (1) and the divided embryos are implanted into different mothers (1). In adult cell cloning, a nucleus from a cell of the animal to be cloned is removed (1) and put into an egg cell that has had the nucleus removed (1). An electric shock makes the egg cell start to divide (1) and when it is a ball of cells/embryo it is implanted in the uterus of a (surrogate) mother (1).

d Any four from: embryo splitting would be better (1); as you could take embryos created by eggs and sperm from known champion cows and bulls (1); split them and put them into many other cows to produce calves (1). It is a method that has been done before (1); and it is likely to be cheaper (1) than adult cell cloning, which is still very new.
(4 marks maximum)

2 a Any four from: nucleus/DNA/chromosomes/genetic material removed (1) from (unfertilised) egg/ovum (1). (Allow empty egg cell for the first two marks: do *not* allow fertilised egg.) Nucleus from the body cell of champion (cow) (1); inserted into egg/ovum (1); electric shock (1); to make cell divide or develop into embryo (1);(embryo) inserted into womb/host/another cow (1) (allow this point if wrong method, e.g., embryo splitting).
(4 marks maximum)

b Any four from: *Pros* – economic benefit, e.g., increased yield/more profit (1); clone calf not genetically engineered (1); genetic material not altered (1); milk safe to drink/same as ordinary milk (1).
(2 marks maximum)

Cons – consumer resistance (1); caused by misunderstanding process (1); not proved that milk is safe (1) (ignore 'God would not like it' or 'it's not natural'); ethical/religious argument (1); reduce gene pool (1).
(2 marks maximum)

Conclusion: sensible conclusion for or against, substantiated by information from the passage and/or own knowledge conclusion at end (1).
(5 marks maximum)

3 a i Any one from: kills insects (which eat crop) (1); increases yield (1).
(1 mark maximum)

ii Any two from: kills insects which may not be pests (1); poisonous to humans (1); expensive (1); pollutes the environment (1); other relevant suggestions, e.g., is not organic (1)   *(2 marks maximum)*

iii Any one from: increases crop yield (1); reduces cost of pesticide use (1).
(1 mark maximum)

iv Any one from: may lead to increased use of pesticides in the long run/description of or reference to last paragraph (1); ethical considerations, e.g., alters genes of crop (1). (*Do not* allow 'not natural' or 'genes may get into wildlife idea' or 'against religion' or 'not organic'.)
(1 mark maximum)

b • There is a clear and detailed scientific description of the sequence of events in genetic engineering. The answer is coherent and in a logical sequence. It contains a range of appropriate or relevant specialist terms used accurately. The answer shows very few errors in spelling, punctuation and grammar.   *(5–6 marks)*

• There is some description of the sequence of events in genetic engineering, but there is a lack of clarity and detail. The answer has some structure and the use of specialist terms has been attempted, but not always accurately. There may be some errors in spelling, punctuation and grammar.   *(3–4 marks)*

• There is a brief description of the genetic engineering, which has little clarity and detail. The answer is poorly constructed with an absence of specialist terms or their use demonstrates a lack of understanding of their meaning. The spelling, punctuation and grammar are weak.   *(1–2 marks)*

• No relevant content.   *(0 marks)*

Examples of biology points made in the response:
• gene from the bacterium
• is cut from the chromosome
• using enzymes
• gene transferred to the cotton
• (cotton) chromosome (allow cell)
• (the gene) controls characteristics
• causes the cotton (cells) to produce the poison.

12 Evolution, adaptations and interdependence

12.1

1 a All the species of living organisms that are alive today (and many more which are now extinct) have evolved from simple life forms, which first developed more than 3 billion years ago.

b Only the animals and plants most suited to their environment – the 'fittest' – will survive to breed and so pass on their characteristics.

2 Any thoughtful point, e.g., Lamarck helped to pave the way for Darwin's ideas; Lamarck's ideas stimulated Darwin's thinking; people had already come to terms with a theory other than the Bible so were more ready to accept Darwin's ideas, debate on the origins of life was opened up; Darwin's theory made more sense than Lamarck's and had an evidence base that Lamarck's did not, which made it easier for it to be accepted.

3 a South American rheas – Darwin found a new species. Two types of the bird living in slightly different areas made Darwin start to think about how they came about.

b Galapagos tortoises, iguanas and finches – these were some of the animals in the Galapagos Islands that varied from island to island, and made Darwin wonder what had brought about the differences.

c The long voyage of HMS *Beagle* – this gave Darwin lots of opportunities to collect specimens and time to think about his theories and ideas.

d The twenty years from his return to the publication of the book *The Origin of Species* gave Darwin time to work out his ideas very carefully and to collect a lot of evidence to support them.

12.2

1 Students can chose any three adaptations eg long legs in cheetahs, strong scent in honeysuckle etc as long as they use the characteristics to explain how a random combination of genes or mutation gives organism a characteristic which increases its chances of surviving and reproducing to pass on the alleles for that characteristic. The advantage may not be apparent until there is a change in the environment. Any valid points

2 a New forms of genes (alleles) result from changes in existing genes known as mutations. Mutations are tiny changes in the long strands of DNA. Mutations occur quite naturally through mistakes made in copying DNA when the cells divide, but the rate of mutation can be increased by radiation or certain chemicals.

b Mutations introduce more variety into the genes of a species and can increase the chances of survival. Many mutations have no effect on the characteristics of an organism, and some mutations are harmful but sometimes a mutation has a good effect by producing an adaptation that makes an organism better suited to its environment. This makes it more likely to survive and breed. The mutant allele will gradually become more common in the population and will cause the species to evolve. If a new form of a gene arises from mutation and this coincides with a change in the environment, there may be a relatively rapid change in a species. If the mutation gives the organism an advantage in the changed environment, it will soon become common. This is why mutations are important in natural selection.

3 a Mutation gave some deer antlers to make them more successful in battles with other stags and more attractive to females. This means that they are more likely to mate and pass on their genes. This process continues until antlers become normal in the population. The stags with the biggest or most effective antlers are the ones which mate most successfully.

b Mutation produced spines instead of leaves. Cactus loses very little water and so survives well and reproduces, passing on advantageous genes until normal in population.

c Mutation gives increased temperature tolerance. These camels have an advantage, so more likely to survive and breed, passing on the mutation until it is normal in the population.

12.3

1 a Geographically by the formation of mountains, rivers, continents breaking apart, etc. Environmentally – climate change in one area and not another or different types of change in different areas.

b Natural selection means organisms best suited to a particular environment will be most likely to survive and breed. So in two different environments, different features will be selected for and the organisms will become more and more different until they cannot interbreed and new species have evolved.

2 a An organism which is found in only one place in the world, where it has evolved.

b Endemic organisms evolve as a result of geographical isolation. When an area of land becomes an island, the plants and animals on it are isolated from the organisms on the mainland and so, as they evolve, they often form new species. These are different from the original species and are unique and endemic to the island.

c Our current model of speciation is that an organism has a wide range of alleles controlling its characteristics as a result of genetic variation from sexual reproduction and mutation. In each population, the alleles that are selected will control characteristics that help the organism to survive and breed successfully. This is natural selection. In the formation of endemic species, part of a population becomes isolated with new environmental conditions. Alleles for characteristics that enable organisms to survive and breed successfully in the new conditions will be selected. These are likely to be different from the alleles that gave success in the original environment. As a result of the selection of these different alleles, the characteristic features of the isolated organisms will change. Eventually, they can no longer interbreed with the original organisms and a new species forms. This is what has happened on islands such as Borneo, Australia and the Galapagos Islands, and it acts as evidence that our model of speciation reflects what happens in the natural world.

3 All populations have natural genetic variation due to sexual reproduction and mutation. This results in a wide variety of alleles in the population. If part of the population becomes isolated and conditions are different from the original population, different alleles are likely to give an advantage. These alleles will be selected for, as the organisms which have them will be most likely to survive and reproduce successfully in the new environment. As a result the characteristics of the organism will change until eventually they can no longer interbreed with the original population and a new species has evolved.

12.4

1 Animals – food and oxygen, water and a mate. Plants: water, carbon dioxide, light energy, minerals from soil. Plants can make their own food using photosynthesis and produce oxygen. Animals have to eat food and need oxygen to break it down to release energy.

2 a An organism that can live in extremely difficult conditions in which most other organisms cannot survive.

b Enzymes which function in very high temperatures; enzymes which function at very low temperatures; ability to get rid of excess salt; ability to respire without oxygen; any other valid point.

3 An adaptation is a feature which makes it possible for an organism to survive in its particular habitat. Award marks for any three good examples of adaptations in animals or plants

12.5

1 a It is very cold, so there is a problem in keeping warm and finding enough food.

b It is very hot, so the main problems are keeping the body cool and finding enough water.

2 Small ears reduce surface area of thin-skinned tissue to reduce heat loss; thick fur acts as an insulating layer to help prevent heat loss; layer of fat/blubber provides insulating layer; any other relevant adaptations, e.g., furry feet to insulate against contact with ice; large size reduces surface area: volume ratio so reduces heat loss, etc.

3 a Large, thin ears for heat loss; loose wrinkled skin to aid heat loss; little or no fur; any other sensible suggestions.

b Animals living in hot dry conditions keep cool without sweating by avoiding the heat of the day and by having large ears, baggy skin, little fur, thin and silky fur and a large surface area to volume ratio to increase heat loss.

c Insulation against cold of water, e.g., blubber/internal fat; thick fur externally; small surface area to volume ratio by being big; small extremities such as ears; ability to take deep breaths, heart rate slows when they dive; any other sensible points.

12.6

1 a For photosynthesis; to support the cells, to support the stems; to transport substances around the plant; any other sensible point.

b Through their roots from the soil.

2 a By evaporation through the stomata.

b Dry places are often hot, so photosynthesis and respiration occur at a faster rate. The stomata are open more, so there is more evaporation. If the air is dry, evaporation occurs at a faster rate.

3 a Small leaves; curled leaves – reduce surface area; thick cuticle – reduces rate of evaporation.

b Any three sensible suggestions, e.g., reduction of leaves to spines; thickened water-filled stems or leaves as a store of water; long, deep roots; rolled leaves to reduce evaporation; reduction of stomata, etc

12.7

1 a If anything happens to their food supply, such as another animal eating it, fire or disease, then they will starve.

b Any suitable examples, such as lions, cheetahs and leopards.

c Any suitable examples, such as rabbits, limpets on a sea shore.

d Members of the same species are all competing for exactly the same things, whereas different species often have similar but slightly different needs. Members of the same species tend to live in the same area.

2 a Fighting: strength, antlers, teeth, etc. Displaying: spectacular appearance, colours, part of body to display (e.g., peacock's tail).

b The answer to this will depend on the method selected for the first part of the answer.
Fighting: advantages – possibility of winning lots of mates, becoming dominant, fathering lots of offspring, females don't usually have any choice, preventing others from mating; disadvantages – the animal could be hurt or killed, needs lots of body resources to grow antlers and to fight.
Display: advantages – don't risk getting hurt, possibility of attracting several mates; disadvantages – uses up lots of resources to grow feathers/carry out displays, females usually choose and may not get noticed, vulnerable to disease or lack of food so don't produce good display, need to be seen. Any other sensible points.

3 a Very good eyesight, good hearing, good sense of smell, binocular vision so they can pounce, any other sensible points.

b Special teeth to grind grass and break open cells, ability to run fast away from predators to avoid being caught, good all round eyesight to detect predators creeping up, good hearing to detect predators, etc.

c Good camouflage, good eyesight, hearing and sense of smell to detect predators, good all-round vision, any other sensible points.

d Teeth and gut adapted to eating plants – crushing the cells to release the cell contents/breaking down cellulose cell walls, ability to reach the top of trees (long neck or good at climbing) to get to the tender leaves, ability to grip on to branches to get to the tender leaves/hold them to pull off the tree, possibly use tail for balance to get to the top of the tree.

12.8

1 a Grow and flower very early before many bigger plants get their leaves; growing tall very fast; growing larger leaves; making more chlorophyll to make the most of any light that arrives; any other sensible points.

b They produce flowers before the oak tree's leaves have grown to full size, so they are not shaded.

2 a To avoid competition between the seedlings and the parent plants and to avoid competition between the seedlings, as far as possible.

b Any three suitable adaptations – look for different ones, for example, fluffy seeds, winged seeds, seeds in berries/fruits which are eaten, explosive seeds, sticky seeds, hooked seeds, seeds that float on water. Any other sensible suggestion.

3 For example, deep taproot (difficult to remove, can regenerate well if severed); low rosette of leaves (avoids blades of lawnmowers and grazing animals); long flowering period, so produces large numbers of seeds; very effective wind dispersal of seed over a large area.

Answers to end of chapter summary questions

1 a D **c** A
b C **d** B

2 a The temperatures are too cold for reactions in the body to work and so for the organisms to survive.

b Problems: overheating in day, too cold at night and early morning to move much, water loss. How they cope with problems: bask in the Sun in the morning to warm up, hide in burrows or shade of rocks to avoid heat of day and cold of night, reduce water loss by behaviour and don't sweat.

c Large surface area: volume ratio allows them to lose heat effectively.

3 a Lots of water loss through the leaves, not much water taken up by roots.
b Most water is lost through the leaves, so less leaf surface area means less water loss.
c Spines, rolled leaves.
d Water storage in stems, roots or leaves; thick waxy cuticle; ability to withstand dehydration.
e They have several different adaptations to enable them to withstand water loss/little water available (spines, water storage in stem, etc.).

4 Pandas feed almost exclusively on bamboo, so if it dies out they have no food and will die out as well. Other animals – bamboo only part of the diet so they simply eat other plants.

5 a Because they are competing for exactly the same things.
b Makes sure there is plenty of food for the animals and their young; advertise their territory to reduce conflict with predators.
c Big advantage is that there is no risk of being hurt or killed if you have a courtship ritual/colouration, whereas fighting can lead to both and/or infection and death after injury. Disadvantages is that it uses up a lot of resources to develop the colouration/carry out the display and also the female gets to choose and you might not be chosen.

6 Students use the bar charts in the practical activity on page 84 to answer these questions.
a First month: crowded seedlings taller than spread out seedlings. Crowded seedlings shade each other so each seedling grows taller to avoid the shade. Spread seedlings don't have that pressure. But over six months, the crowded seedlings do not get light as they shade each other. They photosynthesise less so they cannot grow as tall as the spread out seedlings, which can make as much food as possible.
b i They relied mainly on the food stored in the seed – the crowded ones were taller but the spread ones had thicker stems and bigger leaves.
ii The spread out seedlings get the full effect of the light and grow as well as possible, making lots of new plants (and so wet mass). The crowded plants each get less light therefore less photosynthesis and less wet mass.
c To eliminate as far as possible the effects of genetic variety in the seedlings – the bigger the sample, the more reproducible the results.
d i Any of: light level, amount of water and nutrients available, temperature.
ii So that any differences would be the result of the crowding of the seedlings.

7 Malaria is caused by the single-celled parasite *Plasmodium falciparum*, which has a very complicated life cycle. It spends part of its life cycle in a mosquito and part in the human body. The parasites are passed on to people when the female Anopheles mosquitoes take two blood meals from people before laying their eggs. Once inside the human body, the parasites damage the liver and blood and cause serious disease symptoms including fevers, chills and exhausting sweats.
There are several forms of the malarial parasite, and each form is adapted to survive in different places in different hosts. Gametocytes infect mosquitoes, and this form reproduces sexually. The female mosquito takes them in when she feeds on the blood of someone infected with malaria. The gametocytes make their way to the salivary glands of the mosquito and change into a new form called sporozoites.
Sporozoites are passed on to humans next time the mosquito takes a blood meal. The female injects saliva into the blood vessels of her host to prevent the blood from clotting as she feeds. Sporozoites enter the blood stream with the saliva and are carried in the blood to the liver where they enter the liver cells. In the liver cells, some of the sporozoites divide asexually to form thousands of merozoites, another form of the malaria parasite.
The merozoites are released from the liver into the blood where they enter the red blood cells. Here, hidden from the immune system of the body, some of the merozoites become schizonts. After a time, the schizonts burst out of the red blood cells, destroying them and releasing more merozoites. It is the reaction of the body to this release of schizonts and the destruction of red blood cells that causes the terrible fever attacks that are seen when someone suffers from malaria. Some of the merozoites in the blood go into a stage of sexual reproduction and produce female gametocytes, which can then be transferred to the female mosquito when she bites, and so the whole cycle starts again.

Answers to end of chapter practice questions

1 • There is a clear and detailed scientific explanation of adaptations. The answer is coherent and in a logical sequence. It contains a range of appropriate or relevant specialist terms used accurately. The answer shows very few errors in spelling, punctuation and grammar.
(5–6 marks)
• There is some explanation of adaptations, but there is a lack of clarity and detail. The answer has some structure and the use of specialist terms has been attempted, but not always accurately. There may be some errors in spelling, punctuation and grammar. *(3–4 marks)*

• There is a brief explanation of adaptations, which has little clarity and detail. The answer is poorly constructed with an absence of specialist terms or their use demonstrates a lack of understanding of their meaning. The spelling, punctuation and grammar are weak.
(1–2 marks)
• No relevant content. *(0 marks)*
Examples of biology points made in the response:
• large ears give large surface area to volume ratio, which allows more heat to be lost in hot conditions
• thin ears means blood flows near the surface so there is a shorter distance for heat energy to travel to surroundings
• no fat layer reduces insulation, which would keep heat in body
• bristles do not trap air layer, which would be an insulating layer
• by feeding in early morning, elephants do not create so much heat in the hot middle of the day and their food will have maximum water content collected overnight.

2 a No wings makes it easier to move in fur (1)
b Piercing mouth parts enable flea to suck blood from host as food (1)
c Flattened bodies make it easier to move in fur (1)
d Hard exterior means it is hard to kill/crush (1)
e Long back legs enable flea to jump from host to host (1)
f Bristles and combs secure it in the fur (1)

3 a Herbicides are chemicals that (selectively) kill some plants/weeds. (1)
b The herbicide will kill the weeds but not the vegetable plants (1) and so the vegetable plants will not have to complete (1) so hard for water (1), mineral ions (1), light (1) or space (1) to grow (*allow maximum of 2 marks for factors listed*). This means the vegetables will photosynthesise more (1) and so grow bigger. *(5 marks maximum)*

4 a i percentage of gemsboks feeding (1)
ii Any one from: same area of study (1); same mass of grass analysed (1); same herd of gemsboks (1). *(1 mark maximum)*
b The water content of grass rises during the night (some time after 18.00 hours to a maximum of 25% at 09.00 hours and then falls (more rapidly at first) to a minimum of 5% at 18.00 hours. (1) (Some accurate reference to actual figures in the table is necessary to obtain the mark.)
c Between 24.00 hours and 06.00 hours. (1) (The important words here are 'more than'. Candidates who ignore these words will include the figure of 30% and therefore give a response of 21.00 hours to 09.00 hours.)
d The water content of the grasses that it eats are high over this period (1). It is night and the gemsboks are therefore less easily seen by predators (1). It is cooler and therefore they are less likely to have to sweat and so this helps them conserve precious water (1).

13 Ecology

13.1
1 a Biomass is the mass of material in living organisms
b Because to scale they show the amount of biomass at one trophic level compared with another, therefore shows how much is lost etc.
2 a Draw the pyramid of biomass
b Energy is lost at each stage in the food chain and so there is less available for the next trophic level to grow.
3 a 5%
b 0.6%

13.2
1 a Because not all that an animal eats it is able to digest.
b They use up energy in moving and so it is not used to grow and increase their biomass.
2 Only plants can absorb the energy from the Sun, plants do not cover the whole surface of the Earth, most of the biomass is not transferred from plants to animals.

13.3
1 a Detritivores such as maggots and some types of worms; decomposers, which are microorganisms such as bacteria and fungi.
b Plants remove minerals such as nitrates from the soil through their roots all the time. When animals eat the plants these minerals are passed on through food chains and webs. If plants took minerals out but they were never returned, the soil would soon be infertile as there would be no minerals left. Decay is the process by which the minerals that have become part of the bodies of living organisms are recycled and returned to the environment.

2 a The rate of the chemical reactions in the microorganisms that act as decomposers gets faster with an increase in temperature. Average temperatures are much warmer in the summer than in the winter and so the reactions of decomposition occur much faster – and garden and kitchen waste are turned into compost more rapidly.

b The microorganisms involved in the process of decomposition grow better in moist conditions. If it is too dry they cannot digest their food so easily and they may dry out and stop growing completely.
So if it is particularly dry, even if it is also hot, at least some of the microorganisms involved in the formation of compost from garden and kitchen waste may slow down or stop growing and so in turn compost formation slows down.

c Many of the microorganisms which bring about the decomposition of material in a compost bin are aerobic – they need oxygen to respire and so need aerobic conditions. Turning over the contents of a compost bin increases the air in the mixture and ensures that the microorganisms have plenty of oxygen so they can work as efficiently as possible, increasing the rate at which compost is formed.

13.4

1 a The cycling of carbon between living organisms and the environment.
b Photosynthesis, respiration and combustion.
c Because it prevents all the carbon from getting used up; returns carbon dioxide to the atmosphere to be available for photosynthesis.
2 a Carbon dioxide in the air.
b Students can produce a written description of the carbon cycle or a diagram (e.g., Figure 2, The carbon cycle in nature) to summarise the stages (must cover all points in the carbon cycle).
3 Photosynthesis is the process by which plants use carbon dioxide and water to make glucose and oxygen, using energy from light.
carbon dioxide + water (+ light energy) → glucose + oxygen
Photosynthesis takes carbon dioxide out of the environment.
The glucose made by the plants is used in the plants and by animals (both those that eat plants and those that eat animals which eat plants) in the process of respiration. In respiration, glucose is broken down using oxygen to produce carbon dioxide, water and energy, which can be used in the reactions within cells.
glucose + oxygen → carbon dioxide + water (+ energy)
Respiration returns carbon dioxide to the atmosphere.
Combustion is the burning in oxygen of organic material from living or once-living organisms, e.g., wood, fossil fuels such as oil. It produces carbon dioxide, water and energy – it is the same as respiration but occurs in a rapid, uncontrolled way.
fossil fuel or wood + oxygen → carbon dioxide + water

Answers to end of chapter summary questions

1 a i 10%
ii 8%
iii 12.5%
b The mass of the producers has to support the whole pyramid, relatively little energy is transferred from producers to primary consumers (difficult to digest).
c Relatively little energy is passed up the chain, so not enough to support many carnivores.
d Less energy passed on as warm blooded animals use energy to generate warmth. This is transferred to the environment and so that energy is no longer available to pass on up the chain.
2 The amount of biomass transferred along food chain gets less. Biomass is needed for energy. So by eating plants, the maximum amount of biomass is passed on to people.
Eating meat – plant biomass transferred to animals, animal biomass to people – biomass lost at both stages. Students could draw pyramid of biomass to show plant/person and plant/cow or sheep/person, etc.
3 a Graph plotting, correct scale, labelled axes, axes correct way round, accurate points.
b So that chickens use little energy maintaining their body temperature, so have more energy for growth.
c To reduce movement and thereby reduce energy used in movement, so more energy for growth.
d So they grow fast to a weight when they can be eaten and another set of chickens started up – economic reasons.
e The line should be below the first line. Chickens outside use energy moving around and keeping warm or cool, so convert less biomass from their food for growth.

4 a Low temperatures prevent growth of decay – causing microorganisms.
b Cooking destroys the microorganisms, denatures enzymes so no decay.
c Most decomposers need oxygen to respire – no air, no oxygen, so microbes cannot grow.
d Heat kills microorganisms, no oxygen so no decay.
5 a Photosynthesis.
b Respiration, burning (decay and decomposition).
c Oceans, air (carbonate rocks).
d CO_2 is important for photosynthesis and keeping surface of Earth warm. Excess CO_2 means surface gets warmer; this affects sea levels, living organisms. Less CO_2 means surface cools, affects life.
e A death B burning (combustion) C feeding D respiration E photosynthesis F decay and decomposition.
6 a Higher-temperature means faster reactions. Warm compost means microorganisms digest, grow and reproduce faster. More decomposers results in faster decomposition.
b Makes sure all the decomposing microorganisms have enough oxygen to respire as fast as possible.

Answers to end of chapter practice questions

1 a Three layers (1), each one getting smaller as you go up (1), labelled: trees, caterpillars, birds going up (1).
b Sunlight/radiation from the sun (1)
c Any five from: chloroplasts/chlorophyll (1) in the leaves absorbs/captures the light energy to use in photosynthesis (1) which makes sugar/glucose (1). When the leaves are eaten by the caterpillar they are digested (1) so the glucose /sugar is absorbed (1) into the caterpillar. It is used to make proteins/fats/new cells/new tissues in the caterpillar (1). *(5 marks maximum)*
d i $1000/20\,000 \times 100$ (1) = 5% (1);
ii Any two from: some used for movement/contraction of muscles; some lost as heat to the environment; some lost as waste/faeces; not all caterpillars are eaten; not all parts of caterpillars are digested. *(2 marks maximum)*
e • All scientific points are made using appropriate scientific terms. The account of decay and recycling of carbon dioxide is presented in a clear and logical order. The answer contains a range of appropriate or relevant specialist terms used accurately.
The answer shows very few errors in spelling, punctuation and grammar. *(5–6 marks)*
• There are many correct scientific points presented in a clear manner, although they may not all be in logical order or complete. The answer has some structure and the use of specialist terms has been attempted, though not always accurately. There may be some errors in spelling, punctuation and grammar. *(3–4 marks)*
• There are few scientific points about decay, which may be presented in an unclear or confused way. The answer is poorly constructed with an absence of specialist terms, or their use demonstrates a lack of understanding of their meaning. The spelling, punctuation and grammar are weak. *(1–2 marks)*
• No relevant content. *(0 marks)*
Examples of scientific points are:
• Microorganisms/bacteria/fungi decay/decompose the leaves.
• They use the carbohydrates/sugar/glucose to respire.
• Respiration releases carbon dioxide into the air ready to be used again for photosynthesis.
2 a 10(%) (2) (If incorrect answer, allow 100% – 25 – 35 – 30 for 1 mark)
b 2.5 (megajoules) (2) (If incorrect answer, allow 1 mark for correct working)
c Respiration (1)
d It reduces the calf's movement because it won't walk about, therefore it will use that energy for growth (1). It also reduces the energy transferred by heating/the calf will need to use to keep warm, and that energy can be used for growth (1).
e It is not cost effective as you have to pay for heat OR the calf might catch a disease more easily if it is inside with other animals OR any other sensible suggestion. (1)

Index